SOUTHA

(Uni

The Agricultural Sector of Jordan

Policy & Systems Studies

A. B. Zahlan
Programme Leader and Editor

Subhi Qasem
Team Leader

Published for
The Abdul Hameed Shoman Foundation, Amman
by Ithaca Press London 1985

First published in 1985 by
Ithaca Press, 13 Southwark Street, London SE1 1RQ

Printed in England by
Biddles Ltd., Guildford & King's Lynn

Typeset by Carlton Photosetters, Westcliff-on-Sea

ISBN 0 86372 026 9

British Library Cataloguing in Publication Data

Zahlan, A.B.
The Agricultural sector of Jordan : policy & systems studies.
1. Agriculture——Economic aspects——Jordan
I. Title
338.1'095695 HD2059

ISBN 0-86372-026-9

CONTENTS

ACKNOWLEDGEMENTS

This volume is one of a series of publications prepared in a research programme on a policy-oriented systems analyis of the agricultural sectors of Jordan and Sudan. The programme was financed by the Abdul Hameed Shoman Foundation. Thanks are due to the Foundation for making the programme possible by financing it, and also for unstinting support throughout. Mr Abdul Majeed Shoman and his fellow members on the Board of Trustees of the Foundation gave generously of their time and attention to the progress of the programme. Their support is gratefully acknowledged.

The planning and execution of this research programme has also had the support and participation of a large number of other people.

H.R.H. Crown Prince Hassan graciously took a direct and personal interest in it from its inception. His participation in the Workshop, his observations and encouragement are gratefully acknowledged.

Thanks are due to H.E. Dr Said Nabulsi, Governor of the Central Bank of Jordan, and H.E. Hanna Odeh, Minister of Trade and the Economy, who kindly provided their valuable time to further the progress of the programme.

Dr Ian Carruthers and Mr Mark Mitchell participated in the Workshops in Amman at which the methodologies were finalised. Mr Mitchell also provided detailed comments and observations that were instrumental to the programme's evolution.

Needless to say, the innovative leadership and intellectual guidance of Dean Subhi Qasem were invaluable. Moreover, the cooperation and dedication of all the contributors of papers in this volume facilitated a difficult task.

Thanks are also due to Ms Jana Gough for her thorough and painstaking copy-editing, to Ms Janice Bowen for editorial assistance and to Ms Trish Greenfield for editing the flow charts. Mrs Dorothy Thomson provided cheerful and indefatigable secretarial work. I am greatly indebted to her for the long hours she spent in finalising the typescript.

The assistance of Dr Rosemarie Said Zahlan during all phases of this programme is gratefully acknowledged.

A. B. ZAHLAN
Editor and Programme Leader
June 1984

CONTRIBUTORS

Suleiman Arabiat, Chairman, Department of Agricultural Economics, University of Jordan.

Adib H. Duayfi, Director, Arab Company of Animal Production and Head of the Economics Department, Ministry of Agriculture.

Mahmud Duwayri, Faculty of Agriculture, University of Jordan.

Adeeb K. Haddad, Head, Economic Research Department, Central Bank of Jordan.

Burhan Abu Howayej, Secretary of the Higher Council of Agriculture, Ministry of Agriculture.

M. Haitham el-Hurani, Department of Economics, University of Jordan.

Fahd Salih Natur, Senior Vice-President, Jordan Valley Authority.

Subhi Qasem, Professor of Plant Pathology and Dean of the Faculty of Science, University of Jordan.

Bassam Saket, Royal Scientific Society.

Nureddin A. Taqieddin, Agricultural Credit Corporation.

Schedule of Tables

Schedule of Figures

LIST OF ABBREVIATIONS

AGDP	agricultural gross domestic product
AI	artificial insemination
bn	billion
CBR	crude birth rate
CDR	crude death rate
cm	centimetre
DAP	diammonium phosphate
du	dunum = 1,000 square metres or 0.1 hectares
EC	electric conductivity
GCF	gross capital formation
GDP	gross domestic product
GNP	gross national product
g	gramme
ha	hectare
HO	head office
HYV	high yielding variety
JD	Jordanian dinar = $US 3 approximately
JD/du	Jordanian dinar per dunum
JD/t	Jordanian dinar per tonne
kg	kilogram
km	kilometre
km^2	square kilometre
KWH	kilowatt-hour
kW	kilowatt
l	litre
l/sec	litres per second
LDC	less developed countries
LTO	Lateral-Turn-Out
LL	Lebanese pounds
LS	Syrian pounds
£S	Sudanese pounds
LWG	Live weight gain
m	metre
m^2	square metre
MAP	monoammonium phosphate
mcm	million cubic metres

me/l	milliequivalent per litre
m^3	cubic metre
m^3/sec	cubic metres per second
mg/l	milligrams per litre
mm	millimetre
mmhos/cm	millimhos per centimetre conductivity
mn	million
mn/l	million litres
m^3/c/yr	cubic metres per capita per year
N	nitrogen
NPK	nitrogen phosphate potassium
O&M	operation and maintenance
$P_2 0_5$	phosphate
ppm	parts per million
psi	pounds per square inch
R&D	research and development
SAR	sodium absorption ratio
t	tonne = 1000 kg
TV	television
t/du	tonne per dunum

Key to Flow Charts

Physical activity which does not involve external interaction

Physical activity which involves external interaction

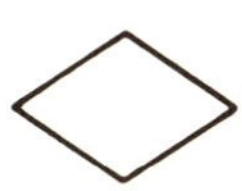

Decision

Stop

LIST OF ACRONYMS

AADAI	Arab Authority for Agricultural Investment and Development (Sudan)
ACSAD	Arab Centre for the Study of Arid Zones and Dry Lands (Syria)
AFESD	Arab Fund for Economic and Social Development (Kuwait)
AGRIS	Agricultural Research Information System (FAO)
AID	Agency for International Development (USA)
AIP	Agricultural Investment Projects
ALAD	Arid Land Agricultural Development Programme (Middle East)
AMO	Agricultural Marketing Organisation (Jordan)
AMREF	African Medical and Research Foundation
ARC	Agricultural Research Corporation (Sudan)
AREAS	Area Resource Analysis Systems
ASAE	American Science Education Administration
AMCCS	Agricultural Marketing and Credit Cooperative Societies (of the Gezira Scheme)
CAP	Common Agricultural Policy (EEC)
CEC	Commission of the European Communities
CIAT	Centro International de Agriculture Tropical (Colombia)
CIMMYT	Centro Internacional de Mejoramiento de Maiz y Trigo (Mexico)
CIP	Centro Internacional de la Papa (Peru)
CMEA	Council for Mutual Economic Assistance
CNEEMA	Centre National d'Etudes et d'Experimentation du Machinisme Agricole
COPAC	Committee for the Promotion of Aid to Cooperatives
DAG	Development Administration Group
DANIDA	Danish International Development Agency
DECARD	Desert Encroachment Programme
DECARP	Desert Encroachment Control and Rehabilitation Programme
DERAP	Development Economics Research and Advisory Service
DLCOAE	Desert Locust Control Organisation for Eastern Africa
DSE	Deutsche Stiftung für Internationale Entwicklung (German Foundation for International Development)
EMASAR	Ecological Management of Arid and Semi-Arid Rangelands

EWLP	Experimental World Literacy Programme
FAO	Food and Agricultural Organisation
GOED	General Organisation of the Euphrates Dam
GRS	Gezira Research Station (Sudan)
GTZ	German Agency for Technical Cooperation
IAC	International Agricultural Centre
IAEA	International Atomic Energy Agency
IAH	International Association of Hydrogeologists
IAHS	International Association of Hydrological Science
IAWPR	International Association on Water Pollution Research
IBP	International Biological Programme
IBPGR	International Board for Plant Genetic Resources (Italy)
ICARDA	International Centre for Agricultural Research in the Dry Areas (Syria)
ICID	International Commission on Irrigation and Drainage
ICOLD	International Commission on Large Dams
ICP	Institute for Comprehensive Planning
ICRISAT	International Crops Research Institute for the Semi-Arid Tropics (India)
IDRC	International Development Research Centre (Canada)
IFPRI	International Food Policy Research Institute (USA)
IFS	International Foundation for Science (Sweden)
IGU	International Geographical Union
IHD	International Hydrological Decade
IHP	International Hydrological Programme
IIEP	International Institute for Educational Planning (UNESCO)
IITA	International Institute of Tropical Agriculture (Nigeria)
ILCA	International Livestock Centre for Africa (Ethiopia)
ILO	International Labour Organisation
ILRAD	International Laboratory for Research or Animal Diseases (Kenya)
IOBS/WPRS	International Organisation for Biological Control of Noxious Animals and Plants, West Paleartic Regional Section
IPPC	International Plant Protection Centre
IPPF	International Planned Parenthood Federation
IRCT	Institut de Recherches du Cotton et des Textiles (France)
IRRI	International Rice Research Institute (Phillipines)
ISNAR	International Service for National Agricultural Research (Netherlands)
ISSS	International Society of Soil Science

IUBS	International Union of Biological Sciences
IUCN	International Union for the Conservation of Nature and Natural Resources
IWRA	International Water Resources Association
JCO	Jordan Cooperative Organisation
JVA	Jordan Valley Authority
JVFA	Jordan Valley Farmers Association
MAB	Man and the Biosphere Programme (UNESCO)
NAS	National Academy of Sciences (USA)
NENARACA	Near East and North Africa Regional Credit Association (FAO)
NIAE	National Institute of Agricultural Engineering (Britain)
NORAD	Norwegian Agency for International Development
NRA	National Resources Authority (Jordan)
NRC	National Research Council (Sudan)
OAPEC	Organisation of Arab Petroleum Exporting Countries
ODA	Overseas Development Administration (Britain)
ODM	Ministry of Overseas Development (Britain)
OEPP	Organisation Européene et Mediterranéenne pour la Protection des Plantes
OHP	Operational Hydrological Programme
OIV	Office de l'Institut du Vin
OPEC	Organisation of Petroleum Exporting Countries
ORSTOM	Office de la Recherche Scientifique et Technique Outre-Mer
SCOPE	Scientific Committee on Problems of Environment
SGB	Sudan Gezira Board
SIDA	Swedish International Development Authority
SIL	International Association of Limnology
SSA	Soil Survey Administration
SSB	Sudanese Savings Bank
STABEX	Export Receipt Stabilisation Scheme under the Lomé Convention
TF	Textile Foundation (USA)
UATI	Union of International Engineering Organisations
UNDP	United Nations Development Programme
UNDP/TA	United Nations Development Programme/Technical Assistance
UNECA	United Nations Economic Commission for Africa

UNECWA	United Nations Economic Commission for Western Asia
UNEP	United Nations Environment Programme
UNESCO	United Nations Educational, Scientific and Cultural Organisation
UNFPA	United Nations Fund for Population Activities
UNIDO	United Nations Industrial Development Organisation
USDA	United States Department of Agriculture
USFGC	United States Feed Grains Council
WCARRD	World Conference on Agrarian Reform and Rural Development
WFP	World Food Programme
WHO	World Health Organisation
WMO	World Meteorological Organisation

INTRODUCTION

A. B. Zahlan

Developing countries have only recently appreciated the importance of agriculture to their societies. After independence, they had relegated it to a secondary role: progress, science and technology, and economic prosperity were then associated with industry. A growing awareness of the role of agriculture as a major employer, of increasing food imports, and of the high rate of advance in agricultural science and technology in industrial countries, has, however, induced a new realism and, with it, a new interest in this sector.

In the Arab world, 'food security' has become of topical concern, particularly because of the rising quantities of food imports. There can be no doubt that the enormous Arab resources—natural, financial and human—could be combined to achieve self-sufficiency in food production. The challenge is to find the way to convert these vast and tangible resources into agricultural produce. The complexity of the agricultural sector and the limited methodological planning capabilities are two of the more important factors that have constrained positive change.

The motivation of the Abdul Hameed Shoman Foundation in initiating this programme is to contribute to the emergence of new insights and a better understanding of the problems of the Third World of which the Arab states are a part. To meet this challenge involves much more than physical and financial resources: the place of the farmer in the organisation of the process of agricultural production has to be reassessed. In many Arab countries, including Jordan, traditional farming is no longer a viable economic or social enterprise.

During the past two decades, agricultural planning in the Arab world has evolved considerably. It has moved away from simplistic projects such as the construction of waterworks without drainage (that invariably led to salinity and soil erosion); and greater attention has been paid to ecological zones, crops, markets and fertilisers. Despite the great strides made to date, however, the widespread belief still exists that the nature of agricultural activity is such that the planner and the decision-maker can modify and shape it at will. The specificity of each plot of land and of each farmer as well as the effectiveness of existing institutional arrangements do not yet appear to be sufficiently significant factors in the planning process. Despite the advances made in planning, the methods remain traditional, both in conception and in spirit.

There is little doubt that prevailing methods have contributed towards the expansion of agricultural production. But the question is not whether present approaches can or cannot make additional contributions, but rather how to grasp those methods that could set the sector on a stable and vigorous course of development and transformation. It must be made clear that we do not subscribe to the view that there exists an easy way to achieve such a transformation. This research programme does not aim to offer simple recipes.

To move away from traditional agriculture and traditional forms of planning is easier said than done. The need for sophistication in the conceptualisation of alternative options, the need for a firm grasp of policy implications and for a working understanding of scientific progress, are essential ingredients to fuel a movement into the desired new era.

This monograph on Jordan's agricultural sector was prepared by a team of senior researchers in the country. Their papers reflect the overall perception and new craftsmanship necessary to pave the way for a transformation of the agricultural sector. In the following, I shall briefly present some of the notions behind this research programme.

Policy Research

Sound planning, decision-making and programme implementation call for considerable policy research and analysis. This type of research spans subjects ranging from the academic to the topical. Research into Ottoman land policies in Jordan and Palestine may be academic at one time and topical at another. Policy analysis relating to the price of wheat or tomatoes in the current season is, however, necessarily topical. Both academic and topical policy research are essential to the establishment of the necessary background for sound decision-making. An established tradition of policy research and forward planning is essential; without it, decisions are made hastily and without adequate analysis.

Policy analysis illuminates events, defines issues, provides early warning of impending crises, and supports the process of policy formation with data, information and alternative courses of action. But these advantages depend heavily on the experience and intellectual imagination of the policy researchers. In other words, the quality of the exercise depends on skills that have to be learned through participation in the decision-making process.

Before independence, the processes of planning and decision-making in the Third World were controlled by the dominant foreign powers, and these activities were carried out partly or wholly in the metropolis (London, Paris, and so on). Thus the institutions created in those days rarely possessed the capabilities to undertake the policy research necessary to support planning and decision-making. After independence, developing countries have been establishing independent capabilities in policy analysis and policy formation. In the interim period, many of these countries resorted to international organisations (for example, the United Nations Food and Agriculture Organisation (FAO), the World Bank, and others), foreign governments and foreign research institutions to meet these needs. Even today, some Arab countries still depend heavily on external agencies for their agricultural policy research.

The difficulty with this arrangement is that policy research is particularly sensitive to national, social, cultural and economic factors. The available data is generally poor, and analysts who are not part of the community and who do not live with its problems are severely handicapped in their advisory roles. Furthermore, the output of policy research is not merely a statement on a particular issue, but rather the experience gained by the analyst in performing the analysis. Unless the policy analyst is a functional element of the system, there is no accumulation of experience, and no collective and institutional learning. Intellectual and professional bonds between research analysts in various public and private institutions are created in the course of this type of research. These bonds are an integral part of the collective capability. Institutional development is a slow and difficult process and every opportunity must be taken to strengthen it.

Dependence on foreign institutions for this type of research has another negative aspect: most of the final reports remain of a classified nature and hence are unpublished in their entirety; many are only released for distribution in an abridged form. Thus the very community that should be aware of these studies sees only sterilised versions, if at all.

The antidote to this state of affairs is for each country to accumulate competence and expertise in policy research. Expertise can only be learned by doing and by the associated public debates on policy issues. Self-reliance in policy research does not presume autarkic behaviour: it is simply the means whereby a society lives in an open world in a manner that protects its national interests intelligently and rationally.

We live in a dynamic and rapidly changing world. Science and technology are constantly altering agricultural practices. It is, of course, possible to wait for advances to trickle in and to allow market forces to disseminate them. Some of the advances may be useless to local agriculture, others may have serious negative side-effects. An alert and wise society pursues a different course: it promotes an awareness of current advances and a critical attitude towards them. A wise, forward-looking public policy attempts to manage change; to encourage the monitoring of new practices; to protect threatened but valuable practices; and to promote realistic new visions of the future.

The formulation of such policies is obviously not a simple matter and calls for considerable experience among a large number of persons deeply rooted in their respective societies. One of the objectives of this research programme funded by the Abdul Hameed Shoman Foundation is to contribute to the development of such expertise in policy research.

Agricultural Systems

Over the past three decades, the agricultural sectors of the Arab countries have received substantial public support in the form of infrastructural and irrigation projects, and have undergone considerable mechanisation. Yet the growth rate of agricultural production has been low, and the changes in yield and labour productivity have been disappointing. The food balance varies from country to country. For the Arab world as a whole, it has been estimated that about 50 per cent of food consumed is being imported. The situation in Jordan is discussed in this volume. This dependence on imports will intensify as the population grows and as *per capita* incomes improve.

Extensive studies on different aspects of the agricultural sectors of developing countries have been undertaken in recent years. But the complexity and the system character of the sector have often been omitted from these studies, thus limiting their usefulness to policy formation. Existing methodologies suffer from a number of constraints. Some emphasise micro-aspects, others focus on macro-economic features.

The farmer, a principal element in the process of agricultural production, is generally relegated to the sidelines. Obviously both the policy-maker and the farmer play crucial roles in determining the overall level of activity. The full power and usefulness of agricultural policies can only be secured when these are firmly grounded in the farmer, for he is crucial to agricultural production. Waterworks, infrastructure, credit and markets do not by themselves create agricultural produce. The farmer's performance depends on the incentives and rewards that such inputs provide. The motivation of farmers varies from country to country, and from region to region within the same country. Each society has to discover its own dynamic optimal solutions. Policy research is a vital tool for the accomplishment of such a discovery.

In recent years, considerable progress has been made in the development of systems analysis and its application to the agricultural sector. Building on such experience, this research programme chose the construction of a realistic model of the system under study as the starting point for analysis. In order to be useful for analysis, a model has to be manageable and capable of lending itself to simulation. Special care was taken to verify that the model-building exercise did not distort the basic issues. Thus we have aimed to construct transparent and realistic models to facilitate verification and testing.

The few experiments on the application of systems analysis to developing countries have shown that the first phase of model construction is the most important. For this reason we have paid considerable attention to this aspect. In the first phase, our effort was concentrated on the organisation of information in a system structure capable of supporting the process of decision-making. The extent and quality of available empirical data were also examined, for an extensive data base is essential to the later phases of model building.

In the present volume, this effort is presented for discussion and evaluation. The comments and reactions to this first approximation will then be utilised to refine it, after which it will become the basis for the second phase of the research programme.

The Model

The methodology pursued is described in detail in Volume 4 of this series. The focii of the model are the farmer and the policy-maker who are linked by a vast and complex network of institutional, infrastructural, technological, economic and social relationship. An important feature of the model is the degree of explicit recognition given to the technology directly utilised by the farmer as opposed to the technology used in the construction of infrastructures.

The various sub-system studies emphasise the technological and institutional aspects underpinning the process of agricultural production. The farmer functions in an ecological environment, in an institutional setting and within a policy framework. Emphasis was thus placed on the package of inputs and the economic rewards involved in the process of agricultural production in Jordan. What the farmer actually produces is a function of the available natural resources (land and water), of the inputs (seeds, fertilisers, machinery), of farming practices and, of course, of the economic rewards generated by his efforts. The policy framework is generally managed by the appropriate public institutions—the Ministry of Agriculture, the Central Bank, the Ministry of Supply, the Ministry of Transport, and so on. On the policy-maker's side, attention was paid to the practices and policies relating to extension services, information, price control, marketing, credit, food balance and water.

In this first phase of model building, the goal was to document the relationships between the farmer and the policy-maker, and to analyse the farmer's output as a function of these factors. Moreover, since both the farmer and the policy-maker are members of a community, they are also supported and/or constrained by broader national, regional and international considerations. Accordingly, each author examined a sub-system according to a methodology that emphasised relationships. The notion of a characteristic farm was utilised to highlight the role of the farmer.

This is the first effort of its type in Jordan—indeed, in the Arab world. The team of researchers brought together a rich and multi-disciplinary experience. At a number of points in the research, information was found to be lacking. One of the objectives of this programme was to identify areas requiring further data collection. Agricultural research in Jordan is of recent origin, and a great deal of work still needs to be undertaken before the full potential of available resources is realised.

This volume also contains valuable information and insights into the practical operations of the Jordanian agricultural sector. The authors have pinpointed a number of essential services that are still lacking for the optimisation of invested labour and capital.

The Series
Each volume of this series is designed to stand on its own, and to be useful to a variety of readers. Yet each volume is also an element within a multi-phase programme. The present volume defines a conceptual model of Jordan's agricultural sector. It identifies a number of key policy issues and it reveals the practical problems facing the expansion of agricultural production.

Instructional Material
Advanced education and research call for special types of instructional material. In many developing countries, a shortage of such relevant literature exists. This shortage is usually met by utilising the textbooks and publications of industrial countries. Much of this literature is of little relevance to developing countries, where the conditions facing agriculture are quite different.

The preparation of topical and useful monographs on the agricultural sector in developing countries should therefore help to improve the standards and quality of education there. The present series has been prepared with this in mind.

The first three volumes, listed below, help the researcher to locate sources:

1 *Agricultural Bibliography of Jordan*
2 *Agricultural Bibliography of Sudan*
3 *Agricultural Bibliography of Syria*

The next three volumes are on methodology and the application of this methodology to two very different countries:

4 *Agriculture & Policy: Methodology for the Analysis of Developing Country Agricultural Sectors*
5 *The Agricultural Sector of Jordan: Policy and Systems Studies*
6 *The Agricultural Sector of Sudan: Policy and Systems Studies*

The final two volumes are the syntheses of the sub-system studies; they provide theory, analysis and applications:

7 *Agricultural Systems and Agricultural Policy in Jordan*
8 *Agricultural Systems and Agricultural Policy in Sudan.*

Visions of the Future
Policy research and systems analysis are useful tools to project visions of the future. One such vision might consider that Jordanian agriculture will always be on a small scale and have a limited impact on the rest of the economy. But when the complementarity of the various Jordanian resources are coherently examined, a different vision may be entertained.

A number of factors have contributed to the economic development of Jordan. A steady demand for manpower in the oil-producing countries has placed a high premium on skills and education. The expansion and improvements in the educational system, and the establishment of the two national universities and the Royal Scientific Society have meant that the population is better equipped to respond to this demand. The combination of formal education and work experience has contributed to the emergence of experienced, successful and competent manpower specialised in a broad range of technical and business fields. Contracting, trade, finance and information are some of the areas in which strong capabilities have emerged. The substantial remittances by Jordanians abroad have in turn made an important contribution to economic expansion.

This demand for manpower has absorbed much of the labour force, reducing the share of agriculture to some 10 per cent, or about 50,000 agricultural workers. Agriculture itself has expanded through mechanisation and the use of improved practices. Mining and industry have also experienced substantial development. The expansion of phosphate production and the establishment of potash and diamonium phosphate industries now enable Jordan to meet total fertiliser demand.

During the past two decades, a business community dedicated to the provision of inputs and the marketing of outputs has evolved. New farming technologies have been successfully introduced. The public sector has instituted important sets of facilities, policies and projects. By intensifying and expanding current developments, agriculture could make substantial progress.

It may appear to some that Jordan's water resources set an upper limit to the scope and future of agriculture. However, know-how in agriculture is itself an important and marketable resource. In the process of developing the technological infrastructure of the sector, it should be possible to utilise existing opportunities in the fertiliser field, in business and in contracting to initiate new and useful services. At the moment, technical and business services to assist the farmer in the scientific application of fertilisers are virtually non-existent throughout the Arab world. The marketing of fertilisers, the level of use and the economic returns to the farmer are all dependent on the existence of such technical services. The introduction of these services in Jordan could be utilised to provide the training and developmental opportunities prior to the export of such services on a regional scale.

Jordanian agriculture could also become a source of selected seeds as well as one of new practices; these could be marketed through local consulting and contracting firms which undertake agricultural projects. In other words, agriculture could become the laboratory in which practices and ideas are applied and tested to serve a technology export industry. The economic success and technical performance of the sector are obviously central to such an exercise. Given Jordan's manpower and institutional assets, such a transformation from the present transitional state of agriculture to one in the front line of quality and research is feasible and within reach. In attaining such a goal, the private and public sectors, farmers and policy-makers, researchers and businessmen all have major roles to play.

The feasibility of such a future depends on continued policy research, and on systematic efforts to encourage and stimulate inter-sectoral and entrepreneurial cooperation. The latent talent in Jordan, and the high demand throughout the region for food supplies and technical services in agricultural science, guarantee high economic returns to a country determined to embark on a sound programme of technological advancement.

Map of Jordan

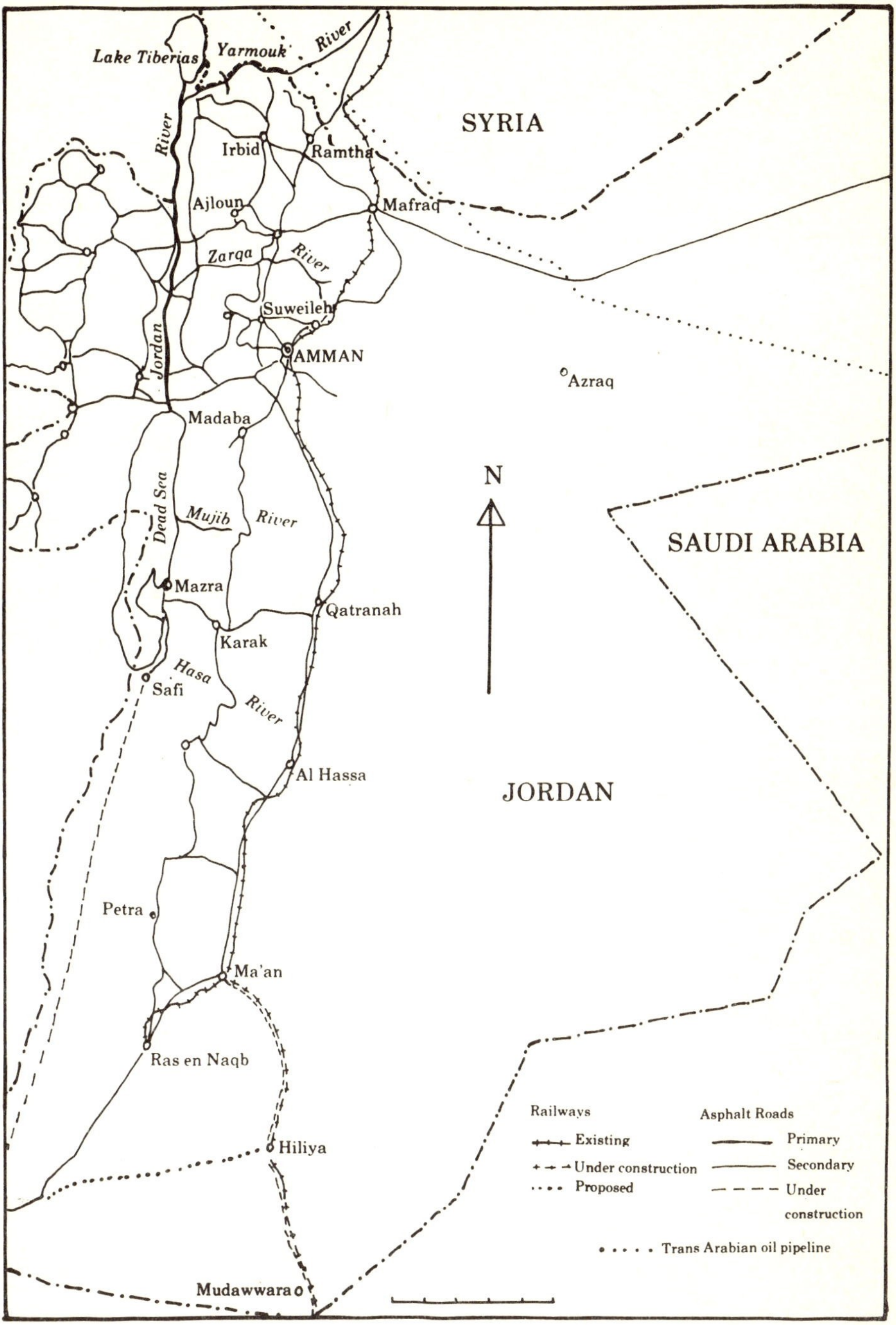

PART I

THE PLACE OF FOOD AND AGRICULTURE IN THE ECONOMY

1 THE JORDANIAN ECONOMY

Bassam Saket

Performance

Political Aspects

Jordan faced enormous problems following the creation of Israel in 1948, which resulted in Jordan's population trebling over a very short period of time with no corresponding measurable rise in material resources. Lines of communication and trade were cut off and the country had to find north-south routes instead of east-west ones almost overnight. At the same time, the Jordanian Government had few financial resources and its ability to raise taxes in sufficient amounts was extremely restricted; this resulted in an increasing dependence upon foreign financial aid.

After a difficult period in the 1950s, a calmer and more suitable climate for investment by Jordanians in the early 1960s allowed the country to proceed with planning. This was disrupted by another major setback in 1967, however, when Israel occupied the West Bank. Overnight, the country lost a large proportion of its resources, agricultural, industrial and historical. In addition large numbers of refugees fled their homes to the East Bank. Subsequent events had adverse effects on the country, and it took about seven years to return to normal. By 1973, however, Jordan was able to embark upon an ambitious Three-Year Development Plan, followed by two more Five-Year Development Plans in 1976 and 1981.

Perhaps the Palestinian question has been, and still is, the most important issue colouring the country's political life. It is also likely that the challenge posed by this issue has stimulated rapid economic and social development. Moreover, there has been little political activity in Jordan since it has not been possible to hold parliamentary elections. In compensation, a National Consultative Council was created in 1978 composed of 60 members appointed by the King; they include members from the East Bank as well as many Palestinians originally from the West Bank. The Council debates new bills and enacts laws.

The country's main targets during the 1980s—as for the past 30 years—are a just solution to the Palestinian problem, and rapid economic, social and regional development. Some analysts would add to that a just solution to the Iran-Iraq War.

Overall, Jordan enjoys a much-envied political stability, greatly enhanced by loyal armed forces and popular approval. Without these factors its economic achievements would not have been possible.

Economic Aspects

Jordan has witnessed a steep rise in the level of economic activity during the past

30 years, as shown by the figures presented in Tables 1.1, 1.2 and 1.3. During the early 1950s its prospects of economic development did not seem very promising, in view of the influx of large numbers of refugees and the meagre resources available for exploitation. In 1957 the World Bank did not expect real growth to exceed 4 per cent per annum—a figure far too low for a country with a *per capita* income of little more than $US 100 per annum. In fact, although no reliable data exist for deflating GNP figures, but considering a widespread belief in general price stability during the 1950s, it is reliably estimated that a real growth rate of 8 per cent or more took place from then until 1966. Growth rates fluctuated thereafter until 1972, reflecting the troubled political climate in the region. Normal healthy growth began again in 1973, however, with the resumption of development planning and the accompanying oil boom in the region.

Per capita income has increased dramatically over the past 30 years, rising from approximately $US 100 during the early 1950s to about $US 900 in 1979 and $US 1,900 in 1982. This was made possible by major improvements in all sectors of the economy, plus an advantageously large inflow of factor income from abroad.

This great improvement was accompanied by a huge increase in the volume of trade and payments with the outside world. In 1952 Jordan's imports amounted to no more than JD 17.3 mn and its exports to only JD 1.5 mn. These figures rose steadily until 1970, when growth began to exceed all expectations, so that in 1976 imports reached JD 338.7 mn and exports JD 68.7 mn. By 1982 imports had shot up to JD 1,142.5 mn and exports to JD 264.5 mn, with a resultant trade deficit of JD 878.8 mn.

Perhaps the most healthy aspect of the ever-increasing trade deficit over the last 30 years has been that Jordan has always been able to turn it into a payments surplus. Indeed, had it not been for this, the country would not have been able to accumulate reserves reaching JD 667.0 mn at the end of 1981, compared to JD 98.0 mn in 1970 and only JD 25.9 mn in 1956. This was made possible by large inflows of funds into the country in the form of remittances from Jordanians working abroad and foreign aid.

Major Factors Determining the Economy's Performance

Natural Resources

Water

Water resources consist primarily of rainfall which replenishes surface and underground water supplies. The amount of rainfall varies according to region, as follows:

Western Hills:	600–800 mm per annum
Jordan Valley:	150–250 mm per annum
Eastern Hills:	500–600 mm per annum
Desert:	less than 100 mm per annum

The rainy seasons runs from mid-October to April. The average annual volume of rain is estimated at over 6,000 mn m^3. Some water is of course lost through evaporation and some in the Dead Sea. The remainder lies underground as a reserve to be tapped through wells. Perhaps the main problem with rainfall, with regard to agriculture, is not that the volume is small, but that distribution is irregular during the season. A very large percentage of rain often falls during one month, leaving others relatively dry. Unfortunately, nothing can be done about this.

Table 1.1: National Accounts, Selected Indicators, 1954-82 (million JD)

	1954	1958	1964	1970[1]	1976	1979	1981	1982
GDP at factor cost, of which:	47.7	69.1	135.5	154.7	387.1	682.8	1,083.1	1,214.5
Agriculture	14.2	12.9	34.1	15.6	37.3	43.6	76.6	85.3
Manufacturing & mining	4.2	7.6	12.5	15.9	72.8	121.6	208.3	216.5
Construction	1.2	2.4	5.5	7.7	26.6	70.5	110.6	119.2
Transportation	4.4	9.0	12.0	14.3	32.5	62.9	102.7	120.4
Commerce & trade[2]	9.3	14.4	28.0	51.5	98.3	182.1	348.4	395.1
Public administration	9.1	15.6	19.7	42.5	81.7	129.1	191.2	225.6
GNP at market prices	52.4	77.1	160.6	187.0	571.1	935.5	1,524.7	1,711.6
Total spending on consumption & gross capital formation	65.0	102.0	181.0	233.6	677.4	1,252.1	1,969.0	2,145.9

Notes: [1] Figures up to 1964 include the West Bank; from 1970, they cover the East Bank only.
[2] From 1970 the figures include the wholesale and retail trade, restaurants, hotels, financing, real estate and business services.

Sources: CBJ b. various issues.
CBJ a. 1982.
Odeh 1972.

Table 1.2: Percentage Shares of Sectors in GDP, 1954-82

	1954	1958	1964	1970	1976	1979	1981	1982
GDP at factor cost (JD mn)	47.7	69.1	135.5	154.7	387.1	682.8	1,083.1	1,214.5
Sectoral shares (%)								
Agriculture	29.8	18.7	25.2	10.1	9.6	6.4	7.1	7.0
Manufacturing & mining	8.8	10.9	9.2	10.3	18.8	17.8	19.2	17.8
Construction	2.5	3.5	4.1	4.9	6.9	10.3	10.2	9.8
Transportation & communications	9.2	13.0	8.9	9.2	8.4	9.2	9.4	9.9
Commerce & trade	19.5	20.8	20.7	33.3	25.4	26.7	32.2	32.5
Public administration	19.1	22.6	14.5	27.5	21.1	18.9	17.7	18.6
Other	11.1	10.5	17.4	4.7	9.8	10.7	4.2	4.4

Ratios calculated from Table 1.1.

Table 1.3: Balance of Payments, Selected Indicators, 1952-82 (million JD)

	1952	1958	1964	1970	1976	1979	1981	1982
Balance of trade	—15.8	—30.5	—41.6	—53.7	—270.7	—468.2	—804.9	—878.0
Exports	1.5	3.5	8.7	12.2	68.8	121.0	242.6	264.5
Imports	17.3	34.0	50.3	65.9	339.2	589.2	1,047.5	1,142.5
Balance on current account	—2.1	—0.9	4.1	—5.9	17.3	—2.1	—13.1	—118.3
Monetary movements	—0.6	—3.6	—16.3	1.5	10.9	—61.2	—61.2	57.9
Selected indicators (net)								
Remittances	n.a.	n.a.	8.0	5.5	129.6	156.4	288.0	319.5
Government transfers	9.1	23.1	26.6	35.5	122.8	318.1	415.3	363.7
Travel	2.2[1]	1.2	4.4	—4.6	28.7	43.6	59.2	52.2

Note: [1] For 1954.
Sources: RSS 1977.
CBJ b. various issues.
Odeh 1972.

The 1981-85 Development Plan estimated water consumption in 1980 to be approximately 434 mn m^3, of which 81.3 per cent was for agricultural irrigation and the remainder for domestic and industrial consumption. The Plan also estimated the daily *per capita* water consumption for both domestic and industrial users at about 104 l, and the annual volume of water used in agriculture at 1,400 m^3 per du. Although it may seem that consumption is considerably less than supply, the major problem is one of geographical distribution.

Land

The greater part of Jordan is desert. The rainfall, as outlined above, divides Jordan into three basic areas with respect to cultivation: the eastern and southern deserts, the hills and the Jordan Valley. The Government has allocated most of the arable land suitable for farming in the Jordan Valley. Taking Jordan as a whole, most landholdings are medium or small. Data from the Department of Statistics (DOS) show that in 1975 there were numerous holdings of 3 ha or less: these accounted for only 7 per cent of the total area, but nearly 50 per cent of the total number of holdings. Large holdings of 200 ha or more were held by little over 100 people, consituting about 0.2 per cent of the total number of holdings, and about 10 per cent of total farm land. The problem of small holdings is now being partially overcome by renting—a phenomenon that is increasing rapidly.

Minerals

It has long been known that Jordan is rich in phosphates, but as the price was fairly low until the early 1970s, revenues from exports were not significant. In 1964 approximately 0.6 mn t were extracted, and exports brought in about JD 2.7 mn. The value of exports had risen to JD 4.5 mn by 1968, JD 19.2 mn in 1976 and about JD 57.2 mn in 1982, with production figures of 1.1 mn t, 1.7 mn t and 4.4 mn t respectively. Proven reserves were estimated at 300 mn t in 1972, rising to 1,200 mn t in 1975. There were also proven reserves of approximately 600 mn t at al-Shidiya.

At Feinan at Kherbit Nohas copper was found in sufficient commercial quantities to support a plant producing 3,000 t annually. The plant is expected to be built in the near future.

Perhaps the most important mineral concern has been the construction of the potash project on the Dead Sea. This has finally been completed at a cost of JD 143 mn.

The Natural Resources Authority (NRA) has undertaken aerial and radioactive surveys of the whole of Jordan. These surveys should prove vitally important to commercial prospecting for minerals in the country. The Authority is still prospecting for oil and has undertaken a re-evaluation of geological data, but to date, no oil in commercial quantities has been found. This fact has cast a shadow over continuous efforts to prospect for oil and other minerals which could prove quite valuable for the country. Studies have revealed reserves of oil shale of 1.3 bn t in the Lajoun area. Economic and feasibility studies are being undertaken with regard to its exploitation.

The Role of Government

Development Planning

The planning era started in 1961, when the Government initiated a seven-year programme of development from 1963 to 1970. This was modified in 1963 as a result of a sudden drop in financial aid. The new programme aimed primarily to reduce drastically the balance of trade deficit, to reduce dependence upon foreign aid, to raise GDP by 7 per cent per annum, and to reduce unemployment.

During this period, much was achieved in transforming productive capacity with the construction of the East Ghor Canal and a number of industrial projects, the most important of which was the expansion of phosphate production. Whilst great headway was made towards achieving targets, the June War of 1967 disrupted development and caused major economic and social difficulties.

In the early days after the War the Government was not in a position to resume planning, and domestic security problems compounded the problem. In 1972, however, the Government prepared the new 1973-75 Development Plan. This Plan was similar to the previous seven-year programme. It aimed to create 70,000 new jobs, raise GDP at a real annual rate of 8 per cent, increase dependence on domestic revenues, lower the rate of increase of the trade deficit and spread development to all regions of the country, especially rural areas. To achieve these ends, the Plan envisaged a total investment figure of JD 179 mn, of which public sector expenditure was to account for about 56 per cent.

Although not all targets were met as planned, great headway was made, and the achievements formed an excellent start. Naturally, the Plan contained a formidable number of projects to be carried out as a means of attaining its targets. Some projects began immediately, but others had to be carried over into the 1976-80 Plan.

This 1976-80 Five-Year Plan had similar targets to those of its predecessor, although it was frequently far more ambitious. To attain its goals the Plan envisaged an expenditure of about JD765 mn as constant 1975 prices, spread over the period of the Plan. In practice this was exceeded by over 10 per cent. However, the other targets (in, for example, GDP, domestic revenues and the balance of trade) were not achieved, although there was considerable progress towards meeting the first. In contrast, the last target was a total failure; whereas the Plan had envisaged a reduction of the trade deficit from JD 184 mn in 1975 to JD 131 mn in 1980 (at constant 1975 prices), the actual figure was JD 290 mn in 1980 (also at constant 1975 prices).

We should stress that this does not mean that the Plan failed. On the contrary, it succeeded in establishing a whole range of policy measures and organisational arrangements which went a long way to ensuring at least partial success for the programme.

The latest 1981-85 Plan is again a continuation of the previous one, for it has adopted similar targets and aims. It emphasises the following policies:

1. Strengthening the free economy while allowing for mixed participation
2. Achieving better policy coordination between different branches of government
3. Adopting policies to achieve a minimum level of income
4. Working to find new energy and water resources
5. Promoting the welfare state
6. Expanding and regulating the labour market
7. Phasing out all supply and petroleum subsidies
8. Promoting Arab cooperation

To implement these policies, a total of JD 2,800 mn will be spent during the period 1981-85, of which 53.4 per cent will be devoted to the productive sectors, including tourism. Spending will be shared equally between the private and public sectors, the latter to rely heavily upon grants and external borrowing.

It should be mentioned in passing that planning institutions have expanded over the years. Planning was initially in the hands of the Jordan Development Board (JDB), which has been restructured and renamed the National Planning Council (NPC). The Council is now responsible for drawing up the major plan targets in cooperation with the various ministries and agencies concerned with the identification of projects for adoption in the Plan. It is also the Council's responsiblity to coordinate the Government's role in executing the Plan.

The Budget and Fiscal Policy

Perhaps the main feature of the data presented in Tables 1.4 and 1.5 pertaining to government finances is the relatively low domestic participation in the revenues and expenditures of the Jordanian Government. It can be argued that domestic participation has in fact grown rather haphazardly during the past years from a low of 35.6 per cent in the fiscal year of 1958/59 to over 55 per cent in 1982, with occasional drops in the rate depending upon particular events. But the trend is clear: that of increasing domestic participation. There has been a much greater emphasis on raising revenue from indirect taxes than from direct taxes. The proportion of direct taxes raised to total revenues has grown only modestly from 5.2 per cent in 1952/53 to 8.5 per cent in 1982. It was only in 1982 that a new income tax law came into effect, replacing the antiquated system which was administered rather inefficiently and raised low revenues. The proportion of indirect taxes to total revenue rose substantially from 20 per cent in 1952/53 to 40 per cent in 1976 but fell to about 32 per cent in 1982. The reason for the high figure in 1976 was that grants fell from over JD 100 mn in 1975 to JD 66 mn, thus causing total revenues to fall slightly that year. With indirect tax revenues rising steadily, their proportion to total revenue seemed rather high at 40 per cent.

One important aspect of the Jordanian Government's revenues is a continuing reliance on foreign grants. In absolute terms, grants rose from below JD 8 mn in 1952/53 to over JD 218 mn in 1982. Despite this, the proportion of total spending financed by grants fell almost continuously from over 65 per cent in 1958/59 to 25 per cent in 1976 (an exceptional year), then rose to 40 per cent in 1979 following and increase in grants sanctioned by the Baghdad Summit, and finally fell to 31 per cent in 1982. While it is desirable for Jordan to continue to receive these grants, the Government should be aware of the unreliable nature of this mode of financing and its possible disruptive effect upon government programmes in years when the grants fall.

Looking at the ratio of revenues to GNP, we notice a healthy rise, the domestic revenue ratio rising from 14.1 per cent in 1958/59 to 21 per cent in 1982. The grants-to-GNP ratio fluctuated widely, reflecting only too well the volatile nature of these grants. The ratio in 1970 demonstrated the extent to which GNP could fall, following the loss of the West Bank's contribution.

On the expenditure side, defence spending requires special attention. Government budget figures indicate that, in proportion to GNP, defence spending stood at 13.4 per cent in 1964/65, peaked at 23 per cent in 1968/69, then fell steadily to 9.1 per cent in 1982. Again, this highlights the extent of the loss of the West Bank and the consequent need to build up the Army. Even when defence spending is measured against total spending, the picture is remarkable: 67 per cent in 1952/53, falling to 49.5 per cent in 1964/65, and to 21.5 per cent in 1982. It should be pointed out that the defence spending figures would be higher in the latter year if physical military assistance were included.

Despite the rise in defence spending, the Government was still able to maintain a capital spending programme after the June War of 1967. The proportion of capital spending to total spending rose steadily from 12.7 per cent in 1952/53 to 20.0 per cent in 1964/65 and to 27 per cent in 1970, to stabilise at 37 per cent in 1982. In relation to GNP, the story is much the same: one of continuous growth. The main reason for this is the increasing size of grants.

The ratio of total government spending to GNP rose after the June War of 1967. While the Government continues to favour a free economy, it still controls expenditures equal to almost half the GNP. This predominance of government

Table 1.4: Government Finances, 1952-82 (million JD)

	1952/3	1958/9	1964/5	1970	1976	1979	1981	1982[1]
Total revenues	13.5	30.6	46.2	68.2	193.7	435.8	598.5	648.1
Total domestic, of which:	5.4	10.9	23.8	30.7	107.6	187.9	309.2	359.2
Direct taxes, of which:	(0.7)	(1.0)	(2.0)	(2.5)	(11.4)	(27.8)	(48.8)	(54.8)
Income tax	n.a.	n.a.	(1.3)	(2.4)	(9.3)	(22.4)	(40.0)	(44.8)
Indirect taxes, of which:	(2.7)	(6.7)	(13.9)	(19.0)	(77.6)	(123.3)	(184.2)	(206.7)
Customs & excise	n.a.	n.a.	(8.6)	(14.2)	(49.1)	(82.6)	(111.0)	(132.0)
Total external	8.1	20.6	22.4	37.5	86.1	247.9	283.0	282.3
Grants	n.a.	(19.5)	(15.4)	(35.4)	(66.2)	(210.3)	(206.3)	(218.1)
Borrowing	n.a.	(1.1)	(7.0)	(2.1)	(19.9)	(37.6)	(75.7)	(64.2)
Total expenditures	13.4	29.3	43.6	80.7	262.5	515.7	647.1	722.3
Current	(11.7)	(23.5)	(34.9)	(59.0)	(185.9)	(321.3)	(391.5)	(455.3)
Defence	(9.0)	(13.5)	(21.6)	(38.2)	(93.3)	(114.9)	(138.0)	(155.2)
Capital	(1.7)	(5.8)	(8.7)	(21.7)	(76.6)	(194.4)	(255.6)	(267.0)
Deficit or surplus	0.1	1.3	2.6	–12.5	–68.8	–79.9	–48.6	–74.2

Note: [1] Preliminary.
Sources: CBJ b. various issues.
Odeh 1972.

Table 1.5: Government Finances, Some Ratios, 1952-82

	1952/3	1958/9	1964/5	1970	1976	1979	1981	1982
Total revenues (JD mn)	13.5	30.6	46.2	68.2	193.7	435.7	598.5	648.1
Total expenditures (JD mn)	13.4	29.3	43.6	80.7	262.5	515.7	647.1	722.3
Percentage rates of:								
Domestic revenues to total revenues	40.0	35.6	51.5	45.0	55.5	43.1	51.6	55.4
Direct taxes to total revenues	5.2	3.3	4.3	3.6	5.9	6.4	8.2	8.5
Indirect taxes to total revenues	20.0	21.9	30.1	27.9	40.0	28.3	30.8	31.9
Grants to total spending	n.a.	66.6	35.3	43.9	25.2	40.8	31.9	30.2
Defence spending to total spending	67.0	46.1	49.5	47.3	35.6	22.3	21.3	21.5
Current spending	46.2	46.4	68.2	52.0	57.9	58.5	78.9	78.9
Capital spending to total spending	12.7	19.8	20.0	26.9	29.2	37.7	39.5	37.0
Percentage ratio of domestic revenues to GNP	n.a.	14.1	14.8	16.4	18.8	20.1	20.3	21.0
Grants to GNP	n.a.	25.3	9.6	18.9	11.6	22.5	13.5	12.7
Total government revenue to GNP	n.a.	39.7	28.8	36.5	33.9	46.6	39.3	37.9
Current spending to GNP	n.a.	30.5	21.7	31.6	32.6	34.3	25.7	26.7
Defence spending to GNP	n.a.	17.5	13.4	20.4	16.3	12.3	9.1	9.1
Capital spending to GNP	n.a.	7.5	5.4	11.6	13.4	20.8	16.8	15.6
Total government spending to GNP	n.a.	38.0	27.1	43.1	46.0	55.1	42.4	42.2

spending has only been possible because of Jordan's continuing access to foreign grants.

Monetary Policy
The Jordan Currency Board (JCB) was established in 1949 as an issuing authority against sterling assets—mainly British government securities and treasury bills invested in London. This provided the currency in circulation with 100 per cent reserves in sterling assets. It is clear that under these conditions the Board was in no position to implement any monetary policy. The decision to establish the Central Bank of Jordan (CBJ) as a separate independent entity was taken in 1959; it started operations in 1964. Its functions were, and still are, the usual duties of the highest monetary authority in the land.

In addition to the CBJ, there were a few commercial banks in 1964, and a relatively small number of specialised credit institutions. Their combined assets amounted to no more than JD 81.5 mn by the end of that year, as shown in Table 1.6, increasing to JD 1,868.8 mn by the end of 1982. Table 1.7 provides the basic monetary data.

It is evident that the currency-holding behaviour of the public has a marked influence on the money supply. Up to 1972 this effect was contractionary, as security problems forced the public to hold more cash in notes, thus depriving the banks of deposits. We therefore see the ratio of currency with the public (MO to M2*) falling from 50.6 per cent in 1958 to 42.9 per cent in 1964, but rising to 63.8 per cent in 1970, then falling steadily to 45 per cent in 1976 and to 33.5 per cent in 1982. By 1972, as the public became prepared to give up cash holdings in favour of deposits, and the banks simultaneously embarked upon an expansionary policy, the money supply was all set to register very high growth rates; in some years these reached 28 per cent.

The CBJ called on other banks to restrain their lending activities and often raised their reserve ratios. As the banks were fairly liquid, these calls did not have a restraining effect, and the growth of bank credit and consequent growth of money supply gave rise to higher inflation by 1974. Together with the vulnerability of the Jordanian economy to the world economy, the maintainence of dinar-dollar parity in 1972 and the oil boom after 1973, this led to an inflation rate which peaked in 1974 at 20 per cent. It fell steadily to 15 per cent in 1976, and settled at about 7 per cent in 1982.

The growth in total credit extended by banks and specialised credit institutions contributed to the great expansion in domestic liquidity during the 1972-82 period. Up to 1970, in fact, there was little growth in total credit: between 1964 and 1970 it amounted to no more than 12 per cent per annum. Thereafter until 1976 the rate accelerated to well over 40 per cent annually, then fell steadily to 24 per cent in 1982.

With regard to the sectoral distribution of credit, the commerce and trade sector remains the highest recipient. This sector's share stood at 42.2 per cent at the end of 1964 and fell continuously to stand at 29 per cent in 1982. Rising in importance is the construction sector, whose share rose from a mere 5.3 per cent in 1964 to a high of 29.8 per cent in 1979, then fell to 22.7 per cent in 1982. The share of the industrial and mining sector rose by only four percentage points during the 20-year period, to stand at 14.3 per cent at the end of 1982. We might mention here that all development plans have called for a structural change in bank credit in favour of the commodity-producing sectors, and this has also been a constant aim of central government policy towards credit.

Other monetary developments include the setting up of the Amman Stock Exchange, which has proved quite a success and is attracting large funds from

* MO: Currency with the public.
* M2: Currency with public plus demand deposits plus quasi-money (time deposits).

Table 1.6: Monetary Survey of the Banking System, 1964-82 (million JD)

Year ending	1964	1970	1976	1979	1981	1982
Foreign assets	52.1	97.4	204.9	450.9	666.9	628.3
Claims on the private sector	27.9	43.6	177.2	444.9	691.1	843.9
Claims on government and semi-government	—	13.2	60.1	120.4	209.4	274.7
Claims on municipalities & public corporations	1.6	2.8	17.2	44.2	78.2	122.8
Assets = Liabilities	81.5	157.0	459.6	1,060.5	1,645.6	1,868.8
Money supply	39.8	105.5	263.6	472.7	701.7	787.5
Quasi-money	13.8	23.7	95.3	300.4	478.2	615.8
Government and semi-government deposits	16.5	11.2	24.2	68.2	116.5	101.9
Foreign liabilities	2.7	3.7	24.1	87.1	178.7	201.9
Others net	8.7	13.0	31.4	58.3	72.9	20.5
Capital and reserves	n.a.	n.a.	20.9	73.9	97.7	141.2

Source: Compiled by author from CBJ a. various issues 1964–82.

Table 1.7: The Monetary Sector, Selected Indicators, 1952-82 (million JD)

Year ending	1954	1958	1964	1970	1976	1979	1981	1982
Currency with public	10.0	13.1	23.0	82.4	161.5	275.4	412.3	469.9
Demand deposits	6.1	9.2	16.7	23.0	102.1	190.2	289.3	317.5
Money supply (M1)	16.1	22.3	39.8	105.5	263.6	465.6	701.6	787.5
Quasi-money	n.a.	3.6	13.8	23.7	95.3	298.3	478.2	615.8
Money supply (M2)	n.a.	25.9	53.6	129.2	358.9	763.9	1,179.9	1,403.3
Currency with public as % of M2	n.a.	50.6	42.9	63.8	45.0	36.1	34.9	33.5
Total credit outstanding[1]	5.8	11.1	34.1	59.6	237.6	514.8	792.8	982.9
Commerce & trade	n.a.	n.a.	14.4	16.9	81.6	134.2	225.8	285.0
Industry & mining	n.a.	n.a.	3.5	7.0	30.0	76.2	112.7	140.1
Agriculture	n.a.	n.a.	5.7	7.3	18.0	36.9	47.3	63.7
Construction	n.a.	n.a.	1.8	11.2	56.5	153.2	205.8	222.8
Transport services	n.a.	n.a.	2.6	7.8	11.1	13.0	23.3	32.9
Gold and foreign exchange reserves	25.9[2]	30.8	55.6	98.1	205.1	451.1	667.0	628.5

Notes: [1] Includes all banks and specialized credit institutions.
[2] This figure is for 1956.

Sources: CBJ b. various issues.
Odeh 1972.

Jordanians for investment in stocks and shares. The CBJ has also licensed several secondary financial institutions which provide investment opportunities for the public.

Other Policies
Besides assuming development planning, economic and civil service functions, the Government has also taken upon itself major responsibility in areas where it felt there had been market failure, in the strictest sense of the word. By this we mean that through its budget and its various agencies and ministries, the Government has gone into the market-place to uphold the prices of certain commodities, and in many cases to compete with the private sector.

As a start, the Government supported prices of wheat and sugar by means of subsidies, and this support assumed considerable importance in 1974/75. Its action followed sharp price rises on the world market for thse products. The Government was able to control prices by monopolising the importation of wheat, sugar and other items through the Ministry of Supply (MOS). The Government then sold flour directly to bakers and fixed the price of bread, allowing a certain mark-up to the bakeries. This subsidy was followed by another on petroleum products, which were then sold domestically at lower than world prices. This constituted a subsidy to both consumers and producers: to consumers in the sense that the Government allowed the public to continue to burn energy at prices lower than those prevailing in the world market; and to producers in the sense that fuel oil as an input was priced below world levels. A substantial part of these subsidies was not reported in the budget because of accounting difficulties.

Population and Manpower

Unfortunately we are limited in this section by the lack of reliable demographic data for Jordan. A census was carried out in 1961 by the DOS on both the East and West Banks, but the next one did not take place until 1979, and then on the East Bank only. Therefore any figures pertaining to years other than 1961 and 1979 must be treated with caution since they are essentially only estimates.

Demographic Characteristics
Because of the unreliability of data except for the census periods, we have limited ourselves to discussing those only. Table 1.8 provides valuable data on population trends for the East Bank, which will facilitate comparisons.

The population of the East Bank stood at 0.9 mn in 1961, compared to about 0.6 mn in 1953 and 2.15 mn in 1979. Reliable estimates put the total number of people who fled the West Bank during 1967/68 at no more than 300,000, and these were naturally incorporated into the 1979 figure.

We can easily gauge the extent of migration from rural to urban areas. The proportion of the rural population to total population (East Bank) was about 49.2 per cent in 1961, falling to 39.6 per cent in 1979. It is clear that a considerable rural-urban migration took place during that period, the motivations obviously being employment and better living conditions.

As regards distribution by sex, the pattern remains much the same, with a 4.5 per cent edge of males over females. The regional distribution shows the Amman region as accommodating 55 per cent of the population in 1979, compared to 48.2 per cent in 1961. In fact, all governorates lost population to the Amman Governorate. This reinforces the fact that most of the country's major economic activity is centred upon the Amman-Zarqa region.

The age distribution pattern shows a shifting of the population below 15 years of age from 45.5 per cent of the total population in 1961 to 50.6 per cent in 1979. Consequently, the 15-64 age group was reduced from just over 50 per cent to 46.6 per cent in 1979. Table 1.9 exhibits the details for 1979.

Table 1.8: Population of the East Bank, Selected Indicators, 1953-79 (thousands)

	1953	1961	1979
Total East Bank population	0.588	0.901	2.153
Urban areas	n.a.	0.458	1.298
Rural areas	n.a.	0.443	0.854
Sex distribution			
Females	0.285	0.431	1.027
Males	0.303	0.469	1.125
Regional distribution			
Amman Governorate	n.a.	0.434	1.188
Irbid Governorate	n.a.	1.274	0.612
Balqa Governorate	n.a.	0.079	0.151
Karak Governorate	n.a.	0.067	0.126
Ma'an Governorate	n.a.	0.047	0.075
Age group (% of total)		100.0	100.0
0-14	n.a.	45.5	50.6
15-64	n.a.	50.7	46.6
65+	n.a.	3.8	2.8
Illiteracy rate (% of population)	n.a.	68.2	34.6
Urban areas	n.a.	53.9	25.4
Rural areas	n.a.	66.5	48.6

Sources: DOS 1979.
NPC 1980.

Table 1.9: Percentage Distribution of Jordanians by Age and Sex, 1979

	Males	Females	Total	Cumulative
0-4	19.3	19.2	19.3	19.3
5-9	17.7	17.4	17.6	36.9
10-14	15.0	14.7	14.9	51.8
15-19	11.4	10.5	11.0	62.8
20-24	7.0	7.6	7.2	70.0
25-29	4.7	5.5	5.1	75.1
30-34	4.8	4.9	4.8	79.9
35-39	4.5	4.5	4.5	84.4
40-44	3.9	4.3	4.1	88.5
45-49	3.2	3.1	3.1	91.6
50-54	2.6	2.2	2.4	94.0
55-59	1.7	1.8	1.8	95.8
60-64	1.5	1.4	1.4	97.2
65+	2.7	2.9	2.8	100.0
TOTAL	100.0	100.0	100.0	100.0

Source: DOS 1979.

Manpower
The estimates provided in the 1961 census show that only 24.2 per cent of the population were economically active. This was due both to the very high proportion of children and to the very low participation by females. If we take the male 15-64 age group, we find a participation rate of 86 per cent. This makes a total labour force in 1961 of 217,800, rising—according to the same source—to only 381,000 in 1975, and representing a participation rate of only 19.6 per cent. According to sample estimates of the 1979 census, we find a participation rate of 38.3 per cent in the 15+ age group, yielding a total labour force of 397,500 in 1979, and representing a rise of 4.3 per cent above the 1975 figure. In other words, we find a participation rate of only 18.5 per cent in 1979. The 1979 estimates are probably far more reliable.

We can summarise the following factors as causes for the low participation rate in the labour force:

1. Over 50 per cent of the total population are below 15 years of age.
2. Much employment is part-time and seasonal, which greatly reduces the significance of any quantitative estimates.
3. The situation is complicated by the problem of the Palestinian refugees. In 1953 a United Nations Relief and Works Agency for Palestine Refugees in the Near East (UNRWA) study showed 17,000 refugees working, of whom only 13.5 per cent were fully employed. The 1960 sample survey of Amman by contrast, showed an unemployment rate among the adult male refugee population of only 10 per cent. This is far more encouraging than previous data and it shows the substantial progress made at that time in the employment of refugees.
4. The fact that females tend not to work after marriage lowers the overall participation rate.
5. There is a large increase in the proportion of working-age males and females enrolled in schools and universities.

In comparing labour force distribution by sector, as shown in Table 1.10, we find that employment in agriculture fell drastically from about 72,975 in 1961, representing 33.5 per cent of the total labour force, to about 41,000 in 1979, or only 10.3 per cent. Employment in public administration, on the other hand rose dramatically from 13.2 per cent to 45.0 per cent. Other data comparisons must be treated with extreme caution, as here we encounter a most astonishing phenomenon—namely, that the industrial sector lost approximately 0.8 per cent of its share. To a certain extent, the same factors which caused a shift in the sectoral distribution of the industrial origin of GDP (see Table 1.2) also apply to the sectoral distribution of the labour force. This is why we recommend that the data presented above be treated with caution.

Census data for 1979 reveal that 40 per cent of the 15+ age group had elementary education, 32.7 per cent had preparatory education and 27.4 per cent had secondary education. These results are based on a sample survey from the census. The survey also indicated that of the 15+ age group, 34.6 per cent were illiterate, 14.4 per cent could read and write and 51.0 per cent had some qualifications.

The 1979 census also produced data on the percentage distribution of Jordanians in the 15+ age group employed in major occupations. About 53.3 per cent were production workers; 13 per cent were professionals, technical and related workers; 10.6 per cent worked in agriculture; 7.9 per cent were sales workers; clerical and service workers constituted 6.9 per cent each; and, finally, 1.4 per cent were administrators and managers.

Table 1.10: Sectoral Distribution of the East Bank Labour Force, 1961 and 1979 (per cent)

	1961[1]	1979[2]
Agriculture	33.5	10.3
Mining & manufacturing	10.2	9.4
Electricity, gas & water	0.4	1.3
Construction	10.2	13.6
Commerce	8.0	9.7
Transport & communications	3.5	8.3
Public administration	13.2	45.0
Other	21.0	—
Financial, insurance, etc.	—	2.4
TOTAL	100.0	100.0
TOTAL LABOUR FORCE (actual numbers)	217,833	397,500

Notes: 1 Results of census, 1961.
2 Figures are based on a sample survey of Jordanians in employment who are 15 years and over.

Science and Technology

It is generally accepted that science and technology play a vital role in a country's development, in the identification of its own particular problems and in setting out to solve them through R and D. A policy towards such activities is therefore necessary. In this section we are concerned with the manpower aspect of such activities—the capability, potential and priorities needed to satisfy and implement policy.

Data available indicate that Jordan's supply of scientists and engineers engaged in scientific and technological activities increased from 142 in 1976 to about 2,800, of whom 44 per cent were employed in the public sector, 12 per cent in the private sector, and the remainder by the universities and the Royal Scientific Society (RSS). In addition, 1,200 technicians were engaged in scientific and technological activities in 1980, as compared to 745 in 1976.

Jordan's supply of scientists, engineers and technicians is expanding rapidly; Ministry of Education (MOE) sources indicate that there were about 80,000 university students, both at home and abroad, during the 1981/82 academic year. This is quite a large pool to be drawn upon by policy-makers. Of course, only a fraction is available, and the students have to receive appropriate training to become useful in Jordan's science and technology effort.

The Migration of Jordanian Workers

The migration of Jordanian workers probably started during the late 1950s and early 1960s, and gathered pace after the June War of 1967. It gained considerable momentum in 1974, following the rise in real oil prices and the rush to spend on development by the Gulf states and other oil-exporting countries in the region.

Kuwait was probably the first country to receive large numbers of Jordanian workers, being one of the first to experience a shortage of labour in view of its desire for development. Others followed shortly, with Saudi Arabia becoming a very large recipient of Jordanian migrant workers.

It should be mentioned that the migration of Jordanian workers is a continuous movement, with its impetus provided by two factors: first, the migration of young men from the West Bank due to the harassment by the Israeli occupying authorities; and second, the coming-of-age of young Jordanians who receive their university degrees or training and apprenticeship in professional jobs.

The NPC estimated in 1980 that there were 240,000 Jordanian workers abroad out of a total Jordanian population of some 500,000 resident abroad.

Table 1.11: Volume of Remittances, 1961-82

Year	Volume (JD mn)
1961	5.25
1962	6.20
1963	6.17
1964	9.30
1965	9.10
1966	10.60
1967	6.60
1968	4.10
1969	6.92
1970	5.54
1971	4.97
1972	7.41
1973	14.70
1974	24.13
1975	53.25
1976	136.41
1977	154.75
1978	159.38
1979	180.42
1980	236.68
1981	340.89
1982	381.87

Sources: CBJ b. various issues.

The official figures for remittances show only those passing through the banking system and the amounts declared and classified as remittances by licensed money-changers. It is widely believed that the 'true' figure of total remittances is much higher than the official published one. Table 1.11 shows clearly the enormous growth in remittances during the two periods 1972-77 and 1980-81. During the early 1960s the volume was fairly low but it later increased slightly, only to fall during and after 1967. It then rose again rather hesitantly, but really took off after 1973. The rate of growth fell substantially after 1976, although absolute levels were maintained. The reasons for this spectacular growth are: first, the sheer number of migrant workers, which had been rising since 1972; second, the increase in oil revenues which enabled the oil-rich states to pay higher wages; and third, Jordan's political stability, which brought about economic improvements which encouraged money to flow in.

Perhaps the most striking effect of remittances was seen in the country's balance of payments. They helped to alleviate the effects of the chronic trade deficit and often caused a payments surplus. One can argue that had it not been for remittances, much of the dramatic rise in imports during the 1976-82 period would not have taken place, as it is generally believed that remittances financed a large amount of conspicuous consumption. We should add that after 1975 the volume of remittances exceeded Jordan's earnings from exports. Remittances have also had a dramatic effect on GNP. The ratio of remittances to GNP stood at 13.8 per cent in 1975, rising to 23.1 per cent in 1976. This had a marked effect on the money supply during 1976 and 1977, when M2 rose by 60 per cent and 24.6 per cent respectively. This happened in two ways. The first was via the demand for credit by businessmen who wanted to finance imports required for families receiving remittances. This was facilitated by the excess liquidity the banking system was enjoying at the time.* The second was through the exchange of foreign currency for Jordanian dinars. To some extent, the rise in the volume of remittances via the rise in credit and money adversely affected Jordan's inflation rate, although it was not the only cause.

A survey into the use of remittances has been carried out recently (RSS 1983). Results show that a significant part of remittances goes on family maintenance, with housing, food and children's education being the main areas of expenditure. The survey showed that property was the most favoured investment in Jordan. Further evidence as to the use of remittances comes from the large rise in non-resident accounts held by commercial banks: their share of total deposits rose from 5.4 per cent in 1975 to 10.6 per cent in 1978, and to 14.7 per cent at the end of 1982. It is possible that non-residents also maintain residents' accounts.

Foreign Workers in Jordan

Foreign workers began to appear in Jordan in 1973. We can summarise the causes of this new phenomenon of importing labour as follows:

1. Jordan has experienced a shortage of labour in certain sectors due to the high emigration of its own labour force (40 per cent of the total in 1980).
2. Two Development Plans were implemented after 1973.
3. Wage rates were higher in Jordan than in the home countries of these foreign workers, and Jordanian firms were looking for cheaper alternatives to local labour.

This does not mean that foreign workers displaced their Jordanian counterparts; Jordanian labour was simply not available in sufficient quantities in certain sectors, especially construction. This situation opened the door to workers from Egypt, Syria, the Sudan, and later South-East Asia, to flow into Jordan to fill the gap. In 1973 the total number of foreign workers amounted to only 376, rising to 18,785 by 1978, and to 93,402 by the end of 1981. The actual number of foreign workers may exceed this total, due to work permit evasion and consequent statistical distortion. (See Table 1.12.) This applies mainly to Arab workers.

Table 1.13 shows that Egyptians constituted approximately 73 per cent of all foreign workers in Jordan during 1981, the majority being production workers or unclassified. Other countries also contributing large numbers were India, Korea, Pakistan and The Philippines. Of the 93,402 workers there were 1,934 females, 60 per cent of whom worked in services.

Important social effects emanate from such a large movement of labour into a small economy. This is demonstrated in the inadaptability of these guest workers

* The dramatic rise in bank deposits during the period 1975-77 can be attributed to increased government spending on development projects, a general economic boom and a rise in remittances.

Table 1.12: Number of Work Permits Issued to Foreign Workers in Jordan, 1973-81

Year	Number
1973	376
1974	519
1975	803
1976	4,790
1977	7,778
1978	18,785
1979	26,415
1980	79,566
1981	93,402

Source: MOL.

Table 1.13: Foreign Workers in Jordan by Nationality, 1979-81

	1979	1980	1981
Arab countries, of which:	18,287	61,967	74,839
Egypt	16,522	55,544	67,796
Asian countries, of which:	5,836	13,954	15,560
India	248	3,737	5,218
Korea	1,427	3,681	2,114
Pakistan	1,206	2,385	1,440
European countries	2,068	3,430	2,637
Others	224	215	366
TOTAL	26,415	79,566	93,402

Source: MOL.

to Jordan's life and culture. It is also possible that these workers, who are by and large unskilled, will introduce some undesirable habits and practices. In addition, the workers need accommodation, which sometimes leads to large numbers living in cramped conditions. In economic terms, the workers represent a flow of remittances out of the country; in 1978 they remitted an estimated total of JD 20.0 mn, rising to JD 46.0 mn in 1980 and to JD 62.4 mn in 1982. However, the amounts remitted fall far below those remitted into Jordan by Jordanians working abroad.

International Economic Relations

Trade

The Government of Jordan has always adopted an open policy with regard to the economy and the world. This is best demonstrated by the sheer size of the foreign trade sector in relation to GNP. The ratio of exports and imports to GNP has risen continuously from about 35 per cent ion 1952 to approximately 42 per cent in 1970, and finally to 82.2 per cent in 1982. This shows how open the economy is and how vulnerable to outside shocks emanating from the world economy. This would have been attenuated had the Government attempted to shield the economy by erecting import barriers or by adjusting the exchange rate in accordance with the direction of shocks. This fact is best demonstrated by events during the early 1970s: when the world economy experienced high inflation rates and the US dollar was devalued; JD-$US parity was maintained. This had the effect of introducing the outside world's inflation into Jordan. Had the Jordanian dinar been revalued against the US dollar, this would not have happened. Of course, the argument against revaluation—that it would harm Jordan's competitive position—is quite pertinent, as is the view that devaluation might act to curb the growth of imports. This did not happen, however; despite the devaluation of the dinar, imports went on rising unabated.

The growth of imports has been overwhelming, and has taken place despite the fact that a large proportion of imports is subject to duties. Jordanian importers have become accustomed to this, however, and simply add duties to their factor costs; this increases the final price to consumers. The demand for certain imports seems to be quite inelastic, so that imports have been growing despite price increases. Table 1.14 shows how crude oil imports have become a major burden, in exceeding total domestic exports. All imports without exception rose substantially during the 1964-82 period. Jordan's food deficit has increased every year since 1964, when it was approximately JD 10.2 mn. In 1982 it stood at JD 151.6 mn. This rise is attributed to the growing demand for food and insufficient domestic production. Exports of phosphates rose substantially between 1979 and 1981, due to increases in volume. Manufacturing and chemicals exports are also rising fairly rapidly.

Table 1.15 shows that Iraq has replaced Saudi Arabia as Jordan's best customers; in 1982 Jordan's domestic exports to Iraq amounted to JD 66.6 mn's worth of goods, most of which were manufactured items. India has become an important market for Jordanian phosphate, along with traditional markets such as Romania, Poland and Yugoslavia. As oil is now the biggest item on Jordan's import bill, Saudi Arabia has become, since 1976, Jordan's major import partner, followed by the US—probably due to the purchase of new aircraft. Other important trading partners are West Germany, Japan, Italy and the UK.

Foreign Aid

Foreign aid has two aspects with respect to Jordan: budgetary grants and foreign borrowing. The former is direct aid to the economy, in effect a straight injection of resources in foreign exchange. It is remarkable that Jordan has been able to attract these funds, usually from sister Arab countries, and they have been invaluable. Their absence would have meant severe economic austerity at home, continuous balance of payments crises, and consequent heavy borrowing on foreign financial markets. It can be argued, of course, that a regime of no grants would have implied large cut-backs on much expenditure by both the Government and the private sector in order to save foreign exchange. Today, although the level of grants reaches well over JD 200 mn, it is still hardly sufficient to pay for Jordan's crude oil bill.

It is generally believed that Jordan will only resort to foreign borrowing to finance development projects. To a certain extent, the country is assured of a degree of flexibility and, at the same time, the ability to pay back these debts, without undue restraint on its foreign exchange reserves. Table 1.16 shows

Table 1.14: Exports and Imports, 1964-82 (million JD)

	1964	1970	1976	1979	1981	1982
Total exports, of which:	7.0	9.3	49.6	82.6	169.0	185.6
Food, live animals, beverages & tobacco	3.7	5.2	17.6	25.1	40.7	45.0
Phosphates	2.4	2.2	19.2	26.3	54.8	57.1
Chemicals	—	0.3	3.5	7.1	17.9	23.1
Manufactured goods	0.1	0.7	3.9	14.1	34.6	32.5
Total re-exports	1.7	2.9	19.2	38.4	73.6	78.9
Total imports, of which:	50.3	65.9	339.5	589.2	1,047.5	1,142.5
Food, live animals, beverages & tobacco	14.1	10.4	83.5	113.3	173.0	196.6
Crude oil	1.9	2.5	34.4	69.0	174.7	230.7
Chemicals	2.3	3.7	16.3	30.3	50.4	54.2
Manufactured goods	12.2	15.5	65.9	142.0	176.6	191.7
Electrical & non-electrical machinery	4.8	6.7	56.8	84.6	171.3	149.1
Transport equipment & spare parts	5.8	3.9	44.7	69.4	166.8	170.3
Balance of trade	—41.6	—53.7	—270.7	—468.2	804.9	—878.0

Sources: CBJ b. various issues.

Table 1.15: Exports and Imports, Major Trading Partners, 1964-82 (million JD)

	1964	1970	1976	1979	1981	1982
Total exports, of which:	7.0	9.3	49.6	82.6	169.0	185.6
Iraq	n.a.	1.3	2.3	12.7	63.5	66.6
Syria	n.a.	1.5	6.4	12.3	10.8	8.4
Saudi Arabia	n.a.	1.3	7.5	19.4	20.9	27.6
Romania	n.a.	n.a.	n.a.	2.5	10.6	13.9
Poland	n.a.	n.a.	n.a.	0.7	4.1	4.9
Yugoslavia	0.7	0.8	0.2	1.0	2.0	4.1
India	0.5	0.3	1.7	6.1	10.3	16.6
Japan	0.1	0.02	1.9	2.9	3.8	3.8
Total imports, of which:	50.3	65.9	339.5	589.5	1,047.5	1,142.5
Syria	n.a.	3.2	7.5	1.9	0.8	1.0
Lebanon	n.a.	4.4	7.3	11.4	12.5	10.4
Saudi Arabia	n.a.	2.5	34.5	69.5	175.8	233.5
West Germany	n.a.	5.9	51.9	68.2	111.6	104.6
Italy	n.a.	2.2	19.5	38.5	61.6	57.2
UK	7.9	8.8	23.7	45.3	57.5	52.6
France	n.a.	1.6	10.5	26.9	45.8	42.5
Spain	n.a.	0.2	4.7	17.5	27.1	20.1
US	7.9	7.4	31.1	44.3	166.7	144.3
Romania	n.a.	1.4	9.5	13.7	24.9	19.4
Japan	2.3	3.9	13.2	37.3	71.5	87.4

Sources: CBJ b. various issues.

Table 1.16: Jordan's Outstanding Public Debt, 1970-82 (million JD)

Year ending	1970	1976	1979	1981	1982
Internal public debt, held by:	14.3	89.3	150.4	231.7	278.2
Central Bank	10.0	38.8	46.7	129.1	175.2
Commercial banks	3.3	22.1	66.5	71.6	71.9
Other	1.0	28.5	37.2	31.0	31.0
Interest paid	n.a.	3.4	6.8	9.4	10.5
External public debt, held by:	41.8	132.6	306.5	528.5	602.1
Foreign Governments	31.5	104.2	194.8	344.7	425.1
International lending institutions	7.3	14.7	34.0	72.2	91.4
Foreign banks	2.1	11.0	74.5	96.0	61.6
Foreign commercial companies	0.8	2.7	3.1	15.7	24.0
Interest	0.3	2.6	36.2	82.2	107.8
TOTAL public debt outstanding	56.1	221.9	456.7	760.2	880.2

Sources: CBJ a. various issues.
CBJ b. various issues.

Jordan's public debt figures for 1970 to 1982. It has been only in recent years that the external debt has increased rapidly; it stood at JD 41.8 mn in 1970, compared to JD 602.1 mn at the end of 1982. Approximately 70 per cent of this debt is held by foreign governments, and 10 per cent by foreign banks. We might add that payment of principal and interest on this debt had risen to stand at JD 70.7 mn in 1982, a figure equal to 38 per cent of Jordan's domestic exports in the same year.

The Exchange Rate of the Dinar
Further to what we have already said about the exchange rate of the Jordanian dinar, the official policy has been one of building up confidence in the currency. The dinar was devalued once in 1972 in line with the US dollar, but since then it has been more or less stable in terms of other currencies. Changes in the rate of the dinar against other currencies have been more a reflection of changes in those currencies than in the dinar. In the present international climate of floating rates, the CBJ publishes a daily bulletin in which the parity of the dinar is fixed according to daily fluctuations on international markets.

The Role of the Private Sector
It is generally considered that Jordan's economy is based upon a healthy free enterprise system with a certain degreee of government involvement. The Government expects the private sector to take an active part in the development process by participating in the implementation of large-scale industrial projects. The Government sees its role *vis-a-vis* the private sector as one of regulating its excesses, and of entering the market-place when it sees fit to correct these excesses. It also views its role as one of providing the basic infrastructural services necessary for the smooth operation of both the private and public sectors.

The 1981-85 Five-Year Plan states that the Government has played a pioneering role, through its participation with the private sector in large-scale industrial projects and through providing incentives and an appropriate economic climate for private investment. The private sector proved its worth by exceeding the investment target set for it by the Plan: of the JD 765 mn planned investment, the private sector was to provide 50 per cent; it actually invested JD 497 mn, representing approximately 60 per cent of total investment expenditure. Overall, the private sector has shown the necessary degree of dynamism for initiating and managing major undertakings.

The major element in any private sector success is the provision of incentives. The Government cannot otherwise expect the private sector to achieve its aim of making money. To this end the Government has provided tax exemptions and other incentives for projects it deems 'economic and approved'. Although the procedure has been cumbersome on some occasions, the private sector and a number of foreign firms have managed to take advantage of these provisions.

But the private sector disagrees with the Government's restrictive supply policies. As mentioned earlier, the Government monopolises the imports of some food items and fixes the prices of others to protect the small consumer. Some private sector participants disagree with the monopoly of certain imports and strongly advocate competition with the MOS.

Sectoral Development of the Economy

The Commodity-Producing Sectors
Perhaps the most outstanding feature of the distribution of GDP by sector (see Table 1.2) is the predominance during the period 1954-82 of the service sector,

which accounted for at least 60 per cent. On the other hand, there has been a remarkable transfer of importance from the agricultural to the industrial sector. The former lost ground continuously, from just under 30 per cent of GDP in 1954 to approximately 7 per cent in 1982. Equally striking is the growth of the industrial and mining sectors from 8.8 per cent in 1954 to approximately 18 per cent in 1982, indicating the rapid progress achieved in the programme of industrialisation.

Agriculture

The main variable in the agricultural sector continues to be rainfall. As mentioned earlier, the amount and distribution of rainfall vary considerably from year to year. Most affected by such fluctuations are the field crops grown in the dryland area. For example, during 1973/74 there was excellent rainfall, but in 1979 rainfall was insufficient. The consequences were striking: wheat production was at its highest (for the East Bank) in 1974 at 244,500 t, compared to 16,500 t in 1979. (See Table 1.17.)

The Jordan Valley has now become the most important irrigated area, with the help of the East Ghor Canal, especially since 1972. Production is focused upon vegetables and citrus trees. Modern methods of irrigation have increased productivity and many products are destined for export.

There are several reasons why fluctuations in agricultural production have not had any major effect upon the economy. To begin with, the share of agriculture in GDP has been falling, as is the case in most other developing countries. Furthermore, the ready availability of food aid, especially wheat and flour, has allowed supplies to be kept at normal levels. When aid was not forthcoming, Jordan was able to buy on the foreign markets by drawing on reserves and because of the high inflow of foreign currencies. Finally, fluctuations have not been of any great importance to the economy because agriculture is not a major supplier of raw materials to manufacturing.

We should mention that the cooperative movement, which started in 1952, was at first effectively limited to credit and savings functions. After 1975 it began to provide a variety of inputs and services, often on subsidised terms: these include fertilisers, seeds, packing, materials and pesticides. The movement has also continued to provide low-credit loans of limited size, but has been fairly inactive as regards marketing.

Although the importance of the agricultural sector to GDP has fallen, value added has increased substantially since 1970. Moreover, the value of agricultural exports has risen steadily over the years, from only JD 5.2 mn in 1970 to reach JD 45.0 mn in 1982. This is mainly due to increasing production from the Jordan Valley.

Industry

The development of the industrial sector during the 1950s was handicapped by the lack of basic infrastructural services, and value added in the sector hardly increased. During the 1960s and up to 1967, the sector experienced a growth rate faster than that of GDP, in which its share rose to 12.6 per cent by 1966. During this period, most industries—soap, detergents, small metal workshops, grain mills, oil presses, clothing, footwear, wood workshops and so on—were producing for domestic consumption. The increase in production was due primarily to an increase in domestic demand and import substitution. The subsequent period, 1967-71, was really one of lost time, during which there was little development because of political events. By 1971 the value of industrial production in the East Bank was the same as that during 1966 for the entire country. After 1972 value added in the sector increased much faster than overall GDP. Phosphate prices rose dramatically but production expanded slowly, despite the strong emphasis laid upon phosphates in the 1975-80 Plan. During the period of this Plan, a new potash works and a fertiliser factory at Aqaba were started; at the same time, the cement plant and the refinery were expanded. The projects proceeded well, and fertiliser production began in 1982. Tables 1.18 provides some basic data on the production of major industries during the 1954-82 period.

To encourage investment in this sector, in 1972 the Government started to provide generous income tax and customs duty exemptions to 'approved economic projects' for a number of years. The exemption period was increased for projects located outside Amman. The Government also followed a protectionist policy with regard to several major projects: this took the form either of tariffs on imports or of prohibition. In some instances, where monopolies were allowed, they were subjected to price control.

It has been consistent government policy to take shares in large undertakings which have a significant impact on the economy. The share taken up by the Government is not the same in all companies, however, and there seems to be no rationale behind this.

With regard to industrial finance, an industrial development bank was set up during the 1960s to provide debt and equity credit to industrialists. The cost of debt credit has not been very concessionary, however, and it is often hardly less than the cost of credit obtained from commercial banks. The latter have increased their lending to industry in recent years, following directives from the CBJ.

Furthermore, government promotion of the industrial sector is evident in the recent establishment of industrial free zones in Zarqa, Amman, Aqaba and on the Syrian-Jordanian border. These zones enable industrialists to import intermediate goods for manufacturing and exporting without payment of duties.

To conclude our discussion of the commodity-producing sectors, we should like to point out that growth here is dramatic and well above that of any other sector, and employment has increased greatly. Investment in this sector rivals all others. We would add that it is vital for Jordan to maintain a strong construction industry, as it facilitates development in general.

Services

As mentioned earlier, the service sector continues to account for over 60 per cent of GDP at factor cost. In the service sector we include retail and wholesale trade, hotels, restaurants, financial services, public administration, defence and transportation. In particular, trade and banking now account for 32.5 per cent of GDP, compared to only 19.5 per cent in 1954. Public administration accounted for approximately 18.6 per cent in 1982 and for most of the period 1954-82, except between 1967 and 1972, when the share was 27.5 per cent, due to heavy defence spending.

Prospects for the Jordanian Economy

Overall Growth

During the past three decades the Jordanian economy has come a long way with regard to growth. Basically, growth in the national product has outspaced that in the domestic product due to the large net factor income from abroad. While this is desirable, more emphasis should be placed upon increasing the domestic product, and policies should be geared towards that end. It is the only way to guarantee a secure long-term future for Jordanians. The past flow of budgetary aid, together with net factor income from abroad, have provided Jordan with a pool of extra resources from which it can draw to finance investment. This process should be

Table 1.17: Agricultural Production, Principal Crops, 1954-82[1] (thousand tonnes)

	1954	1958	1964	1970	1974	1976	1979	1981	1982
Field crops									
Wheat	233.3	65.6	294.7	55.5	244.5	66.6	16.5	50.6	52.3
Barley	104.2	16.8	97.2	5.2	40.2	13.2	4.8	19.2	19.7
Vegetables									
Tomatoes	52.1	74.4	227.8	137.4	133.5	87.9	171.8	204.5	195.3
Eggplants	12.9	22.9	41.7	21.1	32.0	41.9	52.7	62.4	45.0
Cucumbers	125.8	77.3	64.3	6.8	16.0	13.5	22.0	43.7	59.0
Melons[2]	—	—	159.5	22.8	46.0	23.1	15.5	46.9	45.0
Fruit									
Olives	61.4	52.4	98.3	3.0	40.1	22.5	6.8	18.9	40.0
Grapes	52.0	42.6	76.9	6.4	17.8	13.7	23.0	25.8	26.8
Citrus	0.7	3.1	37.1	48.9	33.9	16.5	30.0	53.8	42.2

Notes: [1] Figures up to 1964 include the West Bank; from 1970 they cover the East Bank only.
[2] For 1954 and 1958 these were listed under cucumbers.

Sources: Odeh 1972.
CBJ b. various issues.

Table 1.18: Production of Principal Industries, 1954-82[1]

	1954	1958	1964	1970	1976	1979	1981	1982
Phosphates (mn t)	0.1	0.3	0.6	0.9	1.7	2.8	4.2	4.4
Cement (mn t)	0.1	0.1	0.3	0.4	0.6	0.6	1.0	0.8
Petroleum refining (mn t)	—	—	0.3	0.4	1.1	1.6	2.1	2.5
Electricity (000 mn KWH)	n.a.	n.a.		0.01	0.4	0.8	1.2	1.4
Iron (000 t)	—	—	—	24.1	62.4	81.0	134.9	192.0
Cigarettes (000 mn[2])	—	0.7	1.5	1.6	2.2	3.4	4.7	4.6
Spirits & alcohol (mn l)	—	0.5	1.3	2.2	6.3	7.2	9.0	9.5
Detergents (000 t)	—	1.7	3.0	1.7	5.0	10.6	18.8	15.2
Industrial production indices[3] (1975 = 100)	n.a.	n.a.	n.a.	54.8	125.0	190.7	226.8	273.0
TOTAL								
Pharmaceuticals	n.a.	n.a.	n.a.	n.a.	123.3	161.3	272.1	n.a.
Plastics	n.a.	n.a.	n.a.	n.a.	169.6	341.8	354.5	345.0
Paints	n.a.	n.a.	n.a.	n.a.	221.1	491.0	424.9	440.0

Notes: 1 Figures prior to 1964 include the West Bank; from 1970 they cover the East Bank only.
2 Entries for 1958–70 are in 000 t.
3 Prepared by the CBJ.

Sources: CBJ b. various issues.
Odeh 1972.

developed further and the country should make sure that these resources are not wasted.

We should emphasise that since 1973, when a process of development was implemented, there have been some remarkable achievements, despite the fact that not all expectations have been realised. This should not be seen as discouraging—on the contrary, greater efforts should be exerted towards implementing ambitious targets.

A major factor contributing to recent growth has been the country's relative stability, both political and economic. Without stability there can be no effective development. As mentioned earlier, the 1967—72 period was one of volatile conditions which we characterised as 'time lost' with regard to development.

Under stable conditions, and with continuing efforts by Jordanians, it seems probable that recent trends in development will continue, although perhaps at a lower rate. The following section should provide some insights into the problems which accompany development. Policy-makers would be well advised to take stock and look towards the future.

Areas of Concern

In this section we have surveyed the main determinants of development in the Jordanian economy. The achievements have often appeared to be truly remarkable. We are not suggesting the contrary; we are simply concerned here with describing areas which require special attention but which have escaped mention so far. Putting them together under one heading may prove a useful way of dealing with them. The areas to be discussed will be listed according to sector.

Agriculture

Research should go into methods of solving the problems of dryland farming due to fluctuations in rainfall. A coherent policy is required to alter the basic structure of agriculture with a view to devising new policies to increase production dramatically. Without this, food insecurity will prevail and this cannot be advantageous to the economy. The multiplicity of authorities dealing with agriculture is sometimes confusing. The fragmentation of landholdings often implies inefficient management, so mergers and take-overs should be encouraged. The exodus from rural Jordan should be reduced, and people induced back to the farms.

Water Resources

The demand for water supplies is increasing rapidly as development accelerates. There are three sources of demand: domestic, agricultural and industrial. New policies should take into acount all resources and uses. The public ought to be made aware of the need to conserve water. Recycling could be an area with potential.

Energy Resources (or lack of them!)

As Jordan does not possess many traditional energy resources, conservation should be the main theme until alternatives are found. All possibilities for alternatives should be closely researched and ultimately developed: these include solar power, wind power and oil shale. Exploration for oil should continue while there is still hope. Existing traditional supplies should be diversified.

Industry

Concentration of activity in the Amman region has increased migration to the capital. There has been unregulated importing of inappropriate technologies. This takes place in the absence of R & D suited to Jordanian industry. The lack of efficient and experienced managers has been a cause of low productivity. Government relations with this sector should be simplified and the bureaucracy reduced. Government interference, taking the form of compulsory pricing and occasional protectionism, disrupts the efficient allocation of resources.

Public Finance
The chronic dependence on budgetary aid to finance government expenditure should be reduced. New sources of domestic revenues should be found. Alternatively, spending should simply be reduced, since the public sector has grown too big in recent years. While inefficiency is easily identifiable in government departments, bureaucracy is nevertheless on the increase. There should be more accountability by the Government with regard to public funds, and this area ought to be closely and continuously monitored.

Planning
There should be an important working link between planners, those who implement, and those who follow up each project approved and adopted in the Development Plans. The Plans themselves should effectively link projects with suitable finances.

The Trade Deficit
The severe inbalance between domestic production and domestic consumption should be reduced. The problem is that the economy has become geared to the high level of imports, accompanied by high net factor income from abroad and government transfers. This has contributed to a severe imbalance between the service sector and the commodity-producing sector.

Manpower
The need for trained manpower is manifested by the bottlenecks created following the implementation of the Development Plans. With a low participation rate in the labour force, and the emigration of many skilled and unskilled workers to neighbouring oil-rich states, the shortage of skilled as well as unskilled labour is aggravated. For these reasons, there is a need to strengthen the vocational training schools and enhance their link with development needs. The education system at secondary level can be geared towards providing a choice between an academic higher education or a vocational one. Furthermore, a programme of close cooperation between vocational schools and factories should be developed. Training should also take into account the technology-oriented society in which we are living.

References

CBJ (Central Bank of Jordan), a. *Annual Reports*, various issues, Amman.

—— b. *Monthly Statistical Bulletins*, various issues, Amman.

DOS (Department of Statistics), 1979, *Statistical Yearbook*, Amman.

MOL (Ministry of Labour), *Bulletins*, Amman.

NPC (National Planning Council), 1980, *Study on the Labour Force*, Amman.

—— 1981, *Manpower Study*, Amman.

Odeh, H. 1972, *Economic Development of Jordan*, National Planning Council.

RSS (Royal Scientific Society), 1977, *Numerical Index on Foreign Trade*, Amman.

—— Economic Department, 1983, *Workers' Migration Abroad: Socio-Economic Implications for Households in Jordan*, Amman.

2 THE DEMAND FOR AGRICULTURAL PRODUCTS

Adeeb K. Haddad

Introduction*

The main objective of this chapter is to analyse Jordan's demand for agricultural products of both domestic and foreign origin. The chapter also stresses the extent to which the rising demand for agricultural products is met by domestic production.

The chapter is divided into two parts. In the first part the basic features of Jordanian agricultural external trade are presented as a background to the analysis of demand for agricultural products. The emphasis is placed on the development of agricultural imports, exports and deficit between 1973 and 1981.

The second part deals with major aspects related to the domestic demand for agricultural products. It begins by defining and examining the behaviour of those factors which affect domestic demand for these products. An analysis of the main characteristics of the commodity composition and geographical distribution of domestic consumption throughout the Kingdom is also presented. The section concludes with an examination of the role of international demand in Jordan's agricultural trade and consumption. Finally, a summary of findings is given in the conclusion.

I. External Trade in Agricultural Products

The main task here is to analyse and reveal in detail the major components of the agricultural balance of trade in Jordan over the period 1973-81 as a background to the analysis of demand for agricultural products.

Table 2.1: Development of Agricultural Imports, 1973-81 (million JD)

Year	Agricultural imports	Total imports	Agricultural imports to total imports (%)
1973	30.81	108.20	28.5
1974	42.74	156.50	27.3
1975	49.42	234.01	21.1
1976	81.37	339.53	24.0
1977	75.92	454.41	16.7
1978	87.56	458.82	19.1
1979	108.28	589.52	18.4
1980	118.78	710.97	16.6
1981	167.93	1,047.50	16.0

Source: CBJ b. 1973–81, various issues.

* I am grateful to Mr Mohamad Mougut and Mr Hamad Kasasbeh for their assistance, especially their help in preparing the data needed for this study. Special thanks are also extended to Miss Etikal Arabiat and Miss Amal Sweiss for their meticulous typing.

Agricultural Imports

Jordan's agricultural imports have increased in value over the period 1973-1981, with the exception of 1977 (Table 2.1) Agricultural imports increased from JD 30.8 mn in 1973 to JD 81.4 mn in 1979, but fell to JD 75.9 mn in 1977. They then made a considerable jump in 1978 to reach JD 87.6 mn and continued to rise at a faster rate, reaching a peak of JD 167.9 mn in 1981. The table also shows that average agricultural imports amounted to more than one fifth of Jordan's total imports during the period 1973-81. Although the value of total agricultural imports increased, their percentage of total national imports declined over the same period.

The commodity composition of the value of agricultural imports is presented in Table 2.2. Imports of cereals ranked highest in agricultural imports, showing an increase from JD 10.5 mn in 1973 to JD 37.6 mn in 1980, but fell to JD 35 mn in 1981. The relative importance of cereals in total agricultural imports declined from 34.1 per cent in 1973 to 31.7 per cent in 1980 and to 20.9 per cent in 1981, as shown in Table 2.3. Wheat and flour accounted for most of the total cereal imports, which jumped from JD 1 mn in 1973 to JD 23.4 mn in 1981.

The second most important item was imports of fruit and citrus, which increased from JD 3.3 mn in 1973 to reach a peak of JD 12.5 mn in 1980, but declined in 1981 to reach JD 9.1 mn. The relative share of this category of imports to total agricultural imports decreased from 10.7 per cent in 1973 to 5.4 per cent in 1981.

As for milk and related products, imports increased from JD 3.3 mn in 1973 to JD 12.5 mn in 1979. It should be noted that imports of these items registered a decline from JD 10.7 in 1980 to JD 8.6 mn in 1981. Consequently, their relative share in total agricultural imports decreased from 10.8 per cent in 1979 to 5.2 per cent in 1981.

Jordan is basically a vegetable-producing country, exporting its surplus to nearby Arab states; its vegetable imports are limited to potatoes, onions and other vegetables where domestic production cannot meet local needs. Jordanian imports of vegetables amounted to JD 7 mn in 1981 as compared to JD 2.5 mn in 1973. The relative importance of these vegetables to total agricultural imports fell from 8.3 per cent in 1973 to 4.2 per cent in 1981.

As for meat imports, these have shown a noticeable rise over the last three years. While imports amounted to JD 9.4 mn in 1979, they increased to reach a peak of JD 17.1 mn in 1981 as compared to only JD 0.8 in 1973. The value of sugar imports, on the other hand, has shown irregular fluctuations from one year to another. From JD 1.8 mn in 1973 it increased to JD 20.2 mn in 1976, but fell sharply during the period 1977-80 to stand at JD 8.2 mn at the end of 1980. In 1981 the value of sugar imports witnessed a considerable jump, reaching a peak of JD 21.4 mn, as compared to the previous peak of JD 20.2 mn in 1976. These developments, however, are a result of erratic variations in the international price of sugar during the period 1973-81. More details regarding the impact of international prices on cereals, sugar and meat imports are presented below.

The commodity composition of the volume of agricultural imports is given in Table 2.4. This shows that cereals ranked first in total agricultural imports, followed by fruit and citrus, and then sugar and milk. The volume of cereal imports increased from 131.3 thousand t in 1973 to 429.2 thousand t in 1979. While it declined sharply to 359.2 thousand t in 1980, it increased to 451.2 thousand t in 1981. Wheat and flour accounted for most cereal imports, increasing from 71.3 thousand t in 1973 to 248.0 thousand t in 1981.

Fruit and citrus imports increased from 101.5 thousand t in 1973 to 208.4 thousand t in 1977. It should be noted that imports of these items registered a noticeable decline over the next three years, falling to 168.2 thousand t in 1981.

Table 2.2: Commodity Composition of Agricultural Imports, 1973-81 (million JD)

Item	1973	1974	1975	1976	1977	1978	1979	1980	1981
Cereals	10.48	13.00	11.19	32.51	27.55	31.02	35.23	37.67	35.03
Wheat & flour	7.04	7.14	6.83	14.80	13.87	14.58	19.57	23.00	23.43
Barley	0.22	0.16	0.11	9.54	6.70	8.83	2.40	1.25	0.56
Lentils	—	—	—	—	—	—	—	—	—
Corn	1.25	0.84	1.75	3.63	4.23	4.15	4.92	6.10	5.44
Rice	1.40	4.28	1.87	4.45	2.75	3.46	5.92	6.11	5.41
Chick-peas	—	—	—	—	—	—	1.81	0.82	0.07
Fodder	0.57	0.58	0.63	—	—	—	0.61	0.39	0.12
Other									
Vegetables	2.53	2.49	1.96	3.00	4.13	4.47	6.60	6.35	7.01
Tomatoes	—	—	—	—	—	0.09	—	—	
Eggplants	—	—	—	—	—	—	—	—	—
Potatoes	0.84	0.92	1.50	1.96	2.57	1.51	1.88	2.65	3.81
Onions	0.13	0.29	0.19	0.78	0.46	0.10	1.22	1.13	1.42
Other	1.56	1.28	0.27	0.26	1.10	2.77	3.50	2.57	1.78
Fruit & citrus	3.31	5.26	6.58	10.26	9.38	9.56	11.65	12.48	9.11
Apples	1.41	0.10	0.14	0.20	1.00	0.13	4.79	4.06	3.48
Grapes	0.11	0.10	0.12	0.21	0.12	0.02	0.05	0.11	0.11
Citrus	1.45	3.55	4.29	5.72	6.53	4.84	4.54	4.15	3.18
Melons	—	—	—	0.29	0.20	0.36	1.03	0.47	0.85
Olives	—	—	—	—	—	0.03	0.19	0.32	0.15
Other	0.34	1.51	2.03	3.84	1.53	4.18	1.05	3.37	1.34
Sugar	1.78	6.56	11.53	20.20	6.56	8.37	9.94	8.17	21.40
Meat	0.76	1.90	1.96	2.17	3.15	9.62	9.44	9.89	17.14
Red meat	0.56	1.21	1.49	1.53	2.62	8.33	8.65	7.85	14.85
White meat	0.20	0.69	0.47	0.64	0.53	1.29	6.79	2.04	2.29
Fish	0.32	0.43	0.51	0.54	0.79	0.62	0.54	0.46	1.25
Milk & related products	3.31	3.88	4.82	6.45	7.93	8.23	12.55	10.73	8.60
Eggs	1.12	1.59	1.54	1.47	0.83	1.19	1.25	1.77	1.99
Nuts	0.98	0.70	0.85	1.23	0.69	0.81	1.38	1.68	1.87
Tea, coffee & cocoa	2.52	2.35	2.92	2.99	5.20	6.52	6.12	6.67	7.08
Other	3.70	4.58	5.59	0.56	9.71	7.16	13.58	22.92	57.45
TOTAL	30.81	42.74	49.42	81.38	75.92	87.57	108.28	118.79	167.93

Source: DOS a. 1973–81, various issues.

Table 2.3: Relative Importance of Value of Agricultural Imports, 1973-81 (per cent)

Item	1973	1974	1975	1976	1977	1978	1979	1980	1981
Cereals	34.1	30.4	22.6	39.9	36.3	35.4	32.5	31.7	20.9
Vegetables	8.2	5.8	4.0	3.7	5.4	5.1	6.0	5.3	4.2
Fruit & citrus	10.7	12.3	13.3	12.6	12.5	10.9	10.8	10.5	5.4
Sugar	5.8	15.4	23.3	24.8	8.6	9.6	9.2	6.8	12.7
Meat	2.5	4.4	4.0	2.7	4.1	11.0	8.7	8.3	10.2
Fish	1.0	1.0	1.0	0.7	1.0	0.7	0.5	0.4	0.7
Milk & related products	10.7	9.1	9.8	7.9	10.4	9.4	11.6	9.2	5.2
Eggs	3.6	3.7	3.1	1.8	1.2	1.4	1.2	1.5	1.2
Nuts	3.2	1.7	1.7	1.5	0.9	0.9	1.3	1.4	1.1
Tea, coffee & cocoa	8.2	5.5	5.9	3.7	6.8	7.4	5.7	5.6	4.2
Other	12.0	10.7	11.3	0.7	12.8	8.2	12.5	19.3	34.2
TOTAL	100.0	100.0	100.0	100.0	100.0	100.0	100.0	100.0	100.0

Source: Table 2.2.

Table 2.4: Commodity Composition of Agricultural Imports, 1973-81 (thousand tonnes)

Item	1973	1974	1975	1976	1977	1978	1979	1980	1981
Cereals, of which:	131.3	69.5	94.0	288.3	364.0	389.3	429.2	359.2	451.2
Wheat & flour	71.3	30.5	44.8	132.8	139.0	173.0	201.0	163.0	248.0
Barley	6.0	—	2.6	14.7	67.9	23.4	44.4	20.5	9.8
Lentils	—	—	—	—	—	—	—	—	—
Corn	37.8	15.6	34.3	69.8	86.4	97.1	126.0	109.4	135.0
Rice	16.1	23.3	11.4	69.8	68.4	95.8	34.6	47.3	36.9
Chick-peas	—	—	—	—	—	—	15.5	12.9	21.1
Fodder	0.1	0.1	0.9	1.2	2.3	—	7.7	6.1	0.4
Vegetables, of which:	48.7	42.6	34.5	58.2	44.4	48.5	83.7	64.1	65.0
Tomatoes	—	—	—	—	—	0.9	—	—	—
Eggplants	—	—	—	—	—	—	—	—	—
Potatoes	5.0	7.6	10.2	18.7	29.4	20.1	25.3	31.0	40.1
Onions	2.1	2.2	2.1	11.7	4.9	7.7	16.1	13.6	17.2
Fruit & citrus, of which:	101.5	150.8	168.7	201.5	208.4	181.8	189.3	181.1	168.2
Apples	9.2	14.1	3.6	2.0	1.1	1.8	23.5	25.4	27.3
Grapes	1.0	0.1	0.2	2.6	1.2	1.3	0.3	0.7	0.7
Citrus	77-2	98.8	136.8	160.3	175.7	135.9	143.5	125.8	117.4
Melons	—	—	—	0.3	0.6	0.5	13.7	4.6	8.5
Olives	—	—	—	—	—	1.3	0.7	1.0	0.4
Sugar	19.4	48.3	45.4	108.6	45.0	89.9	96.3	98.8	90.1
Meat	2.8	4.1	5.2	4.4	5.6	14.8	14.8	22.0	25.5
Red meat	0.5	2.0	2.0	2.7	4.3	11.5	12.8	17.8	20.7
White meat	2.3	2.1	3.2	1.7	1.3	3.3	2.0	4.2	4.8
Fish	2.3	2.3	2.9	2.5	3.3	2.3	2.3	9.7	6.7
Milk & related products	34.1	45.2	41.0	47.5	65.9	63.7	97.7	90.1	97.1
Eggs (per mn)	70.0	75.4	89.7	37.4	35.6	24.0	—	—	—
Nuts	1.8	9.1	9.5	0.7	0.8	1.2	1.5	1.6	4.0
Tea, coffee & cocoa	6.0	5.4	5.6	4.9	5.8	7.5	6.1	7.1	7.0

Source: DOS a. 1973-81, various issues.

Sugar imports varied considerably from one year to another. They increased from 19.4 thousand t in 1973 to, 48.3 thousand t in 1974 and subsequently continued to fluctuate, reaching 90.1 thousand t in 1981. By the same token, milk and related products increased from 34.1 thousand t in 1973 to 45.2 thousand t in 1974 and continued to fluctuate, reaching 97.1 thousand t in 1981. Imports of red and white meat, on the other hand, amounted to 2.8 thousand t in 1973 and continued to rise over the period under study, to reach 14.8, 22.0 and 25.5 thousand t in 1979, 1980 and 1981 respectively.

Table 2.5: Geographical Distribution of Agricultural Imports, 1981 (million JD)

Country		Value of imports	Relative importance (%)
Arab countries	Total	*16.87*	*10.1*
Lebanon		4.87	2.9
Egypt		3.44	2.0
Iraq		0.32	0.2
Syria		5.29	3.2
Saudi Arabia		2.16	1.3
Others		0.79	0.5
EEC	Total	*36.16*	*21.5*
France		14.06	8.4
Holland		6.07	3.6
Italy		5.39	3.2
UK		3.84	2.3
Others		6.80	4.0
Other European countries	Total	*8.14*	*4.8*
Greece		4.42	2.6
Spain		1.97	1.2
Others		1.75	1.0
US	Total	*25.19*	*15.0*
Comecon countries	Total	*27.93*	*16.6*
Bulgaria		13.44	8.0
Romania		10.83	6.4
Yugoslavia		1.85	1.1
Others		1.81	1.1
Turkey	Total	*9.55*	*5.7*
Brazil	Total	*7.62*	*4.5*
Other countries	Total	*36.47*	*21.7*
TOTAL		167.93	100.0

Source: DOS a. 1981.

An analysis of agricultural imports by major geographical areas and countries is given in Table 2.5. This shows that more than one fifth of agricultural imports came from European Economic Community (EEC) countries, accounting for JD 36.2 mn, or 21.5 per cent of total agricultural imports in 1981. Imports from Comecon (Council of Mutual Economic Assistance), countries ranked second and accounted for JD 27.9 mn, or 16.6 per cent. This was followed by the US and Arab countries, which accounted for 15 per cent and 10.1 per cent respectively.

It should also be noted that most of the agricultural imports from the US, which accounted for JD 25.2 mn, were for sugar and cereal imports, of which JD 9.8 mn went to sugar and JD 10.9 mn to cereals. Imports from France amounted to JD 14.1 mn, of which JD 7.7 mn went to refined sugar and JD 3.5 mn to wheat. Imports from Bulgaria and Romania reached JD 13.4 mn and JD 10.8 mn respectively, of which JD 11.7 mn and JD 10.1 mn went to meat. Jordan's imports from Turkey, on the other hand, amounted to JD 9.5 mn, of which JD 1.5 mn, JD 2.4 mn and JD 2.6 mn went to flour, meat and pine seeds, respectively. Total imports from Brazil amounted to JD 7.6 mn, of which JD 4.5 mn went to sugar imports.

Agricultural Exports

Jordan's agricultural exports depend primarily on the productivity of the agricultural sector which, due to unpredictable weather conditions, has been unstable over the years. In addition, they depend on the nature of external markets and the prevailing prices at those markets. Table 2.6 shows that agricultural exports witnessed a noticeable growth in their absolute values, whereas their growth rates showed a decline in particular years. Hence, agricultural exports jumped from JD 4.7 mn in 1973 to JD 10.2 mn in 1975 and JD 20.6 mn in 1977, but fell to JD 16.3 mn in 1978. They then jumped again to JD 21.2 mn in 1979 and continued to rise at a higher rate, reaching JD 34.6 mn in 1981.

Table 2.6: Development of Agricultural Exports, 1973-81 (million JD)

Year	Agricutural exports	Total national exports	Agricultural exports to total exports (%)
1973	4.65	14.01	33.2
1974	9.57	39.44	24.3
1975	10.20	40.07	25.5
1976	10.38	49.55	33.1
1977	20.64	60.25	34.3
1978	16.34	64.13	25.5
1979	21.24	82.55	25.7
1980	23.50	120.11	17.9
1981	34.59	169.03	20.2

Source: CBJ b. 1973–81, various issues.

Table 2.7 shows the commodity composition of agricultural exports during the period 1973—81. It reveals that vegetable exports ranked first in total agricultural exports, followed by fruit and citrus, and then lentils. The value of vegetable exports amounted to JD 2.5 mn in 1973 and continued its increase to reach JD 19.5 mn in 1981. It accounted for 57.2 per cent of the value of total agricultural exports in 1981. This relative share is slightly larger than the corresponding percentage in 1973, as shown in Table 2.8.

Table 2.7: Commodity Composition of Agricultural Exports, 1973-81 (million JD)

Item	1973	1974	1975	1976	1977	1978	1979	1980	1981
Cereals	0.48	0.54	0.55	0.24	0.37	0.03	0.45	0.40	0.66
Wheat & flour	—	—	0.05	—	—	—	—	—	—
Barley	—	—	—	—	—	—	—	—	—
Lentils	0.48	0.54	0.50	0.24	0.37	0.03	0.45	0.40	0.66
Corn	—	—	—	—	—	—	—	—	—
Rice	—	—	—	—	—	—	—	—	—
Chick-peas	—	—	—	—	—	—	—	—	—
Fodder	—	—	—	—	—	—	—	—	—
Vegetables	2.42	3.89	3.24	5.34	6.10	8.15	11.95	11.37	19.50
Tomatoes	0.92	1.84	1.28	2.52	2.48	3.70	5.46	5.40	6.67
Eggplants	0.40	0.51	0.48	0.59	0.89	1.11	1.53	1.45	1.61
Potatoes	—	—	—	—	—	—	—	0.04	0.12
Onions	0.07	0.03	0.04	0.10	0.02	0.15	0.14	0.24	0.39
Other	1.03	1.51	1.44	2.13	2.71	3.19	4.82	4.24	10.71
Fruit & citrus	1.57	4.16	6.26	9.07	10.25	6.32	6.63	7.65	9.98
Apples	—	—	—	0.03	—	0.16	—	—	0.04
Grapes	0.10	0.15	0.25	0.25	0.06	0.12	0.36	0.23	0.31
Citrus	1.41	3.73	5.65	8.40	1.00	5.70	5.88	5.95	8.74
Melons	0.03	0.09	0.05	—	—	—	—	—	—
Olives	0.02	0.01	0.03	—	—	—	—	—	—
Other	0.01	0.18	0.28	0.39	9.19	0.34	0.39	1.47	0.89
Sugar	—	—	—	—	—	—	—	—	—
Meat	—	—	—	—	—	—	—	—	—
Red meat	—	—	—	—	—	—	—	—	—
White meat	—	—	—	—	—	—	—	—	—
Fish	—	—	—	—	—	—	—	—	—
Milk & related products	0.06	0.12	0.12	0.16	0.15	0.23	0.16	0.18	0.26
Eggs	—	—	—	—	—	—	0.14	0.82	1.62
Nuts	—	—	—	—	—	—	—	—	—
Tea, coffee & cocoa	—	—	—	—	—	—	—	—	—
Other	0.07	0.86	0.03	1.57	3.77	1.60	1.91	3.08	2.07
TOTAL	4.65	9.57	10.20	16.38	20.64	16.33	21.24	23.50	34.59

Source: DOS a. 1973–81, various issues.

Table 2.8: Relative Importance of Value of Agricultural Exports (per cent)

Item	1973	1974	1975	1976	1977	1978	1979	1980	1981
Lentils	10.3	5.6	4.9	1.5	1.8	0.2	2.1	1.7	1.8
Vegetables	53.1	40.6	31.7	32.5	29.6	49.9	56.3	48.4	57.2
Fruit & citrus	33.8	43.5	61.4	55.4	49.7	38.7	31.2	32.5	29.3
Sugar	—	—	—	—	—	—	—	—	—
Meat	—	—	—	—	—	—	—	—	—
Fish	—	—	—	—	—	—	—	—	—
Milk & related products	1.3	1.3	1.2	1.0	0.7	1.4	0.8	0.8	0.8
Eggs	—	—	—	—	—	—	0.6	3.5	4.8
Nuts	—	—	—	—	—	—	—	—	—
Tea, coffee & cocoa	—	—	—	—	—	—	—	—	—
Other	1.5	9.0	0.3	9.6	18.2	9.8	9.0	13.1	6.1
TOTAL	100	100	100	100	100	100	100	100	100

Source: Table 2.7

Fruit and citrus, next in importance in agricultural exports, ranged between JD 1.6 mn in 1973 and JD 10.0 mn in 1981, with relative shares of about 33.8 per cent and 29.3 per cent respectively. Cereal exports were the third largest, averaging JD 0.41 mn per annum and varying between a low of JD 0.03 mn in 1978 and a high of JD 0.66 mn in 1981.

Table 2.9: Geographical Distribution of Agricultural Exports, 1981 (million JD)

Country	Agricultural exports	Relative importance (%)
Iraq	10.42	30.5
Syria	6.15	17.8
Kuwait	4.91	14.4
Lebanon	1.64	4.5
Saudi Arabia	7.20	21.1
Bahrain	0.25	0.5
Qatar	0.57	1.4
Others	3.45	9.8
Total	34.59	100

Source: DOS a. 1981.

The geographical distribution of agricultural exports is given in Table 2.9, which shows that Arab countries absorbed most of them. Iraq ranked first and accounted for JD 10.4 mn or 30.5 per cent, followed by Saudi Arabia, Syria and Kuwait, with relative shares of JD 7.2 mn or 21.1 per cent, JD 6.2 mn or 17.8 per cent, and JD 4.9 mn or 14.4 per cent, respectively.

Agricultural Balance of Trade

The external trade of agricultural products plays an important role in Jordan's economy. The statistical annexes indicate that both agricultural imports and exports have witnessed a noticeable improvement in terms of value and volume over the period 1973-81. Agricultural exports increased about eightfold in nominal value over the period, from JD 4.6 mn in 1973 to JD 34.6 mn in 1981, or at an average annual growth rate of 22.3 per cent. Agricultural imports rose by about six times during the same period, from JD 30.8 mn in 1973 to JD 167.9 mn in 1981, an increase of 445.1 per cent, representing an average annual growth rate of 23.6 per cent. Moreover, the ratio of exports to imports increased from 13.1 per cent in 1973 to 20.4 per cent in 1981. *Per capita* agricultural imports amounted to JD 73 in 1981, while *per capita* agricultural exports did not exceed JD 15 in the same year.

Although the growth of agricultural exports surpassed that of agricultural imports during the period 1973-81, Jordan's agricultural balance of trade suffered a rising deficit*. This deficit increased from JD 26.2 mn in 1973 to JD

* Agricultural deficit is measured by the net agricultural balance of trade, which represents total agricultural imports minus total agricultural exports. It is also measured by the difference between domestic agricultural consumption and domestic agricultural production.

133.3 mn in 1981, reflecting an increase of JD 107.1 mn or 408.7 per cent during the period, or an average annual growth rate of 22.6 per cent, as shown in Table 2.10. Thus the relative share of the agricultural deficit in the total trade deficit amounted to 31.3 per cent, 18.6 per cent and 16.6 per cent respectively for the years 1973, 1979 and 1981. However, due to the fact that the volume of the agricultural deficit is continuously increasing, we see that the *per capita* deficit is also growing at a noticeable rate, from JD 15.6 in 1973 to JD 40.5 in 1979, and to JD 57.6 in 1981.

An analysis of the commodity composition of the agricultural deficit is given in Table 2.11, which reveals that cereals ranked first and accounted for most of the deficit. This was followed by meat, milk and related products, coffee, tea and cocoa. While the vegetable balance of trade showed a surplus for the whole period, fruit and citrus showed a deficit for some years and a surplus for others.

Table 2.10: Agricultural Deficit and Trade Deficit, 1973-81

Year	Net agricultural deficit	Net trade deficit	Agricultural deficit to trade deficit (%)	Population (000)	*Per capita* deficit (JD)
1973	26.16	83.60	31.3	1,675.1	15.6
1974	33.17	105.90	31.4	1,735.0	19.1
1975	39.22	184.10	21.3	1,810.5	21.6
1976	65.00	270.00	24.1	1,889.3	34.4
1977	55.28	371.10	14.9	1,971.6	28.1
1978	71.24	368.10	19.3	2,057.5	34.6
1979	87.04	467.40	18.6	2,147.9	40.5
1980	95.29	543.40	17.5	2,233.3	42.7
1981	133.34	803.70	16.6	2,322.3	57.7

Sources: Tables 2.1 and 2.6.

More specifically, the agricultural deficit for cereals increased from JD 10.0 mn in 1973 to JD 34.4 mn in 1981. Its relative share in the total agricultural deficit declined from 38.2 per cent in 1973 to 25.7 per cent in 1981, as shown in Table 2.12. The agricultural deficit of red and white meat together rose from JD 0.8 mn in 1973 to JD 17.1 mn in 1981, or by 3.0 per cent and 12.8 per cent, respectively. Sugar varied considerably from one year to another. It increased from JD 1.8 mn in 1973 to JD 20.2 mn in 1976, fell sharply to JD 8.2 mn in 1980, and then jumped to JD 21.4 mn in 1981. The vegetables balance of trade, on the other hand, registered a surplus of JD 1.4 mn in 1974 and continued to rise, reaching JD 12.5 mn in 1981, while the fruit and citrus balance of trade showed a surplus of JD 0.9 mn in 1981 as compared to a deficit of JD 1.7 mn in 1973.

External Trade Indices for Agricultural Products

Available indices for the period 1975-81 indicate that export prices increased by 75.3 per cent in 1976, followed by a drop of 8.7 per cent in 1978 and an increase of 15.8 per cent in 1981. (See Table 2.13.) On the other hand, the quantity index of exports showed a substantial increase, amounting to 24.1 per cent in 1976, 22.3

Table 2.11: Commodity Composition of Agricultural Deficit, 1973-81 (million JD)

Item	1973	1974	1975	1976	1977	1978	1979	1980	1981
Cereals	10.00	12.46	10.64	32.27	27.18	30.99	34.78	37.27	34.37
Wheat & flour	7.04	7.14	6.78	14.80	13.87	14.58	19.57	23.00	23.43
Barley	0.22	0.16	0.11	9.54	6.70	8.83	2.40	1.25	0.56
Lentils	+0.48	+0.54	+0.50	+0.24	+0.37	+0.03	+0.45	+0.40	+0.66
Corn	1.25	0.84	1.75	3.63	4.23	4.15	4.92	6.10	5.44
Rice	1.40	4.28	1.87	4.54	2.75	3.46	5.92	6.11	5.41
Chick-peas	—	—	—	—	—	—	1.81	0.82	0.07
Fodder	0.57	0.58	0.63	—	—	—	0.61	0.39	0.12
Vegetables	0.06	+1.40	+1.28	+2.34	+1.97	+3.68	+5.35	+5.02	+12.49
Tomatoes	+0.92	+1.84	+1.28	+2.52	+2.48	+3.61	+5.46	+5.40	+6.67
Eggplants	+0.40	+0.51	+0.48	+0.59	+0.89	+1.11	+1.53	+1.45	+1.61
Potatoes	0.84	0.92	1.50	1.96	2.57	1.51	1.88	2.61	3.69
Onions	0.06	0.26	0.15	0.68	0.44	0.05	1.08	0.89	1.03
Other	0.48	+0.23	+1.17	+1.87	+1.61	+0.42	+1.32	+1.67	+8.93
Fruit & citrus	1.74	1.00	0.32	1.19	0.87	3.24	5.02	4.83	+ 0.87
Apples	1.41	0.10	0.14	0.17	1.00	0.03	4.79	4.06	3.44
Grapes	0.01	+0.15	+0.12	+0.04	+0.06	+0.10	+0.31	+0.12	+0.20
Citrus	0.04	+0.18	+1.36	+2.68	+5.53	+0.86	+1.34	+1.80	+5.56
Melons	+0.03	+0.09	+0.05	+0.29	+0.20	+0.36	+1.03	+0.47	+0.85
Olives	+0.02	+0.01	+0.03	—	—	+0.03	+0.19	+0.32	+0.15
Other	0.33	1.33	1.75	3.45	+7.66	3.84	0.66	1.90	0.45
Sugar	1.78	6.56	11.53	20.20	6.56	8.37	9.94	8.17	21.40
Meat	0.76	1.90	1.96	2.17	3.15	9.62	9.44	9.89	17.14
Red meat	0.56	1.21	1.49	1.53	2.62	8.33	8.65	7.85	14.85
White meat	0.20	0.69	0.47	0.64	0.53	1.29	0.79	2.04	2.29
Fish	0.32	0.43	0.51	0.54	0.79	0.62	0.54	0.46	1.25
Milk & related products	3.25	3.76	4.70	6.29	7.78	8.00	12.39	10.55	8.34
Eggs	1.12	1.59	1.54	1.47	0.83	1.19	1.11	0.95	0.37
Nuts	0.98	0.70	0.85	1.23	0.69	0.81	1.38	1.68	1.87
Tea, coffee & cocoa	2.52	2.35	2.92	2.99	5.30	6.52	6.12	6.67	7.08
Other	3.63	3.72	5.53	+1.01	5.85	5.56	11.67	19.84	55.38
TOTAL	26.16	33.17	39.22	65.00	55.28	71.24	87.04	95.29	133.34

Note: + means net trade balance surplus of agricultural commodity.

Sources: Tables 2.2 and 2.7.

Table 2.12: Relative Importance of Agricultural Deficit (per cent)

Item	1973	1974	1975	1976	1977	1978	1979	1980	1981
Cereals	38.2	37.6	27.1	49.6	49.2	43.5	40.0	39.1	25.7
Vegetables	0.2	—4.2	—3.2	—3.6	—3.6	—5.25	—6.1	—5.3	—9.3
Fruit & citrus	6.7	3.3	0.8	1.8	—1.6	4.5	5.8	5.1	—0.7
Sugar	6.8	19.8	29.4	31.1	11.9	11.8	11.4	8.6	16.0
Meat	3.0	5.7	5.0	3.3	5.7	13.5	10.8	10.4	12.8
Fish	1.2	1.3	1.3	0.8	1.4	0.9	0.6	0.5	0.9
Milk & related products	12.4	11.2	12.0	9.8	14.1	11.2	14.2	11.1	6.2
Eggs	4.3	4.8	3.9	2.3	1.5	1.7	1.3	1.0	0.3
Nuts	3.7	2.2	2.2	1.9	1.2	1.1	1.6	1.8	1.4
Tea, coffee & cocoa	9.6	7.1	7.4	4.6	9.6	9.2	7.0	7.0	5.3
Other	13.9	11.2	14.1	—1.6	10.6	7.8	13.4	20.7	41.4
Total	100.0	100.0	100.0	100.0	100.0	100.0	100.0	100.0	100.0

Note: — means the relative importance of net trade balance surplus for agricultural commodity.

Source: Table 2.11.

Table 2.13: External Trade Indices for Agricultural Products, 1975-81 (1979 = 100)

	Agricultural imports				Agricultural exports			
Year	Quantity unit	Change in average	Price unit	Change in average	Quantity unit	Change in average	Price unit	Change in average
1975	51.10	—	112.90	—	62.35	—	50.85	—
1976	75.87	48.5	101.10	—10.5	77.38	24.1	89.10	75.3
1977	71.79	—5.3	100.83	—0.3	89.95	16.2	98.46	10.4
1978	84.88	18.2	95.82	—5.0	81.73	—9.1	89.87	—8.7
1979	100.00	17.8	100.00	4.4	100.00	22.3	100.00	11.3
1980	98.44	—1.6	112.97	12.9	110.12	10.1	99.30	—0.7
1981	104.75	6.4	142.10	25.8	126.25	14.6	115.04	15.8

Source: CBJ b. 1976–81, various issues.

per cent in 1979 and 14.6 per cent in 1981. This reflects the fact that the increase in value of agricultural exports resulted from the increase in both the quantity index and the price index.

The price index of agricultural imports showed a decline of 10.5 per cent in 1976, while the quantity index went up by 48.5 per cent. Thus the sharp increase in agricultural imports (by 65 per cent) in 1976 was due mainly to the increase in the quantity index and not the price index. The increase in agricultural imports in 1981 by 41 per cent, however, resulted from the increase in both the price index (by 25.8 per cent) and the quantity index (by 6.4 per cent). Generally speaking, the noticeable fluctuations in import prices were a direct result of the rise and fall in international prices for most of the imported food commodities, especially sugar, meat and cereals.

II. Demand for Agricultural Products

This part deals with the major aspects of domestic demand for agricultural products. The first section explores those factors affecting the demand for agricultural products; the second discusses the commodity composition of domestic consumption and average *per capita* consumption; and the third emphasises for geographical distribution of agricultural consumption throughout the Kingdom. The last section explains the role of international demand in Jordan's agricultural trade and consumption.

Factors Determining Domestic Demand for Agricultural Products

It is hard to differentiate between factors which contributed to the increase in demand for agricultural commodities and those which contributed to the growth in demand for non-agricultural commodities. But it is obvious that increase in population and increase in *per capita* income level have been among the major factors which have led to greater food consumption in Jordan.

Jordan has experienced marked improvements in *per capita* income level. The average *per capita* income at current prices increased from JD 300 in 1976 to JD 640 in 1981. But the average *per capita* income at 1975 prices increased from JD 260 in 1976 to JD 355 in 1981 (or from JD 418 to JD 545 at 1979 prices). As a result, Jordan has witnessed significant changes in food consumption patterns and styles; these have contributed mainly to the expansion in domestic consumption of basic foodstuffs, particularly meat, milk and dairy products. Despite the apparent improvement in general income level throughout the Kingdom, income disparities are still marked in various regions, accounting for the varying rates of *per capita* consumption for various food commodities.

In Jordan, as in most developing countries, population growth is considered to be one of the most important factors contributing to the increase in domestic demand for food commodities. More specifically, Jordan has one of the highest population growth rates in the world, amounting to 4.2 per cent during the period 1973—81. Jordan's fertility rate is also considered to be one of the highest in the world, amounting to 48 per thousand, as compared to 35 per thousand in Egypt and 41 per thousand in Iraq (Haddad 1983: 35-7).

Accordingly, the increase in demand for foodstuffs and consumer commodities throughout the country is clearly related to the increase in population. For instance, if Jordan's *per capita* consumption of food had remained constant over the period under consideration, the population growth rate of 4.2 per cent alone would have been responsible for an annual increase in food consumption of 4.2 per cent.

Since the income earned by any consumer is considered to be one of the major determinants of quantity demanded for various commodities, we would expect that quantity demanded for any given commodity would have some degree of responsiveness to any given income change. This is what in economics is called income elasticity of demand, defined as the degree of responsiveness regarding changes in quantity demanded of food commodities to changes in consumer income. This elasticity differs, however, from one case to another. In Jordan, as in most developing countries, the income elasticity of demand for food commodities amounts to 70 per cent, which means that an increase in consumer income of 1 per cent would result in a 0.7 per cent increase in food demand (Abu-Remailleh 1982). Income elasticity of demand differs from one country to another, due to disparities in income levels and differences in consumption patterns.

In the last few years Jordan has witnessed substantial social and economic changes which have made a major contribution to the expansion of domestic demand for food commodities. For example, female participation in the labour force has meant that most working mothers have been forced to substitute bottle-feeding for breast-feeding, which in turn has contributed to the expansion of domestic consumption of infant food products such as milk.

Customs duties on agricultural imports and exports are also among the major factors directly affecting domestic and external demand for foodstuffs. Imposing tariffs on agricultural exports would definitely put some kind of restriction on

Table 2.14: Tariffs on Agricultural Imports, 1981

Item	Tariff (%)
Flour & wheat	Exempt
Barley	—
Lentils	—
Corn	—
Rice	—
Potatoes	—
Onions	—
Apples	—
Melons	—
Olives	—
Eggs	—
Nuts	—
Sugar	0-35
Canned meat	0-3
Tomatoes	23
Eggplants	23
Grapes	23
Citrus	14
Red meat	10
White meat	11
Fish	11
Dairy products	14
Fruit juice	38
Canned vegetables	33
Tomato paste	53

Source: MOF 1981.

such products. Such tariff policy domestically would cause an increase in the supply and a decrease in the prices of agricultural products. On the other hand, non-restrictive import policies would surely induce new consumption habits and desires in any society and thereby increase the domestic demand for food commodities as well as consumer goods. As stated previously, Jordan's agricultural imports have contributed greatly to the expansion of domestic food consumption. This is due to the fact that most agricultural imports are subject to favourable tariff rates, as indicated in Table 2.14.

Commodity Composition of Domestic Demand for Agricultural Products

Available data concerning the commodity composition of domestic demand for agricultural commodities in Jordan during the period 1973—81 indicate that the domestic consumption of various agricultural products witnessed a sharp increase. As can be seen from Table 2.15, the volume of domestic consumption of cereals in creased from 192.7 thousand t in 1973 to 387.1 thousand t in 1976, and continued to increase at a higher rate, reaching 445.6 thousand t in 1977 and 561.8 thousand t in 1980—with the exception of 1981, when it declined to 537.6 thousand t. Wheat and flour accounted for most of this cereal consumption, increasing from 121.7 thousand t in 1973 to 226.3 thousand t in 1978 and 298.6 thousand t in 1981. Domestic consumption of corn also increased from 38 thousand t in 1973 to 135.2 thousand t in 1981.

Domestic consumption of vegetables during the period under study fluctuated markedly. The volume of vegetable consumption amounted to 137.3 thousand t in 1973, as compared to 184.3 thousand t in 1975 and 98.8 thousand t in 1977. While it showed a sharp increase by reaching 250.4 thousand t in 1978, it fell to 195.0 thousand t in 1981. However, this fluctuation is attributable to the fact that most vegetable products were used for external trading purposes. Thus total exports of vegetables reached 88.7 thousand t in 1973, expanding to 302.0 thousand t in 1980, but falling to 283.5 thousand t in 1981, as shown in Table 2.16. This fluctuation is particularly noticeable in tomatoes, where domestic consumption increased from 50.9 thousand t in 1974 to 91.9 thousand t in 1975; it then declined sharply in 1976 to 20.4 thousand t and continued to fluctuate, reaching 76.0 thousand t in 1981. Exports of tomatoes varied considerably from one year to another. They increased from 38.2 thousand t in 1973 to 82.6 thousand t in 1974 and continued to fluctuate, reaching 128.0 thousand t in 1981. On the other hand, domestic consumption of eggplants reached 37.9 thousand t, 65.1 thousand t and 28.7 thousand t in 1973, 1976 and 1981 respectively. Domestic consumption of onions reached 4.1 thousand t in 1973 and 21.2 thousand t in 1981, and of potatoes 7.6 thousand t and 49.2 thousand t in 1973 and 1981 respectively.

Fruit and citrus domestic consumption increased from 136.5 thousand t in 1973 to 164.2 thousand t in 1981. Domestic consumption of apples, grapes, citrus and melons amounted to 11.4, 20.3, 38.1 and 54.6 thousand t respectively in 1973, as compared to 29.7, 19.3, 31.2 and 55.4 thousand t in 1981.

Domestic consumption of red and white meat amounted to 32.1 thousand t in 1973, and continued to rise over the period under study to reach 53.6, 78.3 and 90.3 thousand t in 1979, 1980 and 1981 respectively. Domestic consumption of fish, on the other hand, reached 2.4 thousand t in 1973, as compared to an increase of 9.8 thousand t in 1980, and declined to 6.8 thousand t in 1981. As a result of increasing efforts on the part of the authorities to develop and improve Jordan's poultry industry, domestic consumption of eggs and poultry rose substantially over the period 1973-81. Domestic consumption of eggs amounted to 109.5 mn in 1973 and rose to 287.7 mn in 1981. By the same token, milk and related products increased from 77.6 thousand t in 1973 to 136.8 thousand t in 1981. In Table 2.17 the commodity composition of domestic production is shown, while in Table 2.18 the commodity composition of the agricultural deficit is computed.

Table 2.15: Commodity Composition of Domestic Consumption, 1973-81 (thousand tonnes)

Item	1973	1974	1975	1976	1977	1978	1979	1980	1981
Cereals, of which:	192.7	392.8	163.4	387.1	445.6	481.6	457.3	561.8	537.6
Wheat & flour	121.7	275.1	94.2	199.5	201.5	226.3	217.5	296.6	298.6
Barley	11.9	40.2	14.4	27.9	79.9	39.0	49.2	58.6	29.0
Lentils	0.1	18.7	1.3	7.5	6.2	7.8	3.5	4.7	4.3
Corn	38.0	16.0	34.7	70.3	86.9	97.2	126.0	109.5	135.2
Rice	16.1	23.3	11.4	69.8	68.4	95.6	34.6	47.3	36.9
Chick-peas	2.2	4.0	4.4	1.9	5.7	2.2	15.5	15.6	24.7
Fodder	2.4	3.9	2.4	2.0	3.8	1.3	8.4	7.6	2.1
Vegetables, of which:	137.3	101.4	184.3	130.0	98.8	250.4	140.2	137.6	195.0
Tomatoes	44.9	50.9	91.9	20.4	23.1	128.6	66.8	52.9	76.0
Eggplants	37.9	9.4	17.6	65.1	1.6	39.4	21.0	25.0	28.7
Potatoes	7.6	10.1	19.3	27.2	37.0	30.3	29.9	37.6	49.2
Onions	4.1	9.4	7.6	14.8	12.1	12.2	17.6	16.4	21.2
Fruit & citrus, of which:	136.5	191.4	113.3	117.2	142.0	197.8	129.0	163.0	164.2
Apples	11.4	17.4	7.4	5.8	4.6	5.2	25.6	27.4	29.7
Grapes	20.3	13.9	5.5	9.6	21.7	27.6	16.7	13.1	19.3
Citrus	38.1	40.7	15.9	26.1	51.6	35.8	36.4	45.3	31.2
Melons	54.6	36.5	38.9	22.8	28.7	40.8	29.2	32.3	55.4
Olives	4.9	38.8	3.6	22.5	8.3	38.2	7.5	41.5	19.3
Sugar	19.4	48.3	45.9	108.6	45.0	89.9	96.3	98.8	90.1
Meat	32.1	29.0	32.2	33.4	40.3	46.0	53.6	78.3	90.3
Red meat	8.9	9.6	9.9	9.2	12.1	19.1	19.9	25.9	29.3
White meat	23.2	19.4	22.3	24.2	28.2	26.9	33.7	52.4	61.1
Fish	2.4	2.4	3.0	2.6	3.4	2.4	2.4	9.8	6.8
Milk & related products	77.6	75.8	80.1	80.7	103.9	105.7	133.9	134.0	136.8
Eggs (per mn)	109.5	125.9	176.1	198.4	249.6	289.0	276.0	285.5	287.7
Nuts	2.9	11.4	11.9	3.3	2.8	7.9	3.3	2.5	5.5
Tea, coffee & cocoa	6.0	5.4	5.6	4.9	5.8	7.5	6.1	7.1	7.0

Notes: Consumption = product ± agricultural deficit.
Agricultural deficit = (imports - exports) = (consumption - production).

Sources: Tables 2.17 and 2.18.

Table 2.16: Commodity Composition of Agricultural Exports, 1973-81 (thousand tonnes)

Item	1973	1974	1975	1976	1978	1979	1980	1981	
Cereals, of which:	4.9	10.8	4.5	1.9	12.2	0.5	4.3	1.6	3.6
Wheat & flour	—	—	0.6	—	—	—	—	—	—
Barley	—	—	—	—	—	—	—	—	—
Lentils	4.9	10.8	3.9	1.9	12.2	0.5	4.3	1.6	3.6
Corn	—	—	—	—	—	—	—	—	—
Rice	—	—	—	—	—	—	—	—	—
Chick-peas	—	—	—	—	—	—	—	—	—
Fodder	88.7	147.3	112.5	130.0	150.2	207.8	296.6	302.0	283.5
Vegetables, of which:	38.2	82.6	53.2	67.5	62.6	81.0	105.0	110.0	128.0
Tomatoes	17.9	23.2	22.0	17.7	23.0	24.7	31.7	27.4	33.7
Eggplants	—	—	—	—	—	—	—	—	—
Potatoes	3.8	1.2	1.8	2.2	0.5	3.5	2.8	5.0	7.7
Onions	69.0	116.1	152.2	167.8	170.0	161.2	180.0	188.0	160.0
Fruit & citrus, of which:	—	—	—	—	—	—	—	—	—
Apples	2.9	4.0	5.8	6.7	1.8	4.4	6.6	5.8	6.7
Grapes	54.5	92.7	133.7	150.7	160.6	138.8	137.0	129.0	140.0
Citrus	1.4	9.5	11.4	—	—	—	—	—	—
Melons	0.3	1.2	1.1	—	—	—	—	—	—
Olives	—	—	—	—	—	—	—	—	—
Sugar	—	—	—	—	—	—	—	—	—
Meat	—	—	—	—	—	—	—	—	—
Red meat	—	—	—	—	—	—	—	—	—
White meat	—	—	—	—	—	—	—	—	—
Fish	1.4	1.7	1.9	1.5	1.7	2.1	2.5	2.1	2.3
Milk & related products	—	—	—	—	—	—	9.0	35.0	62.3
Eggs (per mn)	—	—	—	—	—	—	—	—	—
Nuts	—	—	—	—	—	—	—	—	—
Tea, coffee & cocoa	—	—	—	—	—	—	—	—	—

Source: DOS a. 1973–81, various issues.

Table 2.17: Commodity Composition of Domestic Production, 1973-81 (thousand tonnes)

Item	1973	1974	1975	1976	1977	1978	1979	1980	1981
Cereals, of which:	66.3	334.1	73.9	100.7	93.8	92.8	32.4	204.2	90.0
Wheat & flour	50.4	244.6	50.0	66.7	62.5	53.3	16.5	133.6	50.6
Barley	5.9	40.2	11.8	13.2	12.0	15.6	4.8	38.1	19.2
Lentils	4.8	29.5	5.2	9.4	6.0	8.3	0.8	6.3	7.9
Corn	0.2	0.4	0.4	0.5	0.5	0.1	—	0.1	0.2
Rice	—	—	—	—	—	—	—	—	—
Chick-peas	2.2	4.0	4.4	1.9	5.7	2.2		2.7	3.6
Fodder	2.3	3.8	1.5	0.8	1.5	1.3	0.7	1.5	1.7
Vegetables, of which:	177.2	205.8	262.3	201.8	204.6	410.0	353.5	375.5	413.5
Tomatoes	83.1	133.5	145.1	87.9	185.7	208.7	171.8	162.9	204.0
Eggplants	55.7	32.4	39.6	82.8	24.6	64.1	52.7	52.4	62.4
Potatoes	2.6	2.5	9.1	8.5	7.6	10.2	4.6	6.6	9.1
Onions	5.8	8.4	7.3	5.3	7.7	8.0	4.3	7.8	11.7
Fruit & citrus, of which:	104.0	156.7	96.8	83.4	103.6	177.2	118.7	169.9	156.0
Apples	2.2	3.3	3.8	3.8	3.5	3.4	2.1	1.9	2.4
Grapes	22.2	17.8	11.1	13.7	22.3	30.7	23.0	18.2	25.8
Citrus	15.4	34.6	12.8	16.5	36.5	32.9	29.9	48.5	53.8
Melons	56.0	46.0	50.3	23.1	28.1	40.3	15.5	37.7	46.9
Olives	5.2	40.0	4.7	22.5	8.3	37.0	6.8	44.5	18.9
Sugar	—	—	—	—	—	—	—	—	—
Meat	29.3	24.9	27.0	29.0	34.7	31.2	38.3	56.3	64.8
Red meat	8.4	7.6	7.9	6.5	7.8	7.6	7.1	8.1	8.5
White meat	20.9	17.3	19.1	22.5	27.0	23.6	31.7	48.2	56.3
Fish	0.1	0.1	0.1	0.1	0.1	0.1	0.1	0.1	0.1
Milk & related products	44.9	32.3	41.0	34.7	39.7	44.1	38.7	46.0	42.0
Eggs (per mn)	39.5	50.5	86.4	161.0	214.0	265.0	285.0	320.5	350.0
Nuts	1.1	2.3	2.4	2.6	2.0	6.7	1.8	0.9	1.5
Tea, coffee & cocoa	—	—	—	—	—	—	—	—	—

Source: DOS a. 1973–81, various issues.

Table 2.18: Commodity Composition of Agricultural Deficit, 1973-81 (thousand tonnes)

Item	1973	1974	1975	1976	1977	1978	1979	1980	1981
Cereals, of which:	126.4	58.7	89.5	286.4	351.8	388.8	424.9	357.6	447.6
Wheat & flour	71.3	30.5	44.2	132.8	139.0	173.0	201.0	163.0	248.0
Barley	6.0	—	2.6	14.7	67.9	23.4	44.4	20.5	9.8
Lentils	+4.9	+10.8	+3.9	+1.9	+12.2	+0.5	+4.3	+1.6	+3.6
Corn	37.8	15.6	34.3	69.8	86.4	97.1	126.0	109.4	135.0
Rice	16.1	23.3	11.4	69.8	68.4	95.8	34.6	47.3	36.9
Chick-peas	—	—	—	—	—	—	15.5	12.9	21.1
Fodder	0.1	0.1	0.9	1.2	2.3	—	7.7	6.1	0.4
Vegetables, of which:	440.0	+104.7	+78.0	+71.8	+105.7	+159.3	+212.9	+237.9	+218.5
Tomatoes	+38.2	+82.6	+53.2	+67.5	+62.6	+80.1	+105.0	+110.0	+128.0
Eggplants	+17.9	+23.0	+22.0	+17.7	+23.0	+24.7	+31.7	+27.4	+33.7
Potatoes	5.0	7.6	10.2	18.7	29.4	20.1	25.3	31.0	40.1
Onions	+1.7	1.0	0.3	9.5	4.4	4.2	13.3	8.6	9.5
Fruit & citrus, of which:	32.5	34.7	16.5	33.8	38.4	20.6	9.3	6.9	8.2
Apples	9.2	14.1	3.6	2.0	1.1	1.8	23.5	25.4	27.3
Grapes	+1.9	+3.9	+5.6	+4.1	+0.6	+3.1	+6.3	+5.1	+6.0
Citrus	22.7	6.1	3.1	9.6	15.1	2.9	6.5	+3.2	+22.6
Melons	+1.4	+9.5	+11.4	0.3	0.6	0.5	13.7	4.6	8.5
Olives	+0.3	+1.2	+1.1	—	—	1.2	0.7	1.0	0.4
Sugar	19.4	48.3	45.9	108.0	45.0	89.9	96.3	98.8	90.1
Meat	2.8	4.1	5.2	4.4	5.6	14.8	14.8	22.0	25.5
Red meat	0.5	2.0	2.0	2.7	4.3	11.5	12.8	17.8	20.7
White meat	2.3	2.1	3.2	1.7	1.3	3.3	2.0	4.2	4.8
Fish	2.3	2.3	2.9	2.5	3.3	2.3	2.3	9.7	6.7
Milk & related products	32.7	43.5	39.1	46.0	64.2	61.6	95.2	88.0	94.8
Eggs (per mn)	70.0	75.4	89.7	37.4	35.6	24.0	+9.0	+35.0	+62.3
Nuts	1.8	9.1	9.5	0.7	0.8	1.2	1.5	1.6	4.0
Tea, coffee & cocoa	6.0	5.4	5.6	4.9	5.8	7.5	6.1	7.1	7.0

Note: + means net trade balance surplus of agricultural commodity.
Source: Tables 2.4 and 2.16.

Statistical information regarding *per capita* domestic agricultural consumption is presented in Table 2.19. *Per capita* consumption of cereals increased from 115.2 kg in 1973 to 226.0 kg in 1977 and to 233.2 kg in 1981. *Per capita* consumption of wheat and flour accounted for most *per capita* cereal consumption, and increased from 72.6 kg in 1973 to 128.6 kg in 1981. Consumption of rice *per capita* reached 15.9 kg in 1981, as compared to 9.6 kg in 1973.

As for vegetables, *per capita* domestic consumption rose from 82.0 kg in 1973 to 83.9 kg in 1981. Fruit and citrus *per capita* consumption reached 81.5 kg in 1973 and 71.5 kg in 1981.

Per capita consumption of red and white meat reached 19.2 kg in 1973, of which 5.3 kg went on red meat and 13.9 kg on white meat, as compared to 38.9 kg in 1981, with 12.6 kg for red and 26.3 kg for white meat.

As for milk and related products, it is estimated that *per capita* consumption amounted to 46.3 kg in 1973, as compared to 58.6 kg in 1981. It is also estimated that *per capita* consumption of eggs amounted to 65.4 in 1973 and 79.3 in 1975, increasing to a peak of 139.1 in 1978; but it declined in 1979, 1980 and again in 1981, to reach only 123.9. *Per capita* sugar consumption rose from 11.6 kg in 1973 to 38.8 kg in 1981, while coffee, tea and cocoa consumption decreased from 3.6 kg in 1973 to 3.0 kg in 1981.

Geographical Distribution of Domestic Demand for Agricultural Products

As mentioned earlier, Jordan is incapable of producing foodstuffs in sufficient quantities to meet the increasing needs of its population. Moreover, food is predominant as regards private consumption in Jordan. Thus we find that the average expenditure on foodstuffs constituted about 50 per cent of total private consumption expenditure and about 43 per cent of GDP over the period 1973-81. We may also note that average agricultural imports constituted 34 per cent of consumer commodity imports and 19 per cent of total imports over the same period (CBJ b. 1973-81).

However, statistical information regarding Jordan's geographical distribution of domestic demand for agricultural commodities during the period 1973-81 is not available. Thus we have utilised the results of the *Family Expenditures Study, 1980*, carried out by the Department of Statistics (DOS). This study indicated that there were marked consumption disparities between the various regions of Jordan This is mainly due to income disparities throughout the Kingdom. The summary of this study is presented in Table 2.20. Here we see that meat expenditure ranked highest, accounting for 28.4 per cent of total food expenditure in the Kingdom, followed by eggs and dairy products at 21.2 per cent, vegetables and fruit at 19.3 per cent and cereals at 13 per cent. The study also revealed that Amman Governorate was the leading geographical region in total food expenditures, with a total of JD 195.7 mn in 1980, of which 29.2 per cent went on meat, 21.2 per cent on dairy products, 20.8 per cent on vegetables and fruit and 11.6 per cent on cereals. The second largest region was Irbid Governorate, with total food expenditures of JD 104.5 mn, of which 26.2 per cent went on meat, 23.5 per cent on eggs and dairy products and 15.2 per cent on cereals. Total food expenditures for the rest of the governorates amounted to JD 60.1 mn, of which JD 26.5 mn was accounted for by Balqa Governorate, JD 10.9 mn by Ma'an Governorate and the remainder by Karak Governorate.

Impact of International Demand on Jordan's Agricultural Trade and Consumption

In the above analysis, we discussed the most important economic indicators of

Table 2.19: *Per capita* Agricultural Consumption, 1973-81 (kilogrammes)

Item	1973	1974	1975	1976	1977	1978	1979	1980	1981
Cereals, of which:	115.2	226.4	90.2	204.9	226.0	231.8	212.9	251.6	233.2
Wheat & flour	72.6	158.0	52.2	105.6	101.2	109.0	101.3	132.8	128.6
Barley	7.1	23.1	8.0	14.8	40.5	18.8	22.9	26.3	12.5
Lentils	0.5	10.8	0.7	4.0	3.1	3.8	1.6	2.1	1.9
Corn	22.7	9.2	19.2	37.2	44.1	46.7	58.7	49.0	58.2
Rice	9.6	13.4	6.2	36.9	34.7	46.0	16.1	21.2	15.9
Chick-peas	1.3	2.3	2.4	1.0	2.9	1.1	7.2	7.0	10.6
Fodder	1.4	2.2	1.3	1.1	1.9	0.6	3.9	3.4	1.0
Vegetables, of which:	82.0	58.4	101.8	68.8	50.0	120.5	65.2	61.6	83.9
Tomatoes	26.8	29.3	50.8	10.8	11.7	61.9	31.1	23.7	32.7
Eggplants	22.6	5.4	9.7	34.5	0.8	19.0	9.8	11.2	12.4
Potatoes	4.5	5.8	10.7	14.4	18.8	14.5	13.9	16.8	21.2
Onions	2.5	5.4	4.2	7.8	6.1	5.9	8.2	7.3	9.1
Fruit & citrus, of which:	81.5	110.0	62.5	62.0	72.1	95.2	60.1	73.0	71.3
Apples	6.8	10.0	4.1	3.1	2.3	2.5	11.9	12.3	12.8
Grapes	12.1	8.0	3.0	5.1	11.0	13.3	7.8	5.9	8.3
Citrus	22.7	23.5	8.9	13.8	26.2	14.5	12.9	20.2	13.9
Melon	32.6	21.0	21.5	12.1	14.6	20.0	13.6	14.5	23.9
Olives	2.9	22.4	2.0	11.9	4.2	18.4	3.8	18.9	8.5
Sugar	11.6	27.8	25.4	57.5	22.8	4.3	44.8	44.2	38.8
Meat	19.2	16.9	17.8	17.7	20.4	22.1	25.0	35.1	38.9
Red meat	5.3	5.6	5.5	4.9	6.1	9.2	9.3	11.5	12.6
White meat	13.9	11.2	12.3	12.8	14.3	12.9	15.7	23.5	26.3
Fish	1.4	1.4	1.7	1.4	1.8	1.2	1.1	4.3	3.0
Milk & related products	46.3	43.6	44.2	42.7	52.7	50.8	62.3	60.0	58.6
Eggs (no. of eggs)	65.4	72.6	97.3	105.0	126.7	139.1	128.5	127.9	123.9
Nuts	1.7	6.7	6.7	1.8	1.4	3.8	1.5	1.1	2.4
Tea, coffee & cocoa	3.6	3.1	3.1	2.6	2.9	3.6	2.8	3.2	3.0

Note: Calculations of *per capita* is made by dividing agricultural consumption over population.
Source: Table 2.15.

Table 2.20: Geographical Distribution of Expenditure on Food Commodities, 1980 (million JD)

Commodity	Amman	Irbid	Balqa	Karak	Ma'an	Jordan
Cereals	22.72	15.79	4.03	2.72	1.65	46.95
Meat	57.22	27.62	6.73	7.43	3.11	102.11
Vegetables & fruit	40.67	16.82	5.21	4.42	2.33	69.46
Eggs & dairy products	41.39	24.48	4.82	3.69	2.22	76.40
Other	33.68	19.74	5.74	4.39	1.82	65.37
Total	195.67	104.47	26.52	22.72	10.92	360.30

Source: DOS b. 1982.

Table 2.21: Production, Imports and Consumption of Agricultural Commodities, 1973-81 (thousand tonnes)

Item	Average consumption	Average production	Average imports	Production to consumption (%)	Imports to consumption (%)
Cereals	384.1	121.0	326.1	31.5	93.7
Vegetables	114.5	251.8	49.9	219.9	43.6
Fruits & citrus	145.3	124.1	113.6	85.4	78.2
Sugar	71.4	—	71.4	—	100.0
Meat	48.4	37.3	11.1	77.1	22.9
Fish	3.9	0.1	3.8	2.6	97.4
Milk & related products	103.2	40.4	64.7	39.1	62.7
Eggs (mn)	220.4	236.4	36.9	107.3	16.7
Nuts	5.7	2.4	3.4	42.1	59.6
Tea, coffee & cocoa	6.2	—	6.2	—	100.0

Sources: Tables 2.4, 2.15 and 2.17.

demand for agricultural products in Jordan, such as agricultural exports and imports, agricultural deficit and domestic consumption. Based on these indicators, as can be noted that Jordan suffers from a continuous deficit in its agricultural balance of trade because of a monotonic upward trend in its agricultural consumption.

An exposition of these indicators is given in Table 2.21 which reveals that Jordan imported most of its basic food consumption needs, with the exception of a few commodities, during the period 1973-81. Specifically, cereal imports constituted 93.7 per cent of total average domestic consumption during that period. The relative share of fruit and citrus imports was 78.2 per cent of domestic consumption. Moreover, the relative importance of other agricultural imports to total domestic consumption ranged between 16.7 per cent for eggs and 100 per cent for sugar, coffee, tea and cocoa.

The role of domestic production in meeting domestic consumption needs is also presented in Table 2.21. This role can be defined as the degree of self-sufficiency which measures the ratio of domestic agricultural production to domestic agricultural consumption. The ratios for cereals, vegetables, fruit and citrus, meat, fish, milk and related products, eggs and nuts were 31.5 per cent, 219.9 per cent, 85.4 per cent, 77.1 per cent, 2.6 per cent, 39.1 per cent, 107.3 per cent and 42.1 per cent, respectively.

In addition to the important role of agricultural imports in Jordan's domestic consumption, it is also necessary to underline the impact of international demand upon Jordan's agricultural trade and consumption. In order to understand this influence, one needs to consider the following issues:

1. The external demand for Jordan's agricultural commodities is confined in the first place to Arab countries. This is due mainly to the fact that agricultural products exported by Jordan to neighbouring Arab markets are limited to vegetables and some kinds of fruit and citrus. The demand for Jordanian exports depends on several factors in addition to price. These factors may be of a political and economic nature.
2. Jordan's exports of agricultural products are characterised as being non-strategic commodities (vegetables and fruit) on the international markets. The effect of external demand on Jordan's agricultural exports is therefore very limited.

Conversely, Jordan's agricultural imports from the international markets are characterised as being more strategic (wheat, sugar, meat and rice) than its exports. Thus the impact of international demand on Jordan's imports of such commodities has to be considered. Although it is extremely difficult to identify the impact of various global variables related to international demand on Jordan's imports of basic agricultural commodities, it is still necessary to give a brief exposition of such variables.

As explained earlier, Jordan imports most of its food supplies, particularly cereals, meat, sugar, coffee and rice. The volume of imports is influenced by the level of international demand for such commodities. That said, the international markets of these agricultural commodities have been largely unstable since the beginning of the 1970s. This has not only affected the quantity supplied and demanded of these commodities, but it has also influenced prices.

Despite various collective efforts undertaken by the international community, aimed at eliminating every kind of tariff restriction and barrier that agricultural trade commodities might encounter, we note that agricultural policies designed by major exporting countries have contributed to increasing further the problem of instability on the international markets for agricultural products.

Furthermore, international exports of foodstuffs are basically monopolised by a small number of exporting countries, namely Argentina, Australia, Brazil, Canada, the (EEC), Spain and the US. Their total exports of foodstuffs accounted for more than one quarter of the world's total in 1980 (FAO 1980).

International prices of agricultural products are controlled by various sophisticated and interrelated variables, most of which are the international production quantities, surplus quantities available for exports, and inventories held by exporting countries. In addition, international prices of agricultural products are linked to any changes or developments that may occur in the exporting and producing countries. For instance, a price change on the US wheat market would automatically affect the prices of wheat on the international markets.

Accordingly, the statistical data of the United Nations Food and Agriculture Organisation (FAO) indicate that international prices of agricultural products, especially cereals, meat and sugar, witnessed a considerable increase during the 1970s. As a result, the price index of meat increased from 105 points in 1976 to 166 points in 1980. Similarly, the price index of wheat increased from 119 points in 1976 to 174 points in 1980, while the price index of sugar increased from 185 points in 1976 to 320 points in 1980.

As a result of these developments on the international markets, Jordan's import values of sugar, meat and cereals recorded a substantial increase over the period 1973-81. Jordanian imports of cereals increased from JD 10.5 mn in 1973 to JD 35.0 mn in 1981. As for sugar, the value of its imports increased from JD 1.5 mn in 1973 to JD 21.4 mn in 1981, while the meat value increased from JD 0.7 mn in 1973 to JD 7.1 mn n 1981.

It is worth mentioning, however, that Jordan's imports of sugar, meat and cereals are distributed among five major suppliers for each commodity. Such distribution lessens the impact of international price variations on Jordanian importation of these commodities. Jordan's import values of these commodities increased during the period under study, partly because of an increase in domestic demand and partly because of a substantial increase in the world prices of these commodities.

Conclusion

This study shows that Jordan experienced a substantial increase in its domestic consumption of agricultural commodities during the period 1973-81. This was due mainly to rapid population growth, coupled with an increase in *per capita* income level.

With the exception of a few commodities, most of Jordan's basic foodstuffs are imported from abroad. Indeed, agricultural imports accounted for about a quarter of Jordan's trade deficit during the period under consideration.

In order to cope with this situation the Government, through its agricultural policy, has taken numerous measures aimed at improving the agricultural sector and encouraging agricultural industries. Despite these measures, it has not yet been able to achieve a reasonable degree of self-sufficiency in meeting the increasing domestic need for agricultural commodities.

References and Other Works Consulted

Abdul Hag, Yusif, 1979 *Economic Development and Planning in Jordan*, Cairo.

Abu-Remaillah, Naji, 'Food conditions as a basic link to Arab development', *Tanmiah Magazine* (Amman), No. 112, Nov. 1982.

Aresvik, Oddvar, 1976, *The Agricultural Development of Jordan*, Praeger, New York.

CBJ (Central Bank of Jordan), a. *Annual Reports, 1973-81*, various issues, Amman.

—— b. *Monthly Statistical Bulletins, 1973–81*, various issues, Amman.

Cooper, Charles and S. Alexander, 1972, *Economic Development and Population Growth in the Middle East*, American Elsevier Publishing Company, New York.

DOS (Department of Statistics), a. *Annual External Trade Statistics, 1973-81*, various issues, Amman.

—— b. 1982, *Family Expenditures Study, 1980*, Amman.

FAO (Food and Agriculture Organisation), *Trade Year Books, 1979 and 1980*, Rome.

Haddad, Adeeb, 1983, *The Hashemite Kingdom of Jordan: Regional and Sectoral Study*. Council of Arab Economic Unity, Amman.

Haddad, Mohammed, 1979, *The Agricultural Sector of Jordan*, Royal Scientific Society, Amman.

Khain, Nasir, 1979, *Patterns of Agricultural Development in Arab Countries*, Arab Planning Institute, Kuwait.

Mazur, Michael, 1979, *Economic Growth and Development in Jordan*, Westview Press, Boulder, Colo.

Millham, Charles *et al.*, 1977, *Rationalization of Imports in Jordan*, Royal Scientific Society, Amman.

MOF (Ministry of Finance), *Customs Duties Report, 1981*, Amman.

NPC (National Planning Council), *Three-Year Development Plan, 1973–75*, Amman.

—— *Five-Year Plan, 1976–80*, Amman.

—— Five-Year Plan for Economic and Social Development, 1981—85, Amman.

Qasem, Subhi, 1982, *An Analytical View of the Food Problem in Arab Countries*, Amman.

Valdes, Alberto, 1981, *Food Security for Developing Countries*, Westview Press, Boulder, Colo.

3 THE SUPPLY OF AGRICULTURAL LABOUR

M. Haitham el-Hurani

Introduction*

This chapter aims to analyse the size and characteristics of the agricultural labour force and employment in Jordan. The first part will consist of a brief historical review of the development of the Jordanian population over the past 60 years, discussing demographic conditions and educational status. The second part will analyse the major features of the Jordanian labour force in general, and the third will focus more on its characteristics in the agricultural sector by describing the farming population and agricultural holdings. The final part will deal specifically with certain crucial issues pertaining to the employment and tenancy arrangements prevailing in the Jordan Valley. A brief analysis of employment conditions in rain-fed areas will be a starting-point, followed by a more comprehensive discussion on the renting arrangements for fixed amounts of money, and on share-cropping, which is very widespread in the Jordan Valley.

For the purpose of this research I have depended on two main sources of information, the first being personal interviews with Jordanian farmers, share-croppers and workers. Interviews were also carried out with Egyptian agricultural workers who live in the Jordan Valley. I had meetings with several officials at the Jordan Valley Authority (JVA), who were most cooperative in answering all questions and in providing the requested data and reports. The second source of information was provided by the documents and reports, which were particularly useful in obtaining the data required.

I. Analysis of the Demographic Conditions of the Jordanian Population

The Jordanian population has very distinctive demographic features. It has a high rate of natural growth, which has resulted in a large proportion of the population being below the age of 18. Large flows of people into Jordan from Palestine and the West Bank of Jordan over the last 30 years, the outflow of Jordanian workers to the Gulf states, and finally the influx of workers from Arab and non-Arab countries—all these labour movements have had their effect on the characteristics of Jordan's population, which will be discussed below.

Historical Development

The Emirate of Transjordan was declared a political entity in 1923. The previous

* I wish to extend my gratitude and sincere appreciation to officials at the Department of Statistics, Department of Manpower at the National Planning Council, and the Department of Programming and Planning at the Jordan Valley Authority, who were very kind in providing all the required documents and books. I am also grateful to all farmers and share-croppers and non-Jordanian workers in the Jordan Valley for their kind reception and cooperation.

year, a population count was carried out which estimated total population at about 225,380 persons distributed as shown in Table 3.1. Another population estimate was undertaken in 1928. It put the total population of Transjordan at between 300,000 and 350,000. The Department of Statistics (DOS) was established in 1949 and estimated the population in 1950 as being 600,000.

The West Bank was officially declared part of the Hashemite Kindgom of Jordan in 1950. In 1952 the DOS conducted a housing census which gave a population estimate of 1,329,000 inhabitants; of these, 587,000 were living on the East Bank and 742,000 on the West Bank.

The first comprehensive nation-wide population and housing census was undertaken in 1961. The results indicated that, of the total population of Jordan, 901,000 were living on the East Bank and 742,000 on the West Bank. During the

Table 3.1: Estimated Population of Transjordan, 1922

Type of living	Population	No. of dwelling units
Urban & rural	122,430	20,560
Tribesmen	102,950	20,590 (tents)
TOTAL	225,380	41,150

Source: DOS b. 1981:1.

Table 3.2: Estimated Population of the East Bank by Sex, 1952, 1961 and 1962-82 (thousands)

Year	Female	Male	Total
1952	284.5	301.7	586.2
1961	434.1	469.4	900.8
1962	446.9	484.1	931.0
1963	461.9	500.3	962.2
1964	477.4	517.1	994.5
1965	493.4	534.6	1,028.0
1966	510.0	552.4	1,062.4
1967	653.8	708.2	1,362.0
1968	676.4	732.7	1,409.1
1969	699.7	758.1	1,457.8
1970	723.9	784.3	1,508.2
1971	749.8	812.2	1,562.0
1972	776.4	841.1	1,617.5
1973	804.0	871.1	1,675.1
1974	832.8	902.2	1,735.0
1975	869.0	941.5	1,810.5
1976	906.8	982.5	1,889.3
1977	946.3	1,025.3	1,971.6
1978	987.7	1,069.8	2,057.5
1979	1,025.0	1,122.6	2,147.6
1980	1,065.7	1,167.3	2,233.0
1981	1,114.8	1,207.5	2,322.2
1982	1,152.5	1,226.7	2,415.2

Source: DOS 1982: 2.

period 1952-61 the population growth rate was 2.84 per cent. However, the population increase was concentrated on the East Bank, where the growth rate was 4.27 per cent, while it was only 0.85 per cent on the West Bank. (DOS b. 1981: 2). During the period 1962-66, before the June War of 1967, it is estimated that the rate of population growth was 3.3 per cent annually (DOS 1982: XIV). After the 1967 War, and following the Israeli occupation of the West Bank and Gaza, a forced migration and displacement of about 264,000 people took place from the occupied areas to the East Bank of Jordan. In addition, the rate of natural increase of the population was estimated at 3.4 per cent During the period 1970-74 the rate of population growth was estimated at 3.5 per cent.

Due to the political circumstances following the War of 1967, the DOS was unable to undertake a national population and housing census during the period 1961-79. However, an agricultural census was conducted in 1975 which estimated the population of the East Bank at about 1,952,000. In 1978 another agricultural census was conducted for the Jordan Valley. During the period 1975-78 Jordan witnessed several waves of immigration of Arab and non-Arab workers to the East Bank, and the rate of population growth was estimated at 4.3 per cent annually.

The second comprehensive nation-wide population and housing census was carried out by the DOS in 1979. This census aimed at providing integrated data on the social, economic and demographic characteristics of the population of Jordan, and on buildings and occupied housing units. The results gave the population of the East Bank as 2,152,000. This figure included Jordanians abroad on the census evening for purposes of education, tourism and medical treatment, but it did not include Jordanians who were working abroad and their household members, who were estimated at between 500,000 and 750,000 (DOS b. 1981: 3). Table 3.2 shows the historical development of Jordan's East Bank population from the 1952, 1961, and 1979 censuses as well as the population estimates during the period 1962-82, classified by sex.

The 1982 population estimate reported the population of the East Bank as being 2,415,200 persons, which indicates an increase of 1,829,000 or 312 per cent since the first census in 1952. This high rate, which equals an annual average of 4.7 per cent during the 30-year period 1952-82, is due to migration from the West Bank as well as natural population growth, estimated at 3.5 per cent annually.

Fertility; Mortality and Population Growth Rates

The rate of natural growth of population in Jordan has been quite high. It is estimated at 3.5 per cent annually since 1975, compared to a rate of 2.1 per cent in 1950. According to research work on Jordan's population, it is considered that the country is still undergoing the first state of demographic transition as the fertility rate* has remained quite high (Farooq 1977: 3). The 1976 Jordan Fertility Survey reported a fertility rate of about 7.5. Although the rate has apparently declined from about 9 in the early 1960s, it is still one of the highest in the world. It is over 9 among women in rural areas and 6.5-7.00 among women in cities and towns. Due to these high fertility rates, the CBR* has remained very high in Jordan. It was estimated at 47.3 per thousand in 1960 and increased to about 50 in 1975. The CDR† has been steadily declining due to widespread health services and improved sanitation. It was estimated at 21 per thousand in 1950, 18 in 1961, and

* Fertility rate is defined as the average number of births per woman during her reproductive years.

* The crude birth rate (CBR) is defined as the number of births per thousand population.
† The crude death rate (CDR) is defined as the number of deaths per thousand population.

it dropped to 12 in 1975. It was estimated at 9 in 1981 (Futures Group 1981: 5). Life expectancy at birth has increased from about 49 years in 1961 to 61 years for males and 64 years for females in 1981.

In addition to the principal roles of birth and death rates on the net growth rate of population, the large flows of people into and out of Jordan have been a major factor affecting the size of the population. Between 1961 and 1979 migration from the West Bank to the East Bank added 700,000 to the population. During the same period, the outflow of Jordanian workers to the Gulf states was estimated at 500,000 people. In addition, the influx of workers from Egypt and other countries to take jobs in Jordan has added about 180,000 to the population (Futures Group 1981: 6).

Sex and Age Structure

The distribution of Jordan's population by sex has been very stable since 1952. Table 3.2 above indicates that the ratio of females to the total population was 48 per cent from 1952 jto 1982, and the male ratio was 52 per cent. Table 3.3 shows

Table 3.3: Distribution of Population by Sex and Age, 1979 Census Sample (per cent)

Age	Female	Male	Total
0-4	19.2	18.7	19.0
5-9	17.3	17.1	17.2
10-14	14.6	14.4	14.5
15-19	10.5	11.1	10.8
20-24	7.7	7.4	7.5
25-29	5.6	5.5	5.5
30-34	4.9	5.2	5.1
35-39	4.5	4.8	4.6
40-44	4.3	4.2	4.2
45-49	3.1	3.3	3.2
50-54	2.2	2.5	2.4
55-59	1.8	1.7	1.8
60-64	1.4	1.5	1.4
65+	2.9	2.6	2.8
All ages	100.0	100.0	100.0

Source: DOS 1982: 13.

Table 3.4: Estimated Population of the East Bank by Governorate, 1982

Governorate	Size of population	Per cent
East Bank	2,415,200	100
Amman	1,333,400	55.21
Irbid	685,200	28.37
Balqa	170,00	7.06
Karak	141,800	5.87
Ma'an	84,300	3.49

Source: DOS 1982: 1.

Table 3.5: Distribution of Urban and Rural Population by Governorate, 1979

Governorate	Total population (No.)	(%)	Rural (No.)	(%)	Urban (No.)	(%)
East Bank	2,147,065	100	869,111	40.48	1,277,954	59.52
Amman	1,185,181	100	218,674	18.45	966,507	81.55
Irbid	609,251	100	416,603	68.38	192,648	31.62
Balqa	151,543	100	114,849	75.79	36,694	24.21
Karak	126,114	100	88,757	70.38	37,357	29.62
Ma'an	74,976	100	30,228	40.32	44,748	59.68

Source: DOS b. 1981: 40.

the percentage distribution by age and sex in 1979. Due to the high birth rate, Jordan has a large percentage of children. According to this table, the proportion of those under 15 years of age is 50.7 per cent of the total population. This ratio was 45.5 per cent in 1961. The increase is due to a high birth rate accompanied by a declining death rate as a result of the expansion of health, sanitary and social services and the control of epidemic diseases. Consequently, Jordan has a very high child dependency ratio. There is slightly less than one adult of working age to support every child under 15 years. In a typical industrial country there are 2-3 working-age adults for every dependent child. The proportion of the population in the working age group 15-64 was 50.7 per cent in 1961, and 46.5 per cent in 1979. This decrease has had an adverse effect on the labour force paticipation rate, which is already low.

Distribution and Density

The total area of the East Bank of Jordan is 89,206 km^2, of which 89.2 per cent is semi-desert (*badia*). In 1979 the population density was 2.4 persons per km^2. The density in Jordan's governorates is acutely unbalanced, however, with the population concentrated much more in some areas than others. Tables 3.4 shows the estimated population of the East Bank by governorate in 1982. Amman Governorate accomodates over half (55.21 per cent) the total population of Jordan, while the other four governorates together house only 44.79 per cent. Of course, this maldistribution of population, especially the concentration in Amman Governorate, does not reflect normal regional growth. It is rather the result of an influx of about 264,000 persons from the West Bank and Gaza after 1967, plus the continuing internal migration of people from rural areas to Amman city, seeking better life opportunities. As a result of the unbalanced socio-

Table 3.6: Illiteracy Ratio Among Jordanians Aged 15+ by Governorate, Urban/Rural Area and Sex, 1979 Census

Governorate	Urban/Rural	Total	Male	Female
East Bank	Total	34.6	19.9	45.9
	Urban	25.4	13.3	37.6
	Rural	48.6	29.8	76.4
Amman	Total	30.1	16.9	43.8
	Urban	25.3	13.3	37.7
	Rural	52.2	33.7	70.7
Irbid	Total	37.7	20.5	54.4
	Urban	24.5	13.5	34.7
	Rural	44.5	24.0	64.8
Balqa	Total	43.8	28.4	59.5
	Urban	23.4	13.1	35.0
	Rural	49.0	32.6	65.4
Karak	Total	47.7	31.9	62.8
	Urban	34.5	17.7	49.1
	Rural	52.8	36.9	68.4
Ma'an	Total	38.6	21.9	57.4
	Urban	24.4	10.9	39.4
	Rural	52.3	32.4	75.0

Source: DOS b. 1981: 78.

Table 3.7: Percentage of those Illiterate, Literate and with Educational Qualifications among Jordanians Aged 15+, 1979

Age	Total	Illiterate	Literate (without qualifications)	Holding qualifications
15-19	100	7.5	8.1	84.4
20-24	100	14.6	11.7	73.7
25-29	100	23.2	14.1	62.7
30-34	100	32.0	18.3	49.7
35—39	100	42.6	19.0	38.4
40-44	100	55.6	18.5	25.9
45-49	100	59.4	20.7	19.9
50-54	100	60.3	21.6	18.1
55-59	100	67.5	17.0	15.5
60-64	100	71.9	16.9	11.2
65+	100	84.3	10.7	5.0
All ages	100	34.6	14.4	51.0

Source: DOS b. 1981: 79.

economic development which has occurred in Jordan, Amman region accounts for the principal expansion both of population and of economic projects; its urban population increased from 66.7 per cent in 1952 to a high of 60.0 per cent of total population in 1979. Table 3.5 shows the distribution of urban and rural population of governorate in 1979.

The highest percentage of the urban population clearly resides in Amman Governorate, where the two largest cities, Amman and Zarqa, are located; together they constitute 81 per cent of the total population of the Governorate.

Educational Characteristics

The 1979 census reported a 35 per cent illiteracy rate for the Jordanian population over the age of 15. This high rate is mostly due to illiteracy among older people. Table 3.6 shows the illiteracy ratio among Jordanians aged 15 and above by governorate, urban/rural area and sex. In general, illiteracy among females is more than double that among males, in both urban and rural areas. As for governorates, the highest illiteracy ratio (48) exists in Karak Governorate, followed by Balqa (44), Ma'an (39) and Irbid (38), while the lowest ratio (30) is found in Amman.

The Jordanian Farming Population

The total size of the farming population was 403,721 in 1975, with 82.12 per cent living in rural areas and 17.88 per cent in urban areas. It is estimated that 98 per cent of the total farming population are members of the holder's household, and the remaining 2 per cent are other persons, usually paid agricultural workers, who live on the farm and are mostly non-Jordanians. Table 3.8 shows the distribution of farm population by sex and by urban/rural area. The age structure of the farming population indicates that the age group 15-64 constitutes 41.54 per cent of the total, while those under 10 years of age represent 38.78 per cent. Children usually participate substantially in farming activities, either part-time after school or as full-time unpaid workers. The age group 10-14 constitutes 15.7 per cent. Tables 3.9 shows the distribution of farm population by age both for the holder's household members and for other persons who live on the farm. The educational status of farm populations is very low and reflects the general illiteracy which still prevails in the Jordanian agricultural sector. Those who have a preparatory certificate or higher are estimated at only 14.5 per cent of the total farm population. It is important to note that the rate of illiteracy among females (40 per cent) is much higher than that of males (29 per cent). Tables 3.10 gives a very detailed distribution of the farm poplation according to educational status for both males and females in 1975. The figure of 34.6 per cent for the illiteracy rate (Tables 3.7) is for the age group 15+; the equivalent figure for the entire Jordanian population is 17 per cent.

Tables 3.7 shows the percentage of illiterates, those able to read and write, and those who possess educational qualifications among Jordanians aged 15 and above classified by age group. The illiteracy ratio is directly and positively related to age. The highest ratio (84.3 per cent) is among people of 65 years and above, while the lowest ratio (7.5 per cent) is among the youngest group (15-19). There is a negative relation between age and educational attainments. The highest ratio (84.4 per cent) of those who obtain educational qualifications is found among the youngest group of population (15-19 years). The expectation here is that illiteracy will decline rapidly in the future because of high enrolment rates. Virtually all elementary school-age children and almost 80 per cent of preparatory-age children are presently at school.

II. The Jordanian Labour Force

The Jordanian labour force has played a major role in the remarkable performance of the Jordanian economy, which has had a high rate of growth (El-Hurani 1981:85). It is estimated that Jordan's constant price GDP grew at an average rate of 8 per cent annually during 1954-66. A comprehensive study of Jordan's sources of GDP growth revealed that the rate of growth of employment averaged about 4 per cent per annum during the 1961-66 period and about 5 per cent per annum during the proceding six-year period. This expansion of employment, which is more rapid than in the average developing country, may account partly for Jordan's unusually high rate of GDP growth (Mazur 1979:17).

Prior to the 1970s the Jordanian Government supported the outflow of Jordanian manpower for work abroad because the labour market was somewhat loose and a surplus of labour was evident. The first comprehensive Development Plan (the 1973-75 Three-Year Plan) set as one of its major targets a reduction of the rate of unemployment, which was estimated at 8 per cent, and the creation of 70,000 new job opportunities (NPC a. 1981: 15). By the start of 1974, however, labour market conditions in Jordan had changed due to the dramatic rise in oil prices which enabled Arab oil-producing countries to launch huge investment projects requiring a large Arab labour force. In addition, the Jordanian Government increased its economic role through comprehensive development planning schemes. Ambitious Development Plans, both Arab and Jordanian, generated a large demand for labour, forcing the labour market into a tight corner and leading Jordan to depend upon workers from both Arab and non-Arab countries.

Characteristics

The most salient feature of the Jordanian labour force is its very low participation rate in relation to total population. According to the 1979 census, only 20 per cent were economically active (NPC a. 1981: 285). This rate is very low compared to other developing countries, as well as to developed ones. From a larger sample of countries, the average labour force participation rate for a country of Jordan's level of *per capita* income is about 40 per cent (Kuznets 1966: 402). The following factors are responsible for this low rate:

1. The age structure of Jordan's population. As indicated earlier, 50 per cent are under the working age of 15 years.
2. The large number of professional and skilled Jordanians now working as expatriates in Arab oil countries. It is estimated that about 250,000 Jordanians are working abroad (El-Hurani 1981: 97).
3. The participation rate of males in the age group 15-64. This was estimated in 1961 as 83.2 per cent, but had declined to 75 per cent in 1975 (El-Hurani 1981: 95). This rate is very low even when compared to other Arab countries, with over 90 per cent in Syria, Iraq and Egypt. The low rate might be attributable to the high interest shown by Jordanians in education at all levels.
4. Finally, the low participation of females in the Jordanian labour force. It is estimated that female employment constituted only 7.6 per cent of the total in 1979. The corresponding figure for an industrialised country like Japan is 39 per cent, and for Asia as a whole it is 34.4 per cent (Farooq 1977: 11). One of the contributing factors to this low female participation rate is high fertility. When each female bears and raises seven children, very little time is left for joining the labour force. The female participation rate has been increasing lately, however, because of the high demand for workers. It is estimated that about 15 per cent of females in the age group 15-64 are in the labour force.

Despite the overall low labour force participation rate, its growth rate is a high 4.5 per cent annually as a result of rapid population growth; further, because of the large size of the young population, it is expected that the labour force will continue to grow rapidly in the future.

III. The Jordanian Agricultural Labour Force

Unfortunately, accurate and up-to-date statistical data on the size of the agricultural labour force are not available in Jordan. The lastest general agricultural census was conducted in 1975, with an additional agricultural census of the Jordan Valley in 1978. For the purpose of analysing the characteristics of the agricultural labour force, this study will first discuss the results of the general agricultural census of 1975 with respect to the farming population and agricultural holdings.

Table 3.8: Farm Population by Sex and Urban/Rural Area, 1975

	Female	Male	Total
Total farm population	199,910	203,811	403,721
Urban	35,436	36,692	72,178
Rural	164,424	167,119	331,543
No. of members of the holder's household	196,676	199,435	396,111
No of other persons living on holdings	3,234	4,367	7,610

Source: DOS 1977: 226.

Jordanian Agricultural Holdings*

The agricultural census 1975 reported the total number of agricultural holdings as 55,444. Agricultural landholdings constitute 91.42 per cent, while 8.58 per

* An agricultural holding is defined as a techno-economic unit of agricultural production comprising all livestock and all land wholly or partly for agricultural purposes and operated under the management of one person or more.

Table 3.9: Distribution of Farm Population by Age, 1975

Country total	Total farm population	Under 10 years	10–14 years	15–34 years	35–54 years	55–64 years	65+ years
Farm total	403,721	156,564	63,407	93,648	58,844	15,274	15,984
(%)	(100)	(38.78)	(15.70)	(23.19)	(14.57)	(3.78)	(3.96)
Household members	396,111	153,733	62,536	91,180	57,722	15,085	15,855
(%)	(100)	(38.81)	(15.78)	(23.01)	(14.57)	(3.80)	(4.00)
Other	7,610	2,831	871	2,468	1,122	189	129
(%)	(100)	(37.20)	(11.44)	(32.43)	(14.74)	(2.48)	(1.69)

Source: DOS 1977: 230.

Table 3.10: Distribution of Farm Population According to Educational Status, 1975

	Female	Male	Total
Farm population	199,910	203,811	403,721
(%)	(100)	(100)	(100)
Under 6 years	52,644	53,637	106,281
(%)	(26.33)	(26.31)	(26.32)
Illiterate	79,665	36,751	116,416
(%)	(39.85)	(18.03)	(28.83)
Able to read only	1,535	5,536	7,071
(%)	(7.67)	(2.71)	(1.75)
Elementary	47,127	67,959	115,086
(%)	(23.57)	(33.34)	(28.50)
Preparatory	13,239	23,638	36,877
(%)	(6.62)	(11.59)	(9.13)
Secondary	4,979	12,146	17,125
(%)	(2.79)	(5.95)	(4.24)
Academic	4,797	10,926	15,723
Agricultural	39	554	593
Other	143	666	809
College or institute	721	4,144	4,865
(%)		(2.03)	(1.2)
Agricultural	94	772	866
Non-agricultural	627	3,372	3,999

Source: DOS 1977: 231.

Table 3.11: Distribution of Agricultural Holdings by Rural/Urban Area and Land/Animal Category, 1975

Rural/Urban	No. of holdings		
Total agricultural holdings			55,444
Rural area		46,195	
Landholdings	42,327		
Animal holdings	3,868		
Urban area		9,249	
Landholdings	8,364		
Animal holdings	885		

Source: DOS 1977: 206.

cent are animal holdings. It is reported that 83.3 per cent of the total agricultural holdings are located in rural areas, while the remaining 16.7 per cent are in urban areas. Tables 3.11 summarises these general characteristics.

Table 3.12 shows the distribution of agricultural holdings by governorate. It indicates that Irbid Governorate ranks first, with more than half the total holdings (57.4 per cent), followed by the Governorates of Amman (16.3 per cent), Karak (11.9 per cent), Balqa (11.1 per cent), and lastly Ma'an with only 3.3 per cent.

Table 3.12: Distribution of Agricultural Holdings by Governorate, 1975

Governorate	Number	Per cent
Total agricultural holdings	55,444	100
Irbid Governorate	31,830	57.4
Amman Governorate	9,040	16.3
Karak Governorate	6,608	11.9
Balqa Governorate	6,130	11.1
Ma'an Governorate	1,836	3.3

Source: DOS 1977: 208.

The majority of agricultural holdings (97 per cent) are owned by individuals or families, with the remaining 3 per cent owned by companies, cooperative association or government institutions. The census of 1975 also reported that 99 per cent of total holdings are operated by the holders,* with 1 per cent operated by hired managers. About 65 per cent of the total holders declared that their principal occupation is farming. With respect to educational status, the census reported that 52 per cent of total holders are illiterate. Almost three-quarters of the agricultural holdings are entirely owned by the operator. Of the remaining 25 per cent some are wholly rented, others are partly owned and partly rented. By 1975 Jordan agriculture had become largely commercialised. However, 48 per cent of the total landholdings were producing mainly for family consumption.

With respect to the use of hired labour, the agricultural census reported that 64.7 per cent of total agricultural holdings depend completely on the members of the holder's household for all agricultural work, while 31.13 per cent depend on both members of the holder's household and hired agricultural workers, with only 4 per cent entirely dependent on hired agricultural workers. Tables 3.13 shows the distribution of agricultural holdings by governorate and according to the source of workers.

Size and Characteristics of the Agricultural Labour Force

As indicated before, accurate data on the present size of the agricultural labour force are not available. According to the 1979 housing and population census, Jordan's agricultural labour force amounted to 46,200, or 10.3 per cent of the total labour force (DOS b. 1981: 85). The 1961 census reported a comparable figure of 69,600.* The 1981-85 Five-Year Social and Economic Development Plan, however, estimated the agricultural labour force in 1980 at 63,000, or 14 per cent of the total labour force (NPC b. 1981: 283).

Despite the imprecise estimates, it can be safely stated that there was virtually no change in the general size of the agricultural labour force between

* A holder is the person who exercises management control over the operations of the agricultural holdings and makes the major decisions regarding the utilisation of the available resources. He has the technical responsibility for the holding, which he may operate directly as owner or tenant or through a manager to whom he has delegated the responsibility for day-to-day management.

There have been several estimates of the Jordanian agricultural labour force. In addition to the general agricultural census of 1975, the World Bank estimated the size at 73,000 in 1975, while the National Planning Council (NPC) estimated 73,200 permanent workers and 28,000 temporary workers. Also, the NPC forecasts that the permanent agricultural labour force in 1985 will amount to 74,000 and the temporary labour force to 76,000.

Table 3.13: Distribution of Agricultural Holdings by Governorate and Source of Workers, 1975

District	Total no. of holdings	Number of holdings on which:					
		All agricultural work is done members of holder's household		Most agricultural work is done by members of holder's household with the assistance of waged workers		Most agricultural work is done by waged workers	
		(No.)	(%)	(No)	(%)	(No.)	(%)
Total holdings	55,444						
Amman	9,040	7,187	79.50	1,262	13.96	591	6,53
Irbid	31,830	18,398	57.81	12,251	38.48	1,182	3.17
Balqa	6,130	3,251	53.03	2,622	42.77	257	4.19
Karak	6,608	5,271	79.76	1,199	18.14	138	2.08
Ma'an	1,836	1,772	96.51	23	1.25	41	2.23

1961 and 1980. Given the high natural rate of growth of the rural population, this implies substantial rural/urban migration. Naturally, this emigration of labour is from the rain-fed farming areas. The irrigated Jordan Valley is attractive to the agricultural labour force, with seasonal workers and paid farm employment growing by an average 4 per cent annually (El-Hurani 1981: 97).

As the general agricultural census of 1975 was still the most detailed and comprehensive survey available at the time of completing this research, it is worth analysing its results with respect to agricultural labour. It estimated the total permanent agricultural labour force as 77,456, with 94.6 per cent being paid workers.* Interestingly enough, this census reported that 63.8 per cent of these permanent labourers are male and 36.2 per cent are female. This indicates a very high rate of female participation in agricultural labour. (A more detailed analysis of female agricultural work follows.) The age structure of permanent workers indicates that three-quarters of the total permanent agricultural labour force are between 15 and 54 years of age, while one fifth are above 55 years, as shown below.

Age	% of total permanent agricultural labour force
10–14	5.7
15–34	34.7
35–54	40.0
55+	19.6

Source: DOS 1977: IV. IV.

The temporary waged agricultural labour force† is an estimated 15,519, of which 66.1 per cent are male and 33.9 per cent are female. About 85 per cent of temporary workers are between 15 and 54 years of age. The age structure of these agricultural workers is shown below:

Age	% of total temporary agricultural labour force
10–14	4.0
15–34	46.6
35–54	38.0
55+	11.4

Source: DOS 1977: IV. IV.

The occasional agricultural labour force* is estimated at 36,443, all of them being wage-paid workers. Of these workers, 73.8 per cent are male and 26.2 per cent are female. About 92 per cent are between 15 and 54 years of age. The following is the age structure of these occasional workers:

* Permanent workers are defined as those who work more than two-thirds of the agricultural season.

† Temporary workers are defined as those who work less than two-thirds but more than one third of the agricultural season.

* Occasional workers are defined as those who work less than one third of the agricultural season.

Age	% of total occasional agricultural labour force
10–14	2.9
15–34	49.5
35–54	42.4
55+	5.2

Source: DOS 1977: IV. IV.

IV. Characteristics of Employment and Tenant Farms

Agricultural Employment in the Rain-Fed Areas

The typical life-style of farmers in the rain-fed areas centres partially around farming, and partially around other activities between the planting and harvesting seasons. Jordanian wheat farmers are often described as input minimisers because of the traditional farming methods they generally practise. As increased demands for labour have arisen in Jordan's urban areas and in neighbouring countries, potential farm labour has been attracted away; this has increased incentives to use labour-saving equipment, primarily tractors and combines in the cultivation of rain-fed grain crops. A previous study which surveyed the area around Irbid Governorate, Jordan's principal grain-growing region (El-Hurani 1975: 326), revealed that 92.5 per cent of farmers interviewed hired custom tractor operators, and about 90 per cent hired custom combine harvesters; and because most wheat farmers are old and illiterate and cannot read labels or directions, they showed an increasing desire to hire custom sprayers. Many wheat farmers expressed a wish to use grain drills because they face increasing difficulty in hiring broadcast seeding men (El-Hurani 1975: 391). Because of the general trend of decreasing employment in rain-fed farming areas, more attention is directed towards analysing labour and tenancy arrangements in the irrigated Jordan Valley.

Agricultural Employment in Irrigated Farming Areas

As a result of the developing irrigation systems in the Jordan Valley, the cultivated area has been enlarged. Land has been cropped more frequently by reduced fallowing of rain-fed land and double-cropping of irrigated land; and the cropping pattern has shifted away from cereal cultivation towards more labour-intensive fruit and vegetables. It is estimated that between 1975 and 1978 paid farm employment increased by an average of 4 per cent per annum, and non-farm employment by about 12 per cent. It has been noted that paid employment was on the increase for both permanent and temporary workers (Dajani *et al.* 1980: 116). The life-style of a Jordan Valley family centres around agriculture more intensively than in the rain-fed areas. Work starts in about June of every year, when the farmer ploughs the land, prepares the irrigation furrow and plants a nursery. Within 60 days he is ready to plant his vegetables, which will be ready for harvesting in November or December, and which will continue to produce until the end of March (Dajani 1979: 12).

The most recent agricultural census in the Jordan Valley was conducted in 1978. The results reported the total number of agricultural land holdings as 6,930, with only 32.5 per cent depending completely on the members of the holder's household, and 67.5 per cent taking on paid agricultural workers (DOS a. 1981: 17). The results of the census also reported the total number of permanent workers as 11,662, the temporary workers as 10,574 and the occasional workers as 15,703.

A recent study reported that female paid employment in the Jordan Valley is growing faster than male paid employment (Dajani *et al.* 1980: 206). This shift suggests that some females who were previously unpaid are now receiving payment, perhaps because they no longer work on their family holdings. Most female participation in agricultural work starts in September for the planting season; women usually come from the same area, and the average wage rate is JD 2 per day. Alternatively, women sometimes work on an hourly basis at the rate of 300 fils (JD 0.3) per hour.

Because of the availability of relatively cheap imported labour, many Jordanian farmers have become more interested in working as share-croppers and land renters, as will be discussed later. It is estimated that 4,000 non-Jordanian agricultural labourers are at present working in the Jordan Valley. A comprehensive survey of these workers was conducted recently; it showed that 97 per cent are Egyptians (JVA 1981: 39). A few Pakistanis also work in the Jordan Valley. Egyptian workers interviewed reported that they work nine months on average and then return home, some of them coming back to the Valley for a further season. The wage rate ranges from JD 60 to JD 75 per month, or JD 2 to JD 2.5 per day. Labourers usually work from six in the morning until twelve noon. In addition to cash wages, employers also provide accommodation on the farm together with a free meal, usually breakfast, for workers hired by the day, and three meals for workers who are paid monthly. The survey reported that more than three-quarters of the Egyptian workers (76 per cent) are illiterate (JVA 1981: 40).

Tenancy Arrangements in the Jordan Valley

It is estimated that there were about 3,800 share-croppers or land renters working in the Jordan Valley in 1978. It is believed that most of them used to be agricultural workers, but with the influx of non-Jordanian labour to the Valley, they shifted their occupational status (Dajani *et al.* 1980: 126).

The renting arrangement for a fixed payment usually covers a period of five years. The reason is that the plastic houses and drip irrigation equipment have an average life-span of five years; and the agricultural companies usually give a five-year guarantee and maintenance service. The market rate for renting a dunum of land without a drip irrigation system ranges between JD 20 and JD 25, based on the quality of the land; the rent for a dunum of land with a complete drip irrigation system is from JD 30 to JD 40 annually.

A share-cropping arrangement is the most widely accepted form of tenancy between Jordanian landowners and tenants. Where a drip irrigation system is installed, the landowner pays all the expenses of purchasing the equipment for it, but the total costs of preparing the land and the variable inputs—namely seeds, fertilisers and pesticides—are shared equally between landowner and tenant. However, the tenant pays a lump sum of JD 20 per du to the landowner as compensation for the use of a drip irrigation system. The tenant also pays the wages of the labour he hires. The net returns are shared equally between landowner and tenant.

Where a plastic house and a drip irrigation system are in use, the landowner pays all the capital costs for the plastic houses and irrigation equipment. The cost of variable inputs—namely seeds, fertilisers and pesticides—are shared equally between landowner and tenant. The tenant pays the landowner a lump sum of JD 15 per du for the use of the plastic houses and drip irrigation system, and the tenant also pays the wages for hired labour. The net returns are divided: 65-70 per cent of the total net returns to the landowner, and 30-35 per cent to the tenant.

With respect to income levels, it is believed that there are wide variations in share-croppers' annual incomes. Assuming the typical share-cropper farms about 25 du and splits the net returns of JD 50-100 per du with the landowner on a 50-50 basis, the share-cropper earns about JD 1,200 per year. Each share-cropper's income supports about 5.2 persons, on average. This means that *per capita* income is an estimated JD 231 per year. The national average *per capita* income is between JD 250 and JD 350, which puts the share-cropper at slightly below the national average (Dajani *et al.* 1980: 126).

A recent study reported that there were about 2,100 landowning farmers in the Valley in 1978. Although there is an extremely wide range of farm size, technology and income level, if we assume a farm size of 40 du and an average net return of JD 40-80 per du, this implies an annual *per capita* income of about JD 462, which is considerably higher than the national average (Dajani *et al.* 1980: 126).

With the rapid influx of foreign farm workers, Jordanian labour is less intensively used on the farm, as indicated earlier. If the wage rate for Jordanian workers ranges between JD 90 and JD 120 per month, or about JD 1,080-1,440 per year, assuming that each worker's income supports about 4.2 dependents, this suggests an annual *per capita* farm income of about JD 257-342, a level which is equal to or very slightly below the national average (Dajani *et al.* 1980: 125).

Appendix A
Selected Statistics on the Jordanian Labour Force

Table 3.A1: Distribution of the Labour Force by Economic Activity and Nationality, 1980

Economic activity	Jordanians (000)	Non-Jordanians (000)	Total (000)	Total (%)
Agriculture	37.7	5.6	43.3	9.2
Industry	36.1	29.1	65.2	13.9
Electricity & water	5.4	0.8	6.2	1.3
Construction	54.2	18.2	72.4	15.5
Trade	32.8	7.0	39.8	8.5
Transportation & communications	33.9	1.4	35.3	7.5
Public administration & defence	157.1	6.8	163.9	35.0
Other services	31.8	10.6	42.4	9.1
TOTAL	389.0	79.5	468.5	100.0

Source: Sareh 1982:42.

Table 3.A2: Distribution of the Labour Force by Economic Activity and Sex, 1961 and 1979 (thousands)

	1961			1979		
Economic activity	Male	Female	Total	Male	Female	Total
Agriculture	69.3	0.3	69.6	41.7	0.4	42.1
Industry	18.6	2.4	21.0	49.3	1.7	51.0
Electricity & water	0.9	—	0.9	5.3	0.1	5.4
Construction	21.2	—	21.2	61.4	0.1	61.5
Trade	16.5	0.2	16.7	39.7	0.9	40.6
Transportation & communications	7.5	0.1	7.6	32.0	0.2	32.2
Public administration & defence	45.3	0.4	45.7	130.7	20.3	151.0
Other services	24.7	4.0	28.7	29.0	8.3	37.3
TOTAL	204	7.4	211.4	389.1	32.0	421.1

Source: Sareh 1982: 42.

Table 3.A3: Distribution of Jordanians Aged 15+ by Age and Type of Activity, 1979 Census Sample (per cent)

Age	Employed	Other[1]	Total
15-19	15.7	84.3	100
20-24	44.5	55.5	100
25-29	51.3	84.7	100
30-34	51.8	48.2	100
35-39	50.6	49.4	100
40-44	47.1	52.9	100
45-49	38.3	51.7	100
50-54	47.3	52.7	100
55-59	37.3	62.7	100
60-64	34.0	66.0	100
65+	14.5	85.5	100
All ages	38.3	61.7	100

Note: [1] This category includes those unemployed and those economically inactive, (i.e., students, housewives and others).

Source: DOS b. 1981: 81.

Table 3.A4: Distribution of Employed Jordanians Aged 15+ on the East Bank by Governorate and Employment Status, 1979 (per cent)

Governorate	Total	Worker on own account	Employee	Other[1]
East Bank	100	22.4	72.0	5.6
Amman	100	17.4	74.8	7.8
Irbid	100	30.6	67.5	1.9
Balqa	100	25.4	68.4	6.2
Karak	100	31.5	66.6	1.9
Ma'an	100	23.9	73.4	2.7

Note: [1] This category includes those employers family workers and unspecified cases.
Source: DOS b. 1981: 86.

Table 3.A5: Distribution of Employed Jordanians Aged 15+ on the East Bank by Industry, 1979 (per cent)

Industry	% employed
Agriculture, hunting & fishing	10.3
Mining	1.0
Manufacturing	8.4
Electricity, gas & water	1.3
Construction	13.6
Trade, restaurants & hotels	9.7
Transport, storage & communications	8.3
Financial, insurance & real estate services	2.4
Public administration, defence & other services	45.0
TOTAL	100.0

Source: DOS b. 1981: 85.

Table 3.A6: Distribution of Employed Jordanians Aged 15+ on the East Bank by Major Occupation, 1979 (per cent)

Major occupation	% employed
Professional, technical & related workers	13.0
Administrative & managerial workers	1.4
Clerical & related workers	6.9
Sales workers	7.9
Service workers	6.9
Agriculture, animal husbandry & forestry workers	10.6
Production & related workers, transport equipment operators & labourers	53.3
TOTAL	100.0

Source: DOS b. 1981: 84.

References

Dajani, Jarir, 1979, 'A baseline socio-economic study of the Southern Ghors and Wadi Araba', Report to the US Agency for International Development, Amman.

——, Jared Hazleton *et al.*, 1980, 'An interim evaluation of the Jordan Valley development effort, 1973-80', Report to the US Agency for International Development, Amman.

DOS (Department of Statistics) 1977, *General Results of the Agricultural Census in 1975*, Amman.

—— a. 1981, *Results of the Agricultural Census of Jordan Valley in 1978*, Amman.

—— b. 1981, *Main Findings of Advance Tabulations of Housing and Population Census in 1979*, Amman.

—— 1982, *Statistical Yearbook*, Amman.

Farooq, Ghazi, 1977, 'Population growth, manpower, migration and development with special reference to Jordan, Iraq, and Syria', paper presented to Seminar on Population, Employment and Development by the National Planning Council and ILO, Amman.

Futures Group, The, 1981, *The Effects of Population on the Economic and Social Development of Jordan*, Washington, DC.

El-Hurani, M. Haitham, 1975, 'Economic analysis of the development of the wheat subsector of Jordan', unpublished PhD dissertation, Iowa State University, Ames, Iowa.

—— 1981, 'Analysis of Jordan labor force estimations,' *Journal of Economics and Administration*, Research and Development Center, King Abdul-Aziz University, Jeddah.

JVA (Jordan Valley Authority), 1981, *Jordan Valley Magazine*, Amman.

Kuznets, Simon, 1966, *Modern Economic Growth: Rates, Structure and Spread*, Yale University Press, New Haven, Conn.

Mazur, Michael, 1979, *Economic Growth and Development in Jordan*, Westview Press, Boulder, Colo.

NPC (National Planning Council) a. 1981, *Three-Year Plan for Social and Economic Development, 1973-75*, Amman.

—— b. 1981, *Five-Year Plan for Social and Economic Development, 1981–85*, Amman.

Sareh, Yasir, 1982, 'Labor force and employment in Jordan until 1985', Report to the National Planning Council, Amman.

PART II

FARM SYSTEMS

4 FARM SYSTEMS IN IRRIGATED AREAS

Burhan Abu Howayej

Introduction

Until the beginning of the 1960s irrigated land in Jordan was devoted to growing vegetables and field crops. Later on farmers gradually began to concentrate on fruit trees, particularly citrus. Farmers in the northern part of the Jordan Valley were more active in this respect and consequently citrus trees are grown mainly in that region. Farmers in the aquifer regions concentrate more on olive and stone-fruit trees and summer vegetables.

Farming was carried out in the open until the middle of the last decade, when farmers started using plastic covers for growing vegetables. This plastic culture* has been practised at an increasing rate since then and the total area under plastic reached 15,000 du by 1981.

Farmers in irrigated areas differ in their cropping patterns; some specialise in growing a single crop, while others diversify. Consequently, and for the purposes of this chapter, irrigated farms can be grouped as follows:

1. Farms fully devoted to growing cucumbers in plastic houses with a drip irrigation system
2. Farms fully devoted to growing tomatoes in the open fields with a drip irrigation system
3. Farms fully devoted to growing mixed vegetables with a flood system
4. Farms fully devoted to growing only citrus trees

Each of the four groups above is considered a production sub-system and has been analysed accordingly.

* Plastic culture in Jordan takes the form of large steel-frame hothouses up to 3 m high and 5 m long, or small tunnels about 0.5 m high, supported by steel wires. It also involves pegging black plastic sheets over ploughed land and punching holes at intervals through which the seeds can grow. This is called plastic mulch. Usually plastic culture is used with drip irrigation systems.

Environments and Activities Related to Irrigated Farm Systems

Our objective in analysing farm systems is to establish the relationships between farm inputs and outputs and agricultural policies. To achieve this, we must be aware of all on-farm and off-farm conditions, such as physical environment, relevant institutions and availability of information. We must also identify physical human activities and decisions relevant to farm systems.

Inspection of farming activities showed that we can consider each activity as a sequence of simpler ones. Those activities, described as follows, proved adequate for satisfactory analytical results.

To identify activities and conditions relevant to each of the four production sub-systems, a typical farm has been visualised and a flow chart of relevant activities has been prepared. In what follows we record all relevant conditions and farming activities.

Analysis of a Cucumber Plastic Farm System

Environments

Environments in this study include everything related to a cucumber plastic farm other than human activities. Consequently, they cover all the on- and off-farm environmental/physical conditions, institutional arrangements and information available which interrelate with human activities. The following is a detailed description:

On-Farm Environments

Physical

Our farm is located in the Jordan Valley. It consists of 35 du, including 5 du under plastic, and has water rights sufficient to irrigate the whole area from the main canal of the East Ghor Irrigation Project. (Not all farms in the Valley have the same rights, and some of them get water from artesian wells.)

The Jordan Valley climate prevails and favours the growing of off-season winter vegetables and any other crops that need sub-tropical weather conditions. High temperatures in summer do not encourage growing vegetables during that season. Winter vegetables are subject to drops in temperature and frost, which have an adverse effect on growing vegetables in early winter. To protect the crop from such unfavourable conditions, the farmer has adopted the use of plastic houses. He has constructed 10 plastic houses and installed a drip irrigation system. He also uses black plastic sheets to cover the ground beneath these houses during the growing period.

The plastic houses are used to grow cucumbers and the rest of the farm to grow other vegetables. Before adopting plastic culture the farmer used to grow mixed vegetables by traditional methods, obtaining average yields and paying average costs, as shown in Tables 4.1 and 4.2. The majority of farmers in the Jordan Valley use plastic houses for growing cucumbers, and some of them use them for tomatoes or peppers (Tables 4.3-4.5). The farmer has to pay the average costs of production shown in Table 4.3 to grow cucumbers and to get the average yield of 8 t per du. Detailed levels of input, prices and costs, as well as methods of application, are shown in the record of each activity undertaken on the farm.

Permanent constructions other than the plastic houses are a dwelling house of three rooms and a wooden farm shed.

Electrical power is obtained from a small generator attached to the drip irrigation system. Work is done by the family and hired labour. The farmer can get electricity off the Valley's main supply.

Total assets are estimated at JD 15,900, and total liabilities are estimated at JD 19,500 just before the harvest of last season's crop. These liabilities are likely to decrease if the farmer uses the returns from harvesting to repay loans. Total assets include the following items:

Table 4.1: Production Costs of Mixed Vegetables Grown on a Mixed Vegetable Farm as at 1982 (JD per dunum)

Cost item	Tomatoes	Eggplants	Squashes	Beans	Cauliflowers
Material inputs					
Machinery & tools	2.5	2.5	2.5	2.0	2.5
Seeds	3.0	4.0	6.0	1.5	9.0
Chemicals	48.0	2.7	2.1	2.1	0.3
Manure	25.7	25.7	25.7	25.7	25.7
Triple phosphate	4.0	4.0	4.0	4.0	4.0
Ammonium sulphate	4.2	4.2	4.2	4.2	4.2
Fuel	1.0	1.0	1.0	1.0	1.0
Water	3.6	6.0	2.4	2.0	4.2
Sub-total	92.0	50.1	47.9	42.5	46.9
Labour inputs					
Preparing soil	0.5	0.5	0.5	0.5	0.5
Planting	0.8	1.0	2.0	0.5	1.0
Irrigating	3.3	5.0	1.3	6.3	3.8
Applying chemicals	1.3	2.0	1.5	1.0	0.5
Applying fertilisers	1.0	1.0	1.0	0.5	0.5
Tending the crop	3.8	4.3	0.5	3.5	3.8
Harvesting	8.8	4.5	6.0	3.8	1.5
Sub-total	19.5	18.3	12.8	16.1	11.6
GRAND TOTAL	111.5	68.4	60.7	58.6	58.5

Source: Steitieh 1978.

The above estimates are based on one year's ownership of plastic houses. They do not include the value of the farm land or the interest on borrowed capital.

Institutional

The farm was inherited from the farmer's father. His heirs must operate the farm as one unit to maintain their ownership and right to farm. If they want to sell the farm, they have to sell it to the Jordan Valley Authority (JVA), at a price determined by the Government. To keep up the minimum size of landholdings in the Jordan Valley, the JVA does not allow fragmentation to less than 30-35 du. Not all farms are in ownership as above, however. Some were acquired by

* Agricultural Credit Corporation.

† Jordan Valley Farmers' Association

Table 4.2: Labour Requirements and Frequencies for Traditionally Grown Vegetables in Irrigated Areas (man-hours per dunum)

Crop	Planting		Cultivating		Fertilising		Irrigating		Spraying		Harvesting		Total
	H	F	H	F	H	F	H	F	H	F	H	F	H
Potatoes	3	1	6	2	4	4	7	7	4	4	17	1	41
Green beans	2	1	14	2	2	2	25	16	4	4	15	6	57
Broad beans	1	1	12	3	0.5	1	6	3	4	4	15	6	38.5
Squashes	8	1	2	3	4	2	5	6	6	6	24	12	53
Cucumbers	3	1	20	5	4	4	6	5	6	6	24	12	63
Onions	3	1	11	2	2	2	6	5	2	2	30	1	52
Tomatoes	3	1	15	5	4	5	13	9	5	5	35	13	75
Eggplants	4	1	17	4	4	4	20	15	8	8	18	9	71
Peppers	4	1	10	3	3	1	20	12	3	3	25	15	65
Cauliflowers	4	1	15	2	2	2	15	10	2	2	6	1	44
Melons	2	1	8	2	2	2	4	5	2	2	4	2	22

H = Man-hours
F = Frequency

Source: Steitieh 1978.

Table 4.3: Production Costs of Cucumbers Grown under Plastic Houses as at 1982 (JD)

Cost item	Cost per du	Total cost of 5 du
Material inputs		
Machinery & tools	17	85
Seedlings	320	1,600
Manure	80	400
Triple phosphate	4	20
Compound NPK	17.4	87
Chemicals	110	550
Water	2	10
Fuel	8	40
Depreciation	480	2,400
Sub-total	1,039.4	5,192
Labour inputs		
Preparing soil	26	130
Planting	4.4	22
Irrigating	9.2	46
Applying chemicals	28.2	141
Applying fertilisers	3	15
Tending the crop	35.6	178
Harvesting	74.4	372
Sub-total	180.8	904
GRAND TOTAL	1,220.2	6,096

Sources: Physical requirements obtained from estimates in Steitieh and Musa 1980; prices obtained from market prices collected by the researcher in 1982.

farmers as a consequence of the 1980 land reform programme in the Jordan Valley.

The family consists of six members: father, mother, two sons and two daughters. The father and mother work full-time on the farm and the children go to school and work on the farm during holidays. The father was schooled to preparatory level.

Informational

The farmer has a radio and TV set and tunes in occasionally to agricultural programmes. He has sales receipts for his past crops, which include wholesale prices. He has some hand-outs received from salesmen, which include information about agricultural inputs. He also has samples of some seeds and chemicals which he has not yet tried. The JVFA extension agent visited him three times last year to offer advice, and salesmen visited him several times to promote sales of their products. He also received visits from ACC men supervising their loans. The driver who transports his crops to market gives the farmer verbal reports of sales there. It is noticeable that the Ministry of Agriculture (MOA) extension agent is not in contact with this farmer, and the staff of the MOA

Table 4.4: Production Costs of Tomatoes Grown on a Specialised Tomato Farm as at 1982 (JD)

Cost item	Cost per du	Total cost of 35 du
Material inputs		
Machinery & tools	4.6	161.0
Seedlings	6.6	231.0
Manure	25.7	899.5
Triple phosphate	4.0	140.0
Compound NPK	11.6	406.0
Chemicals	9.6	336.0
Water	2.0	70.0
Fuel	1.2	42.0
Mulch	20	700.0
Depreciation	20	700.0
Sub-total	105.3	3,685.5
Labour inputs		
Preparing soil	0.6	21
Planting	4.3	150.5
Irrigating	4.0	140
Applying chemicals	2	70
Applying fertilisers	2.5	87.5
Tending the crop	3	105
Harvesting	28.7	1,004.5
Sub-total	45.1	1,578.5
GRAND TOTAL	150.4	5,264.0

Source: Physical requirements obtained from Steitieh and Musa 1980.

have not contacted him even though the farm is only a few kilometres from their offices.

Off-Farm Environments

The part of the Jordan Valley where the farm is located has been the recipient of an integrated development programme for a water supply to irrigate 272,000 du, the establishment of health, social, educational and agricultural services, the construction of paved and farm roads, the supply of electrical power, and land distribution and reform. These improvements in the off-farm environment have facilitated the farmer's efforts to grow cucumbers.

Human Activities

The human activities related to a cucumber plastic farm are all those decisions and physical actions which, together with the environment, go into producing the crop for marketing. As each activity is made up of a sequence of simpler ones, we selected and recorded those shown below. Although the farmer started using plastic houses a year previously, the record of activities starts with those undertaken when he first considered using these houses. Activities related to growing other crops outside plastic houses are ignored because we are analysing 10 plastic houses as a farm system in itself.

Table 4.5: Average Yield of Main Crops Grown in Irrigated Areas (tonnes per dunum)

Crop	Yield	Crop	Yield
Tomatoes[1]	6.0	Broad beans	0.59
Tomatoes[2]	3.4	Green beans	0.55
Tomatoes[3]	2.0	Dry onions	1.1
Cucumbers[4]	8.5	Potatoes	1.16
Cucumbers[5]	6.0	Watermelons	0.5
Cucumbers[6]	2.2	Sweet melons(f)	1.5
Sweet melons[6]	1.7	Wheat	0.3
Squashes	0.73	Barley	0.3
Peppers	0.58	Clover (hay)	2.0
Eggplants	1.53	Alfalfa (hay)	2.0
Cauliflowers	1.4	Citrus	3.0
		Grapes	3.0

Notes: [1] Plastic houses and drip irrigation.
[2] Open field and drip irrigation.
[3] Open field and surface irrigation (traditional method).
[4] Plastic houses and surface irrigation.
[5] Plastic tunnels and drip irrigation.
[6] Plastic tunnels and surface irrigation.

Sources: Steitieh 1978.
Steitieh and Musa 1980.

Item	Unit price (JD)	Total value (JD)
Plastic houses	742	7,420
Drip system	3,500	3,500
House	600	600
Pick-up	2,000	2,000
Tractor	2,000	2,400
TOTAL		15,920

Liabilities include the following items:

Item	Value (JD)
Mid-term loan from ACC*	11,500
Seasonal loan from ACC	2,500
Seasonal loan from JVFA†	1,000
Loan from a middleman	2,000
Value of unpaid input	500
Loans from other sources	1,000
TOTAL	19,500

*Agricultural Credit Corporation.
†Jordan Valley Farmers' Association.

The sequence of both physical activities and decisions related to the farm as listed above is shown in Figure 4.1. As far as the time sequence is concerned, the chart does not include the possibility of moving from the completion of one job to the start of another. Thus it does not show up the possibility of buying chemicals before preparing the soil. The time sequence is partially covered by recording for each activity the preceding ones.

Activity Analysis

Physical Activities

1. ASSESS ASSETS
Purpose: To develop a financial statement for the farm.
Inputs required: Price and value estimates of outstanding fixed assets, used machinery, unsold crops, stocks of chemicals and fertilisers, case in hand, debits and credits.
Determining factors: Lack of farm records and information about prices of used materials.
Preceding activity: None.
Comments: Farmers carry out their assessements one way or another, while they are thinking about what to do during the next season, without putting them on paper.

2. DECIDE TO FARM VEGETABLES
Purpose: A sensible farmer is always thinking of increasing his income and consequently asks whether his occupation is the most he can achieve. Therefore at the close of every season, he has to decide whether to continue farming or to move into some other business. The man we are studying chose farming as his career and decided to continue the following season.
Inputs required: Information about alternative careers.
Alternatives: To farm or to go into some other business.
Determining factors: Potential opportunities available to the farmer, income from farming, social attitudes towards farming, risk involvement, farmer's attitude towards risk-taking, and farmer's expectations of the impact of change upon agricultural conditions in the future and changes in farm management and/or the means of production.
Latest time to decide: July, when the growing season starts.
Preceding activity: 1.
Comments: The strength of each factor and the impact of the combination of factors on a farmer's decision have not yet been explored to reveal how the farmer makes his decision.

3. DECIDE ON CROPPING PATTERN
Purpose: To select the most economical pattern.
Inputs required: Information on the economics of alternative cropping patterns and their requirements.
Alternatives: To grow one single seasonal crop, or mixed crops, or fruit trees only, or vegetables on one part of the farm and fruit trees on the other.
Determining factors: Lack of recommendations at regional and farm level, and lack of official advice on what to grow; also lack of information on projected costs of production and market prices of agricultural crops.

Figure 4.1: Activities Related to a Cucumber Plastic Farm System

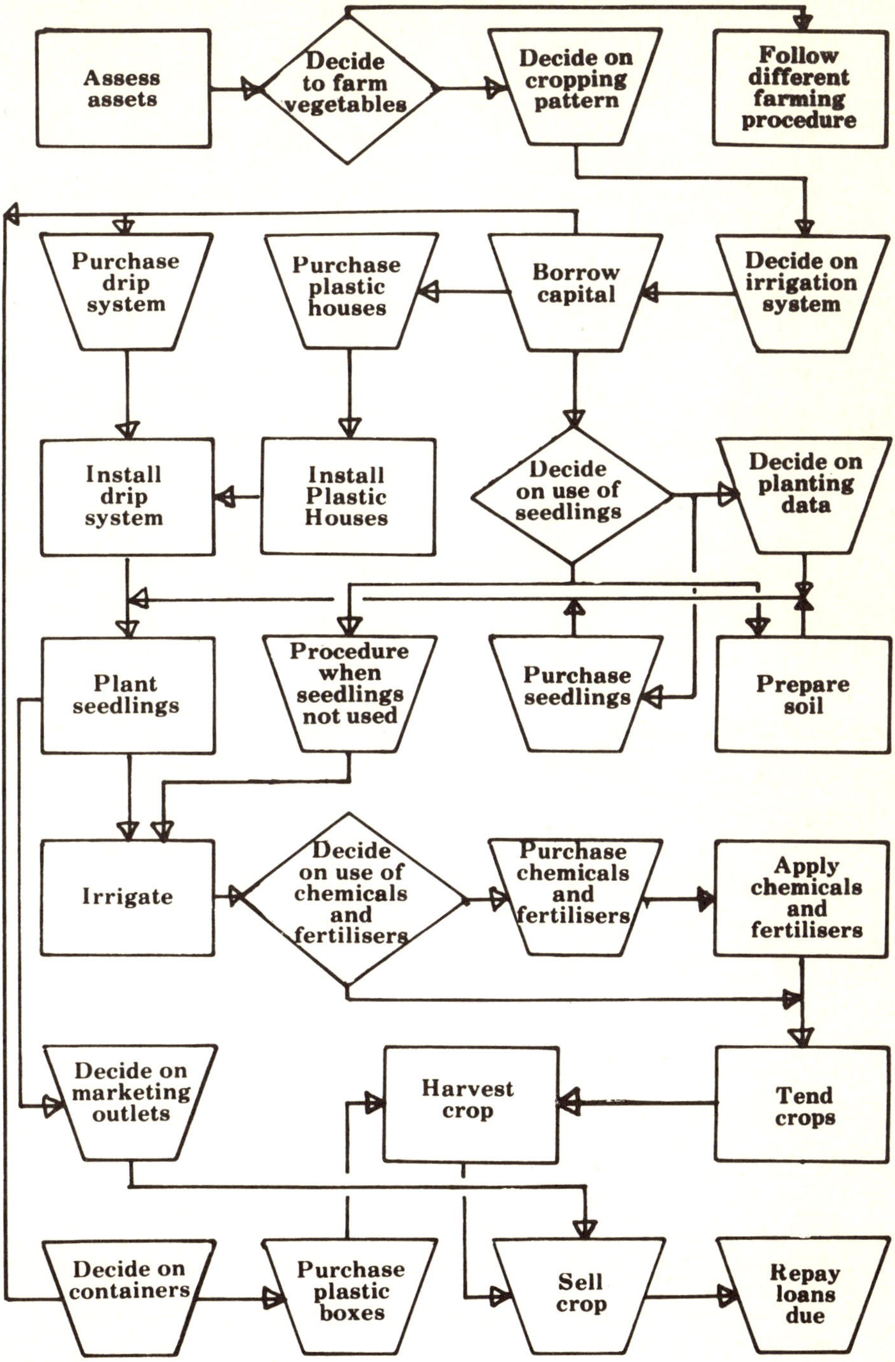

Statistics available from official sources show that the first and third alternatives are capital-intensive. The first needs capital investment if the farmer wants to use plastic culture and a drip irrigation system. The third and fourth need capital investment to establish a fruit-tree farm. Each dunum needs JD 1,500 for plastic houses and JD 120 for a drip irrigation system. Fruit trees need JD 477 per du to establish a citrus farm of 30 du. The first alternative yields a high income in a short space of time, while fruit trees yield a high income over a relatively longer period. The economics of each alternative are included in Tables 4.3-4.6.

Latest time to decide: Late July. Our farmer selected the first alternative, to grow only cucumbers.

Preceding activities: 1 and 2.

Table 4.6: Annual Costs of Growing Citrus Trees (JD per dunum)

Item	Year 1	2	3	4	5	6	7
Material inputs							
Machinery	3	3	3	3.5	4	4	4.3
Seedlings	80	8	—	—	—	—	—
Fencing	19.5	—	—	—	—	—	—
Manure	4	—	—	4	—	—	4
Compound NPK	4.6	5	5.5	6	6.5	8	9.3
Ammonium sulphate	1.8	1.8	3.6	3.6	10.2	10.2	10.2
Triple phosphate	4	—	—	—	—	—	—
Chemicals	0.4	1	1.6	4.5	5	5.3	5.3
Water	1.8	1.8	2.1	3.6	3.6	3.6	3.6
Sand	6	—	—	—	—	—	—
Sub-total	125.1	20.6	15.8	25.2	29.3	31.1	36.7
Labour inputs							
Fencing	6	—	—	—	—	—	—
Planting	30	3	—	—	—	—	—
Cultivating	14	14	14	14	14	14	14
Irrigating	24	24	24	24	24	28	28
Applying chemicals	1.2	1.2	1.5	2	2	2.5	2.5
Applying fertilisers	1.5	1.5	1.5	2.0	2.0	2.2	2.2
Maintaining the trees	—	2.3	2.3	2.3	2.0	3.0	3.0
Harvesting	—	—	—	—	—	14	14
Sub-total	76.7	46	43.3	44.3	45	49.7	63.7
GRAND TOTAL	201.8	66.6	59.1	69.5	74.3	80.8	100.6

Source: Data collected and averaged by author through formal interviews with a selected number of citrus farmers in Jordan in 1983.

4. DECIDE ON METHOD OF CULTIVATION

Purpose: To select the most economic and suitable method of cultivating cucumbers.

Inputs required: Information on costs, returns and requirements of each alternative.

Alternatives: Plastic houses, plastic tunnels and open fields. Each of these three alternatives contains further choices. Plastic houses can be used either with a drip system or a flood system for irrigation. They can be used with or without mulch. The same applies to the other two alternatives. Our farmer decided to use plastic houses with a drip irrigation system and mulch.
Determining factors: The farmer's financial capability to borrow capital. The motivation is to secure a higher income from growing an off-season cucumber crop by plastic culture. In a study conducted by Steitieh and Musa (1980), a comparison was made between various alternatives. The findings were as follows:

Item	Plastic houses & drip system	Plastic houses & flood system	Plastic tunnels & drip system
Price (JD/t)	256	266	238
Yield (t/du)	8.52	6	2.2
Gross return (JD/t)	2,176	1,599	672
Production costs (JD/du)	337	296	114
Marketing costs (JD/du)	313	225	85
Variable costs (JD/du)	651	522	196
Net returns (JD/du)	1,525	1,076	425
Initial capital investment (JD/du)	1,998 2	1,718	366

Selection of the first alternative entails commitment to farming for at least five years to recover capital investment.
Latest time to decide: Late July.
Preceding activities: 1, 2 and 3.

5. DECIDE ON IRRIGATION SYSTEM
Purpose: To select the most economic and suitable irrigation system.
Inputs required: Information on the economics and requirements, as well as merits, of each alternative.
Alternatives: Drip system and flood system. The first alternative saves water and increases efficiency of water use from 30 to 80-90 per cent, but requires high capital investment. The average cost of putting 1 du under a drip system on a farm of 30 du is JD 120. The flood system is less expensive but needs more water and labour. It also introduces diseases from outside the farm which create soil problems. The full long-term implications of the drip system on soil have not been explored under local conditions. This system is not recommended yet by any official agency in Jordan, and farmers use it at their own risk. In the meantime, agencies are not discouraging farmers from adopting it. Our farmer selected the first alternative.
Latest time to decide: Late July.
Preceding activities: 3 and 4.
Comments: Plastic houses in Jordan are used together with the drip system, as observed by the researcher.

6. BORROW CAPITAL
Purpose: To obtain enough capital to purchase and operate 10 plastic houses.
Inputs required: Property as collateral, visits to and contacts with credit sources to apply for loans, and arranging for the extension of loans.
Determining factors: ACC and Jordan Cooperative Organisation (JCO) services, which facilitate mid-term loans. However, their terms, which include 30 per cent capital investment by the farmer, force him to borrow from more than one source. Furthermore, they include a ceiling on the amount of loans for the purpose of buying plastic houses. The ACC and JCO do not issue loans to cover more than the cost of 10 plastic houses for each landholder, even though the farmer possesses the required collateral. The JVFA, as well as being dealers in agricultural supplies, are sources of seasonal loans. If the farmer has no collateral, he cannot get a mid-term loan from institutional sources.
Timing: June-September. Time required: 3 months.
Preceding activities: 1, 2, 3, 4 and 5.
Comments: The farmer has to visit the ACC and/or JCO several times, starting in June, to get the loan finalised, and he has to accept supervision by these institutions.

7 and 8. PURCHASE PLASTIC HOUSES AND DRIP SYSTEM
Purpose: Upon extension of the mid-term loan, the farmer sets out to purchase the plastic houses and the drip irrigation system he has decided upon.
Inputs required: Capital equal to the total value of both items, estimated at JD 11,500 for the plastic houses and JD 3,300 for the drip system. The farmer needs a mediator to secure a good deal and the best service from the dealers. In their supervisory capacity, the ACC and JCO deduct the cost of the plastic houses from the approved loan and pay it to the dealer, who then delivers the houses to the farmer.
Determining factors: Choice of dealers. Some offer the service of supplying plastic houses and constructing them on the farm; some have demonstration farms to show farmers how to assemble the house frames and to instruct farmers in the merits of their products. Some of them sell on a credit basis to their clients against co-guarantor. The same applies for the drip irrigation systems.
Timing: June-September. Time required: 4 days.
Preceding activity: 6.

9 and 10. INSTALL PLASTIC HOUSES AND DRIP SYSTEM
Purpose: to get the plastic houses and drip system ready for use on time.
Inputs required: Installation costs are included in the price, and the agency assists in the execution of this operation. The plastic houses can be assembled by the farmer with trained hired labour, or by the agency if specially contracted. If the farmer decides to construct the houses on his own, he needs 3 days training and 35 man-days to do the work.
Determining factors: There are specialists in the construction of plastic houses who offer a service to farmers at JD 50 per du and there are trained labourers available to the farmer at JD 5 per working day. these speed up the work. Our farmer lacks experience in installing a drip system and has to rely upon the specialists.
Timing: September-October. Time required: 2 weeks.
Preceding activity: 7.

11. DECIDE ON USE OF SEEDLINGS

Purpose: To decide whether to use seedlings or seeds for growing cucumbers.
Inputs required: Information on the economics and merits of each alternative.
Alternatives: Use of seedlings produced by a specialist grower or use of seeds. The first alternative, which our farmer selected, saves time and work, helps get healthy seedlings and better planting. It is more costly because it includes the profits of the specialist grower. The second choice is cheaper and helps use the resources available on the farm and family labour, but it involves a high risk of failure in tending the seeds after sowing, and reliance upon the relative inexperience of the farmer.
Determining factors: Availability of and access to services of a specialist seedling grower. Risks as stated above. The farmer should be encouraged to select the first alternative.
Latest time to decide: September.
Preceding activity: 6.

12. PURCHASE SEEDLINGS

Purpose: Having decided to use seedlings, the farmer has to purchase the required quantity.
Inputs required: Capital for the total cost of seedlings, estimated at JD 1,600 for 20,000 plants. The farmer has to contact specialists to get a good deal, and to arrange for delivery of the plants to the farm gate.
Determining factors: Late contact with seedling growers leads to delay in delivery and/or inability to obtain the required quantity. This will result in late planting, or even missing the growing season.
Timing: September-October. Time required: 2 days.
Preceding activity: 6.

13. DECIDE ON PLANTING DATE

Purpose: To select the best planting date.
Inputs required: Information on probable marketing conditions during the harvesting period, and information on probable weather conditions during the growing period.
Alternatives: To plant the whole area early, or late, or to plant intermittently and have three main plots with three planting dates. The first alternative yields a high income but is very risky. The second is least risky but gives the lowest returns. The third is a kind of diversification to minimise risk and increase returns. This last alternative involves greater management effort.
Determining factors: A major constraint is the lack of reliable data on the market and weather conditions. Fluctuation of prices in the past caused our farmer to select the third alternative. Market prices in 1982 showed that the wholesale price reached JD 1 per kg early in the season but dropped to JD 0.25 per kg late in the season.
Latest time to decide: Late October.
Preceding activity: 11.

14. PREPARE SOIL

Purpose: To render the soil suitable for growing cucumbers.
Inputs required:

— Water	100 m^3 at 3 fils per m^3 and a total cost of JD 0.3
— Chicken manure	20 t at JD 20 per t and a total cost of JD 400

— Triple phosphate	0.5 t at JD 40 per t and a total cost of JD 20
— Machinery (tractor for ploughing)	2.5 working hours at JD 6 per hour and a total cost of JD 15
— Soil fumigants	5 barrels at a total cost of JD 150
— Labour	65 man-days at JD 2 per day and a total cost of JD 130

Determining factors: Some farmers put off fumigation until the second year if weeds are not too much of a problem. Soil fumigation is a relatively long-term investment, as is the use of manure. Consequently, the application of manure and soil fumigants is not repeated every year. Manure is applied every three years and fumigants when necessary. There are agents specialised in manure application who provide a service to farmers. They charge JD 17 per house, excluding the cost of chemicals. Chicken manure is not always obtainable on the market. Tractors can be rented, so they need not be owned.
Timing: October-November. Time required: 20 days.
Preceding activity: 9.
Comments: This activity includes several simpler activities such as the purchase of manure, triple phosphate and soil fumigants. It also includes contact and arrangements with the tractor agent and fumigating agent to obtain their services on time. Finally, it includes developing an irrigating schedule to be approved by the JVA for water supply.

15. PLANT SEEDLINGS
Purpose: Upon delivery of seedlings, the farmer has to plant them.
Inputs required: In addition to the seedlings, the farmer needs labour (females are preferred to do the work) of 11 man-days at JD 2 per day, and a total cost of JD 22.
Determining factors: Poor handling of seedlings and delay in transplanting leads to a high rate of wastage and additional work to replace dead plants.
Timing: October-December. Time required: 3 days, intermittently.
Preceding activities: 11 and 12.

16. IRRIGATE
Purpose: To give the plants the required quantity of water at the right time.
Inputs required: A drip irrigation system waters the plants mechanically as long as the system is running. Labour is required to turn the system on and to check for clogged drippers. Labour requirements approximate 23 man-days at JD 2 per day and a total cost of JD 46. The drip system requires water to be pumped into the piplines. total water requirements are 1,500 m^3 at 3 fils per m^3 and a total cost of JD 4.5. It also needs fuel oil at an average of 1,000 l, costing JD 40.
Timing: Throughout the season, starting from the first day of planting and ending a few days before the last harvest.

Preceding activity: 13.

17. DECIDE ON USE OF CHEMICALS
Purpose: To decide whether or not to use chemicals for the control of plant diseases and insects.
Inputs required: Relevant information on different kinds and forms of chemicals: cost, function, rate and method of application, and precautions against toxicity.

Alternatives: To use chemicals or not. To use them gives a better chance of growing healthy plants and of obtaining higher yields but involves higher production costs. Our farmer selected the first alternative.
Determining factors: Lack of information on the best use of each chemical based on local tests. However, the Department of Agricultural Research and Extension (DARE) at the MOA has conducted limited research into the best application of some chemicals for the control of certain insects and diseases, and the results have been published and are known to farmers.

Merchants import several kinds of chemicals for different plant protection purposes, and their services are available to farmers. They provide the farmer with information about the best method and rate of application and precautions. Farmers differ in their rates of application of the same chemical, and they rely heavily on suppliers for information.

In using chemicals, the farmer wants them either for control or for protective purposes, and he develops his schedule of application accordingly.
Latest time to decide: Just after planting the seedlings.
Preceding activities: 3 and 6.
Comments: Some farmers use leaf nutrient together with chemicals.

18. DECIDE ON USE OF FERTILISERS
Purpose: To decide whether to use fertilisers or not.
Inputs required: Information on functions, costs, output, rate and method of application for each kind of fertiliser.
Alternatives: To use fertilisers or not to use them. There are various reports, and farmers believe that fertilisers increase yield but also increase costs.
Determining factors: The decision is oriented more towards selection of the best kinds of fertiliser rather than whether to use them or not. Information about the best method and rate of application for each specific area is lacking and the response of crops to different rates of application has not yet been investigated. Recommended rates and methods of application are based on manufacturers' advice and farmers' experience. Our farmer selected the first alternative.
Latest time to decide: One month after planting the seedlings.
Preceding activity: 6.
Comments: The farmer must simultaneously select the kind of fertiliser to be used.

19 and 20. PURCHASE CHEMICALS AND FERTILISERS
Purpose: To obtain the kinds and quantities of chemicals and fertilisers chosen.
Inputs required: Capital to cover the cost of 40 kg chemicals and 750 kg NPK fertilisers; the average price of the former is JD 6 per kg, and of the latter JD 0.116 per kg. Total capital required is JD 487.
Determining factors: There are several agents dealing with each of the above-mentioned inputs who provide a service to farmers. In the absence of recommended rates of application, the farmer has to estimate his needs for each input. Consequently, he may buy more or less than he actually needs, and may use more than an economical amount.
Timing: October. Time required: 3 days.
Preceding activities: 17 and 18.

21. APPLY CHEMICALS
Purpose: To protect the crop against insects and diseases.
Inputs required: In addition to chemicals, the farmer needs labour of 16

man-days at JD 2 per day and a total cost of JD 32, and he also needs machinery for spraying the chemicals. The cost of machinery is JD 0.5 per working hour and the cost for the total hours required is JD 16.5.
Determining factors: In the absence of effective extension programmes and the lack of research on the best use of chemicals, there is a tendency for farmers to over-use them.
Timing: October-June. Time required: 16 days, intermittently.
Preceding activities: 15 and 20.

22. APPLY FERTILISERS
Purpose: To increase yield.
Inputs required: In addition to the fertilisers, the farmer needs labour to apply them. He needs 7.5 man-days at JD 2 per day and a total cost of JD 15.
Determining factors: If using the drip system, the farmer can add soluble NPK to the water and apply it through the pipelines. However, not all farmers are in this situation. Those who use the flood system have to use dry NPK and apply it to the soil twice during the season. Furthermore, farmers who use the drip system obtain higher efficiency in the application of fertilisers. The farmer usually applies fertilisers one month after planting, and repeats the application one month thereafter.
Timing: November-January. Time required: 6 days, intermittently.
Preceding activity: 15 and 19.

23. TEND CROP
Purpose: This activity is made up of several different activities which include mounding the plants, pruning and trellising, ventilating, installing drippers and plastic mulch, and inspecting for diseases and insects.
Inputs required: Labour of 89 man-days at JD 2 per day and a total cost of JD 178. Hand tools are also required at a total cost of JD 20.
Determining factors: Skilled labour for these jobs is not always available on a daily basis and the farmer has to hire workers by the month at a relatively higher cost. The monthly wage per labourer is JD 60.
Timing: October-June. Time required: 35 days, intermittently.
Preceding activities: 3 and 15.

24. DECIDE ON CONTAINERS
Purpose: To select the best kind of containers for marketing the crop.
Inputs required: Information about different kinds of containers available.
Alternatives: Plastic containers, wooden containers or no containers at all.
Determining factors: If the farmer decides to sell the crop directly to a contractor, the latter also takes care of purchasing the containers and harvesting the crop, and the farmer need not worry about it. There are several factories in Jordan which manufacture boxes for agricultural marketing use. Second-hand boxes are also available on the market at cheaper prices. Our farmer decided to use new plastic boxes. It is worth mentioning that some farmers told the researcher that exporters prefer crops to bc in plastic boxes and will pay higher prices for them. However, plastic boxes are the most expensive. There are agents who charge commission on selling boxes and offer a delivery service to farms. The Government is controlling specifications of containers manufactured by local factories.
Latest time to decide: January, or just a few days before harvesting begins.
Preceding activity: 6.

25. PURCHASE PLASTIC BOXES
Purpose: To purchase boxes to permit transport/delivery of product.
Inputs required: Capital of JD 2,000 to cover the cost of containers. Arrangements are made with agents to take delivery of boxes on time.
Determining factors: The farmer can buy boxes direct from the factory but then he has to do the loading, transporting and unloading. This involves several visits to the factory and use of his pick-up, but saves the commission fee. There are several agents who offer a delivery service and sell both new and second-hand boxes on credit. Their service is available to farmers on request. Some farmers deal with them to avoid paying cash for boxes bought direct from the factory.
Timing: December-June. Time required: 10 days, intermittently.
Preceding activity: 24.

26. DECIDE ON MARKETING OUTLET
Purpose: To select the marketing outlet with the most convenient service and best possible prices.
Inputs required: Information on alternative outlets.
Alternatives: Amman Central Wholesale Market (ACWM), wholesale markets in other towns, and Me'adi Assembly Market in the Jordan Valley.

The former offers, in general, the greatest possibility of obtaining the best prices but involves the highest marketing costs. Auctions in the market are held both for local consumption and for export. The second alternative carries the same marketing costs but the lowest number of bidders and thus the least possibility of getting high prices. The third alternative carries the lowest marketing costs and is mainly devoted to export sales at auction; exporters are almost the only bidders in this market. Our farmer selected ACWM to sell his crop.
Determining factors: Selection of any market does not involve permanent obligation, and the farmer can move from one market to another at any time. An exception to this is where the farmer agrees to sell through a middleman in the market against credit services provided by the middleman. The selection of ACWM involves the necessity to harvest at night, which in turn further increases costs. Each market has agents who sell the crop on behalf of the farmer and charge a 5 per cent commission. This fee is subject to negotiation and may fall to 3 per cent. The ACWM charges 2 per cent to the buyer and 2 per cent to the seller for use of its services. This is the case with other wholesale markets, but not with Me'adi Assembly Market.
Latest time to decide: December, just a few days before harvest.
Preceding activity: 15.

27. HARVEST CROP
Purpose: To get the crop ready for delivery to the market.
Inputs required:
— Labour of 186 man-days at JD 2 per day and a total cost of JD 372
— Plastic boxes (2,000) at JD 1 per box and a total cost of JD 2,000
Determining factors: The farmer has to harvest at night and consequently he cannot hire women; women are preferred for harvest work and he has to pay higher rates for night work. Rainy days may delay the harvest and affect the quality of the fruit. Harvesting has to be done several times because of the kind of plant and the staggered planting dates.
Timing: January-June. Time required: 40 days, intermittently.
Preceding activities: 25 and 26.

28. SELL CROP
Purpose: To obtain a cash income from the cucumber crop.
Inputs required: Negotiations with sale agents to secure a good deal and the best service, and arrangements for delivery of output. The sale agent usually receives the shipment in his store in the central market, supervises auction of this shipment, receives the value of the shipment from the highest bidder, deducts marketing commission and his commission from the value of the sale, and pays the remainder to the farmer as agreed between them.
This activity also requires contacts and arrangements with a transport agent if the farmer's pick-up is not available to carry the crop to market. Such agents charge JD 3 per ton.
Determining factors: Oligopsony, poor marketing service, and marketing costs affect the farmer's success and may force him to move to another market.
Timing: January-June. Time taken: 40 days, intermittently.
Preceding activity: 27.

29. REPAY LOANS DUE
Purpose: To maintain the source of credit for next season and to fulfill commitments to lenders.
Inputs required: Enough capital to cover loans due and interest. Visits to lenders for payment of all amounts due to reschedule repayments.
Determining factors: If the farmer does not make enough income to cover all loans due, he has to make the rescheduling of repayments a priority. Studies on delinquency show that farmers give low priority to repaying the loans from institutions.
Timing: July-September following the harvesting period. Time required: 3 days.
Preceding activity: 28.

Yield and Cost of Production
The interrelating conditions and activities described above produced a total of 40 t of cucumbers. This output cost a total of JD 6,096; the cost of material inputs and labour inputs per du for each item is shown in detail in Table 4.3. Data on the response of the cucumber crop to different amounts of each input are lacking and this needs investigation.

Analysis of a Specialised Tomato Farm System

Environments
As with the cucumber plastic farm, environments here cover everything other than human activities.

On-Farm Environments
Physical
The farm is located in the Jordan Valley and has 35 du of land. It is fully devoted to growing tomatoes during the winter season in natural conditions with drip irrigation and mulch. It has water rights to irrigate the whole area from the East Ghor Irrigation Project.

The climate on the farm is identical to that of the cucumber plastic farm previously described.

Electrical power is needed for running the drip system and is obtained from the main supply which services the whole Valley. Manpower is provided by family members and hired labour. Machinery is rented.

Total assets are estimated at JD 2,400, excluding the value of farmland, and total liabilities are estimated just before harvest at JD 4,000.

Institutional

The farm is owned and run by the farmer. This is the case with 52 per cent of farms in the Jordan Valley. The rest are run by tenant farmers. The family consists of seven members: the farmer, his wife, three sons and two daughters. One boy and girl, in addition to the farmer and his wife, work permanently on the farm; the rest are still young and go to school. The farmer who is 35, was educated to preparatory level.

The rest of the institutional arrangements on the farm are the same as those on the cucumber farm.

On- and Off-Farm Informational Environment

As the farm is located in the same region as the cucumber farm, it has the same conditions as those previously described.

Human Activities

The main activities selected for analysis are described below. Time sequences of both the physical activities and the decisions involved are shown in Figure 4.2. This chart, however, does not depict working at two activities simultaneously or moving from one completed activity to a second, or others. The actual sequence is fully explained in the record of each activity.

Activity Analysis

Physical Activities

1. ASSESS ASSETS

This activity has the same record as the similar one described above on the cucumber farm.

2-8 DECIDE TO FARM VEGETABLES: DECICDE ON CROPPING PATTERN AND IRRIGATION SYSTEM BORROW CAPITAL; PURCHASE AND INSTALL DRIP SYSTEM; DECIDE ON USE OF SEEDLINGS

All these activities and decision follow the same pattern as those previously described for the cucumber plastic farm, except that:

— The farmer decided to grow tomatoes.

— The farmer needed JD 4,000 to purchase a drip irrigation system.

— The farmer decided to use tomato seedlings.

9. PURCHASE SEEDLINGS

As for the cucumber farm, except that the farmer needed to purchase 57,800 seedlings at a total cost of JD 231.

10. DECIDE ON PLANTING DATE

Comments: As for the cucumber farm, except that the farmer decided to plant in late winter.

The date of planting tomatoes in the Jordan Valley is strongly affected by the danger of leaf-curl virus attack and by prevailing weather conditions. The location of the farm determines to a large extent the most suitable time

Figure 4.2: Activities Related to a Tomato Farm System

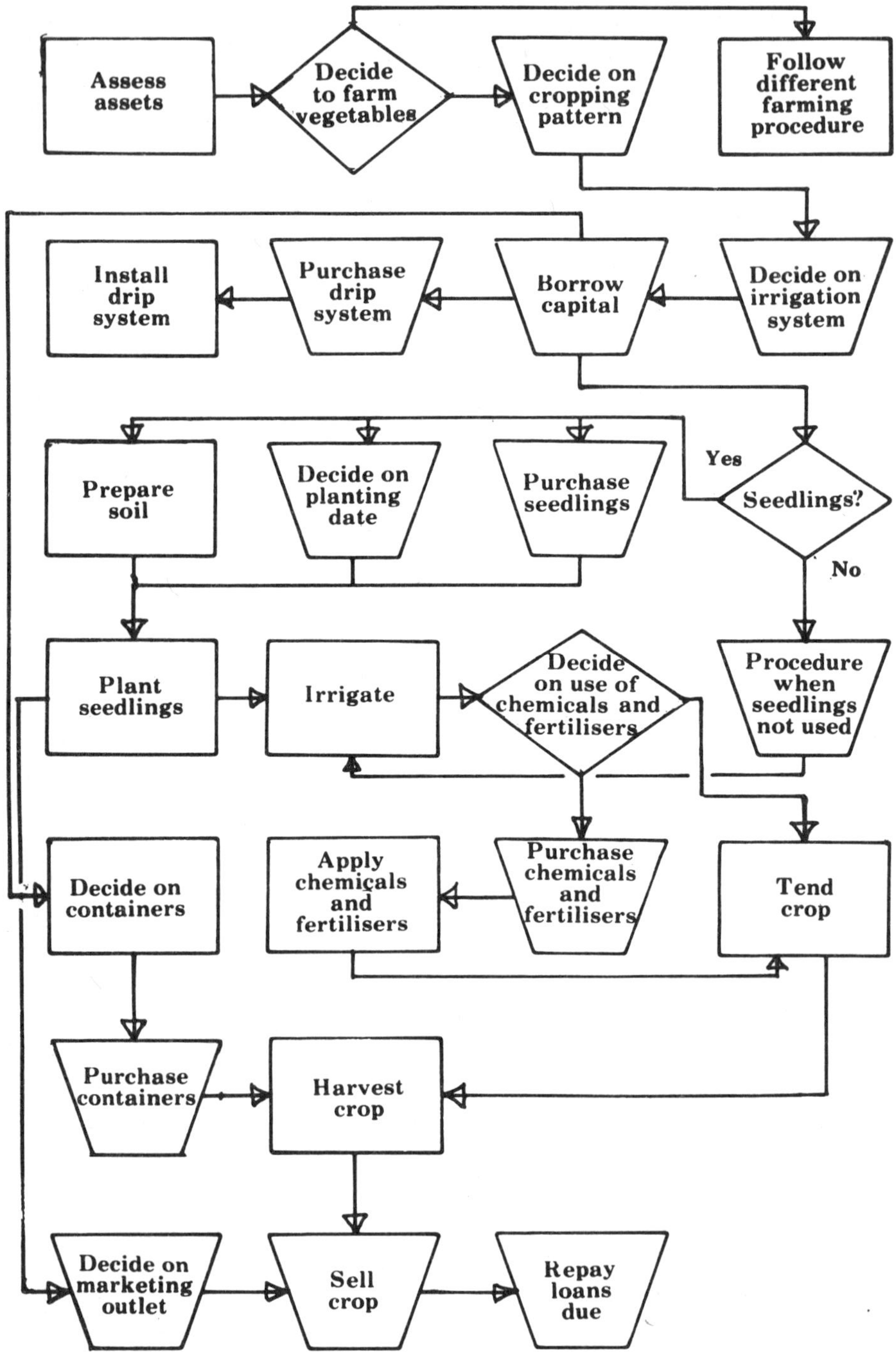

for planting. Farms in the southern part of the Valley are better able to grow early tomatoes, with this possibility gradually decreasing as the farms extend northwards. Late planting has been found to be a way of avoiding leaf-curl virus attack on the tomato crop.

JVFA statistics on yields and the wholesale market price of tomatoes grown in the Jordan Valley show that early tomatoes yield 1.5 t per du and late tomatoes yield 2.5 t. The price of early tomatoes is 190 fils per kg, and for the late ones it is 110 fils per kg.

11. PREPARE SOIL

Purpose: To render the soil suitable for growing a winter tomato crop.

Inputs required:

— Labour: 10.5 man-days	at JD 2 per day and a total cost of JD 21
— Machinery for ploughing: 14 working hours	at JD 10 per hour and a total cost of JD 140
— Chicken manure: 45 t	at JD 20 per t and a total cost of JD 900. (This input is added every three years.)
— Triple phosphate: 3.6 t	at JD 40 per t and a total cost of JD 140
— Water: 700 m^3	at 3 fils per m^3 and a total cost of JD 2.4

Determining factors: Rented farms, and farms operated on a share-cropping basis rarely use manure.

Timing: September. Time required: 5 days.

Preceding activities: 1, 2 and 3.

12. PLANT SEEDLINGS

Purpose: Upon delivery of the seedlings the farmer has to plant them.

Inputs required:

— Labour of 60 man-days	at JD 2 per day and a total cost of
— Mulch: 8.75 rolls	JD 120 at JD 20 per du

Timing: September-January. Time required: 3 days, intermittently.

Preceding activities: 9 and 11.

13. IRRIGATE

Purpose: To give the plants their water requirements.

Inputs required:

— 117,000 m^3 of water	at 3 fils per m^3 and a total cost of JD 35
— Labour of 120 man-days	at JD 2 per day and a total cost of JD 240
— Fuel oil to run the drip system	1,000 l at a total cost of JD 42

Timing: September-June; every day of the growing period.

Preceding activity: 12.

14 and 15. DECIDE ON USE OF CHEMICALS AND FERTILISERS

These two activities have similar records to those for the cucumber farm.

16 and 17. PURCHASE CHEMICALS AND FERTILISERS

These two activities have similar records to those of the cucumber farm system, except that quantities to be purchased and capital requirements are different, as follows:

Item	*Quantity (t)*	*Price (JD/t)*	*Total cost (JD)*
Compound NPK	3.5	116	406
Chemicals	0.112	3	336

18. APPLY CHEMICALS
Purpose: To protect the plants against insects and diseases.
Inputs required: In addition to chemicals, the farmer needs labour of 35 man-days at JD 2 per day and a total cost of JD 70. He needs machinery at a total cost of JD 55.
Determining factors: Farmers use the same chemical for several years running, usually at higher rates than required. Lack of research on the best method of application of chemicals under local conditions is the main constraint on the optimum use of chemicals.

19. APPLY FERTILISERS
This activity is the same as that for the cucumber farm, except that the quantity used is 3.5 t of compound NPK and the labour requirement is 3 man days at a total cost of JD 6.

20. TEND CROP
Purpose: This activity is made up several different activities, such as inspection for insects and diseases, weeding, maintaining the drippers, replacement of dead plants, removal of mulch and drip pipelines, and removal of plants after harvest.
Inputs required: Labour of 52.5 man-days at a total cost of JD 105, and hand tools at a total cost of JD 20.
Determining factors: Farmers, as well as most Jordanian hired labourers, have been growing tomatoes for many years and are experienced in their cultivation.
Timing: Throughout the growing season, starting in September and ending in June. Time required: 53 days, intermittently.
Preceding activity: 12.

21. DECIDE ON CONTAINERS
This activity has a similar record to that of the cucumber farm. The farmer decided to use plastic boxes whenever possible, of medium size, i.e. boxes holding 20 kg.

22. PURCHASE CONTAINERS
This activity is similar to that of the cucumber farm, except that the farmer purchased 2,625 boxes at a total cost of JD 1,050.

23. HARVEST CROP
Purpose: To have the crop ready for delivering to market.
Inputs required: Labour (females are preferred): 502 man-days at a total cost of JD 1,004. Containers: 2,625 boxes at a total cost of JD 1,050.
Determining factors: The varieties of tomatoes grown are not suitable for mechanised harvesting and produce fruit which ripen gradually. Therefore, the farmer has to harvest several times and use manual labour.

The harvesting period falls during winter, so rain affects the quality of the produce and the harvesting schedule.
Timing: December-June. Time required: 13 days, intermittently.

24 and 25. DECIDE ON MARKETING OUTLET AND SELL CROP
These records are similar to those of the cucumber farm, except that the farmer decided to sell his produce at Me'adi Assembly Market.
Comments: The Government takes a special interest in tomatoes and enters into the market as a buyer to maintain a minimum price for growers when prices fall below average costs, and where there are marketing bottlenecks. (The Government did this in 1983 to maintain a minimum price for tomatoes of 35 fils per kg.)

26. REPAY LOANS DUE
This activity is exactly the same as that for the cucumber farm.

Yield and Cost of Production

The activities and decisions recorded above, together with the conditions described, produced a total amount of 122.5 t of tomatoes. Total cost of production, cost per du and detailed cost per item of material inputs and labour inputs are shown in Table 4.4. This farmer is facing the same problem as the cucumber farm owner as regards lack of data on response of yield to the rate and quantity of each input.

Analysis of a Mixed Vegetable Farm System

Environments

This farm is located in the Jordan Valley with the same conditions as the tomato farm. Consequently, identification of the environment as stated above applies here also.

Human Activities

The main activities selected for analysis are described below. The activities and decisions involved are shown in Figure 4.3 with the time sequence relevant to each. However, the chart does not allow for simultaneous activities, nor for moving from one completed activity to the beginning of another, or several others, at the same time. This is explained in the record of each activity.

Activity Analysis

Physical Activities

1 and 2. ASSESS ASSETS AND DECIDE TO FARM VEGETABLES
These two activities involve the same activities and decisions as for the cucumber farm system.

3. DECIDE ON CROPPING PATTERN
This activity is similar to the relevant one for the cucumber farm, except that our farmer decided to grow mixed vegetables by the traditional method.

Figure 4.3: Activities Related to a Mixed Vegetable Farm System

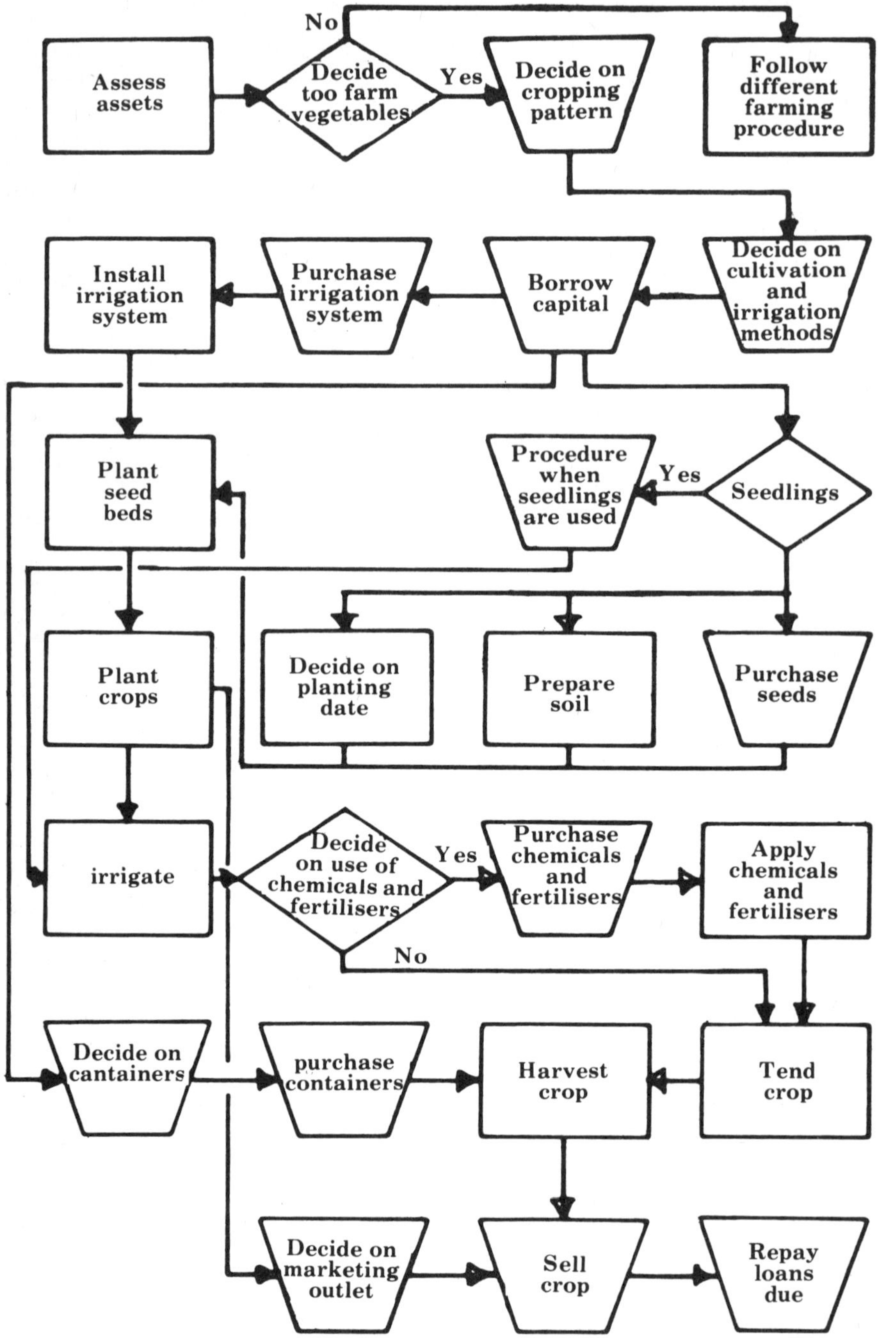

Thus he decided to plant 15 du of tomatoes, and 5 du each of eggplants, squashes, broad beans and cauliflowers.*

4 and 5. DECIDE ON IRRIGATION SYSTEM AND BORROW CAPITAL
These two activities are as recorded for the cucumber farm, except that the farmer decided to use the flood system for irrigation, and to borrow seasonal loans only.

6. DECIDE ON USE OF SEEDLINGS
As for the cucumber farm, except that the farmer decided to use plant seeds where possible, and to transplant his own seedlings from nursery beds when necessary.

7. PURCHASE SEEDS
Purpose: to get the required quantity of each kind of vegetable seeds needed.
Inputs required: Contact with seed merchants to get the best varieties, best deal and best services. The capital required to purchase seeds is JD 146.5.
Determining factors: In addition to the farmer's own experience of the best varieties, there are suppliers who provide seeds with service to farmers. These suppliers are highly competitive and try to stay in the market by supplying the best they have. Research on the best varieties for particular regions is still lacking and there is no continuity with past research. This kind of research is left to private firms who try to test out each variety in the field before releasing it.
Timing: Late August. Time taken: 3 days.
Preceding activity: 6.

8. PREPARE SOIL
This activity is similar to that for the tomato farm.

9. DECIDE ON PLANTING DATE
Purpose: To select the best time for planting each kind of vegetable to be grown.
Inputs required: Information on probable weather and marketing conditions related to each crop.
Alternatives: There are two real alternatives for each crop: either to plant early, or late. The third possibility of planting at intervals is not relevant because of the small areas devoted to each crop. Our farmer decided to plant late, to minimise his risks.
Determining factors: Our farmer selected different crops with different requirements for successful cultivation. Consequently, there were different best planting times. In the absence of weather and marketing forecasts, the farmer has to rely upon his own judgement in selecting best times. His only guidance was past experience of weather and marketing conditions.
Latest time to decide: Late August.
Preceding activity: 3.

10. PLANT SEED-BEDS
Purpose: To produce the required seedlings on time.
Inputs required: In addition to seeds for all the selected crops, except broad beans, the farmer needs small quantities of chemicals and fertilisers. He also needs labour to prepare the soil and tend the plants for one month. The

* This is the most commonly used cropping pattern in mixed vegetable farms.

cost of seedling production on the farm is estimated at 2 fils per plant and total costs at JD 120.

Determining factors: With the increasing incidence of plant diseases in the Jordan Valley, farmers are facing greater problems in getting healthy seeds by traditional methods of production. Specialist seed growers are expanding their services to help farmers get healthy seedlings, and reliance upon them is accelerating. Some farmers are trying to improve their method of seedling production by using soil fumigants, but they still risk uncontrolled weather conditions. The prices of seedling agents encourage farmers to change over and rely on them. These prices are 50 per cent up on farmers' costs.
Timing: August-September. Time taken: 1 month.
Preceding activity: 7.

11. PLANT CROP
Similar to the tomato farm, except that broad bean seeds can be planted direct.

12. IRRIGATE
Purpose: To give the plants their water requirements on time.
Inputs required:

— Labour of 65 man-days	at JD 2 per day and a total cost of JD 130
— Water: 21,000 m^3	at fils per m^3 and a total cost of JD 63

Determining factors: The farm enjoys water rights and the farmer can get his requirements from the main canal of the East Ghor Irrigation Project upon request and also his approved irrigation schedule. The flood system is water-and labour-intensive and may carry new diseases to the farm through water supply. It does not water each plant equally and it is only 30 per cent efficient. Compared to other systems, it needs the least capital investment.
Timing: September-June. Time require: 9 days, intermittently.
Preceding activity: 11.
Comments: When water supply has not met total demand due to drought, the JVA has in the past given priority to tree growers. Vegetable growers had to leave part of their land uncultivated.

13 and 14. DECIDE ON USE OF CHEMICALS AND FERTILISERS
These two activities produced the same records as those for the cucumber farm.

15 and 16. PURCHASE CHEMICALS AND FERTILISERS
As for the tomato farm, except that the farmer purchased 225 kg chemicals at JD 3 per kg and a total cost of JD 675. He also purchased 2.16 t ammonium sulphate at JD 68 and a total cost of JD 147.

17 and 18. APPLY CHEMICALS AND FERTILISERS
Purpose: To protect the plants and increase yield.
Inputs required:
— Labour for applying chemicals (123 man-days at JD 2 per day) = JD 246
— Labour for applying fertilisers (56 man-days at JD 2 per day) = JD 112

Detailed requirements and costs per dunum were shown in Tables 4.1 and 4.2.

— Chemicals (225 kg) and fertilisers (2.16 t)

— Sprayers at a cost of JD 17.5

Determining factors: Same as for the cucumber farm activities regarding application of chemicals and fertilisers.

Timing: September to May. Time required: 7 days, intermittently, as detailed for each crop in Table 4.2.

Preceding activity: 15.

19-25. TEND CROP; DECIDE ON AND PURCHASE CONTAINERS; DECIDE ON MARKETING OUTLET; HARVEST AND SELL CROP; REPAY LOANS DUE

These activities are similar to the corresponding ones for the tomato farm, except that the total labour requirement for tending the crops is 58 man-days at JD 2 and a total cost of JD 116. The farmer needs to purchase 2,000 boxes at JD 0.4 each and a total cost of JD 800. He needs workers to bring in the harvest requires 105.5 man-days at JD 2 and a total cost of JD 211. Labour requirements, frequencies and costs of each crop were shown in Tables 4.1 and 4.2.

Yield and Cost of Production

The previously recorded activities and decisions, in conjunction with the conditions described, produced the following quantities of each crop:

Crop	Tonnes
Tomatoes	30.0
Eggplants	7.7
Squashes	3.7
Broad beans	3.0
Cauliflowers	7.0
TOTAL	51.4

Total cost of production and detailed cost items of material inputs and labour inputs per du were shown in detail in Table 4.1.

Analysis of a Citrus Farm System

Environments

As with the cucumber plastic farm, environments here cover everything other than human activities.

On-Farm Environments

Physical

The farm is located in the Jordan Valley and consists of 35 du. It has enough water rights to irrigate the entire farm all year round from the main canal of the East Ghor Irrigation Project. The farmer can therefore tap water from the canal once he submits his irrigation schedule and this is endorsed by the JVA. Some citrus farms in the Valley do not enjoy such rights; they have their own artesian wells and draw upon them as needed. Such farms are concentrated in the southern part of the Valley at Rameh and Kafrein, and some of them eventually be served by the East Ghor Canal extension, which is under construction.

The weather on the farm favours growing citrus trees, as can be seen from the excellent performance of citrus farms. A working paper submitted by a group of Jordanian experts to the Higher Agricultural Council (HAC) in 1982 stressed the limited area of suitable land for growing citrus.

Our typical farm was established in 1973 to grow several varieties of citrus trees. Most citrus farms in the Jordan Valley grow a mixture of varieties and this is preferred both by the farmers and by the contractors who purchase the crop, as the author found when discussing the subject with farmers and contractors.

Permanent construction on the farm is limited to a dwelling house of three rooms, a store and a farm shed, plus a metal fence. Electrical power is supplied off the main source which serves the whole Valley. Its use is confined to the home.

Total assets are estimated at JD 123,800, excluding the value of the farm land, and total liabilities just before harvesting the last crop in 1982 were estimated at JD 3,000. Assets include JD 18,000 worth of citrus trees, JD 600 as the value of the house, JD 1,200 for the fence, JD 4,000 for machinery, plus a dwelling house in Amman, valued at JD 10,000. Our farmer could repay the loans borrowed to establish his farm by relying on non-agricultural resources. (Estimates of assets and liabilities are based on market prices in 1982.)

Institutional

The farm is owned soley by the farmer, and he and his family live away; he is running it as an absentee landlord. He has a full-time manager (*wakil*), who takes care of the farm as a permanent employee. This manager receives instructions from the landlord and reports back verbally to him upon request. The landlord visits the farm occasionally. Absentee ownership of citrus farms is common in the Valley and represents 0 per cent of total land-holders according to the 1978 agricultural census.

A large number of farmers do not do the harvesting themselves but make arrangements with contractors, who purchase the fruit while it is still on the tree, and then do the harvesting and sell the crop in the market later on. The census of 1978 showed that 38.5 per cent of holdings follow this pattern.

The family consists of six members, none of whom lives on the farm or is involved in agricultural work. Some landlords do live on their farms but still behave like absentee landlords.

In running a citrus farm, the owner usually hires permanent labour on a monthly basis and takes on seasonal labour when necessary.

Informational

The farmer has a secondary-level education. Living away in the capital, he can easily obtain information on resources and extension programmes. As a consequence, the off-farm information described in the background to this study is available to this farmer.

Off-Farm Environments

These are similar to the situation on the cucumber plastic farm.

Human Activities

These include all physical activities and decisions undertaken on the farm which, in conjunction with conditions, produce output for the market. However, these activities are not repeated in the same manner every year, as is the case with seasonal crops, because of the slow growth of trees and the changing requirements from one year to the next. Timewise, activities can be divided into two groups. The first includes those related to the establishment period, while the second includes those related to the mature period. We should add that the first group of activities takes placc simultaneously with other activities for growing vegetables among the trees while they are still very young.

Each activity related to a citrus farm can be seen as a sequence of simpler ones. We have therefore identified and selected the following main activities for analysis. Although the farm under consideration is 10 years old, the analysis covers those activities which have taken place since the farmer started thinking of setting up the farm. Costs and inputs are estimated at actual market prices of 1982 rather than the prices of previous years.

The time sequence of activities related to a citrus farm is shown in Figure 4.4. The sequence is explained in the record of each activity.

Activities and Decisions Related to the Establishment Period
These are set out in the activity analysis below.

Activities and Decisions Related to the Post-Establishment Period

1. Decide on use of chemicals
2. Purchase chemicals
3. Apply chemicals
4. Decide on use of fertilisers
5. Purchase fertilisers
6. Apply fertilisers
7. Irrigate
8. Cultivate
9. Maintain trees
10. Decide on marketing outlet
11. Market crop
12. Repay loans due

Activity Analysis

Physical Activities

1 and 2. ASSESS ASSETS AND DECIDE TO FARM FRUIT
These two activities have similar records to those for the cucumber farm system previously described.

3. DECIDE ON CROPPING PATTERN
As for the cucumber farm, except that this farmer decided to grow citrus on all his land.

4. BORROW CAPITAL
Purpose: To obtain sufficient capital to establish and operate a 35 du citrus farm.
Inputs required: Contacting and making arrangements with the ACC for a mid-term loan, and contacting other lending agencies for obtaining seasonal loans when needed.
Determining factors: The farmer has non-agricultural income to repay any loans due during the establishment period. He possesses the collateral required to secure a mid-term loan. The ACC is specialised in extending agricultural mid-term loans for tree planting. The farmer can afford to live on non-farm income.
Timing: July-September. Time required: 1 month.

5. FENCE THE FARM
Purpose: To protect the farm from human and animal damage.
Inputs required:

Figure 4.4: Activities Related to a Citrus Farm System

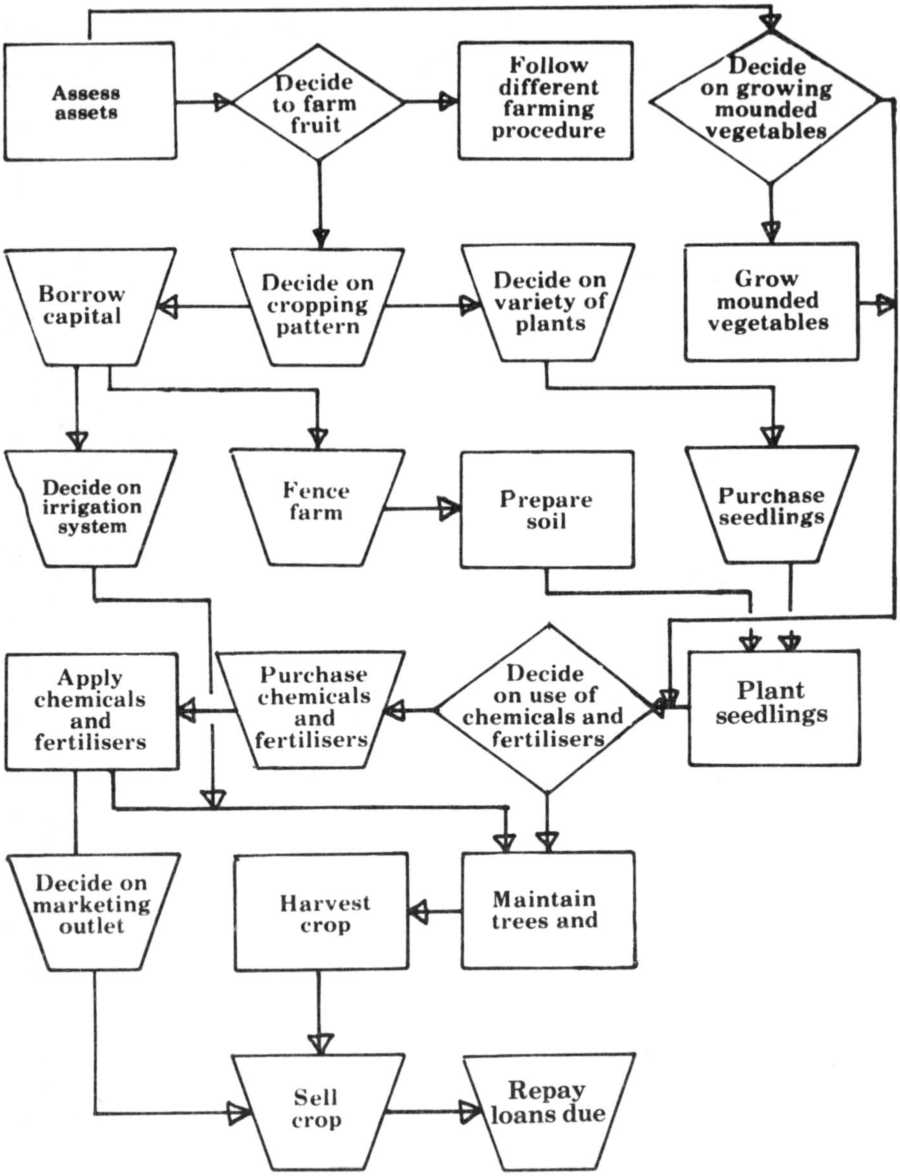

Input	Quantity	Total cost (JD)
Fence wire	6,400 m	231
Fine fence wire	20 kg	6
Iron bars	160	160
Iron poles	18	104
Iron farm gate	1	50
Cement	1 t	35
Sand	6 m^3	12
Labour	30 man-days at JD 20	600
Total		1,198

Determining factors: Not all farmers use the same materials but cheaper fences need more maintenance and their life-time is much shorter. Some farmers put wind-breakers along the fence. Contractors for this work are available.
Timing: Any time during the year before starting activities inside the farm. Time required: 10 days.
Preceding activity: 4.

6. PREPARE SOIL
Purpose: To render the soil suitable for planting citrus trees.
Inputs required:

	(JD)
— Machinery for ploughing for 20 hours at JD 3 per hour	60
— Labour of 93 man-days	186
— Manure:* 35 t at JD 20 per t	700
— Super phosphate: 3.5 t at JD 40 per t	140
— Water: 700 m^3 at 3 fils per m^3	2.1
— Farm design, including the location for each tree	

* Chicken manure is preferred, followed by sheep manure.

Determining factors: The farmer has to follow a time sequence to complete this activity. First he applies manure and triple phosphate, then water. After a few days he has to plough twice and mark out the land to fix the location for each tree. Then he has to dig holes of 0.5 × 0.5 × 0.5 m for each tree. While doing all this, the farmer must keep in mind that he is going to grow vegetables in between the trees and make adjustments accordingly.
Timing: September. Time required: 1 week.
Preceding activities: 3, 4 and 5.

7. DECIDE ON VARIETY OF PLANTS
Purpose: To select the most suitable and economic varieties of citrus plants to be grown.
Inputs required: Information on the merits of different varieties of citrus trees, including production and marketing aspects.
Alternatives: One variety or several varieties. Each alternative contains further choices; the single variety may be just oranges or just lemons.
Determining factors: As mentioned earlier, the contractors prefer to purchase the produce of a mixed citrus farm and will pay more for such farm produce. Most of the farms in the Valley are indeed mixed citrus farms. Market prices in Jordan are unstable and diversifying is one way of

minimising the risk of price fluctuations. Our farmer decided to grow mixed varieties of citrus plants.
Latest time to decide: September, just before preparing the land. This is vital, because the distance between trees depends upon the variety to be grown..
Preceding activity: 6.

8. PURCHASE SEEDLINGS
Purpose: To choose the seedlings on time.
Inputs required: Information about seedling growers and their varieties, plus cost and quality of each variety. Contact to arrange a good deal and delivery to the farm gate on time. Capital for the total cost of seedlings, estimated at JD 2 per plant and a total of JD 2,800.
Determining factors: There are several nurseries which specialise in growing citrus seedlings. They grow and sell grafted and ungrafted seedlings, as well as providing a service to farmers. Some nurseries import seedlings from abroad. The farmer therefore has to choose as far as grafting is concerned. the MOA is producing citrus seedlings to meet some of the local demand and it sells them at cost price. The farmer can get part of his needs from the MOA if he applies it in time. The MOA is controlling technical production of seedlings to minimise the risk of growing unhealthy specimens.
Timing: late September. Time required: 3 days.
Preceding activity: 7

9. PLANT SEEDLINGS
Purpose: The farmer plants seedlings once these are delivered to the farm.
Inputs required: In addition to the seedlings, the farmer needs labour of 350 man-days at a total cost of JD 700. Trained labour is preferred.
Determining factors: There are trained labourers who offer their services for planting seedlings and who charge JD 0.5 per tree to farmers. If the farmer chooses to grow one single variety, the distance between trees will be different and the number of plants and labour requirements will change accordingly.
Timing: Late September. Time required: 3 days.
Preceding activities: 6 and 8.

10. DECIDE ON GROWING MOUNDED VEGETABLES
Purpose: The farmer may need to use his land while the citrus trees are still very young and occupy only small areas. He therefore, has to decide about growing crops in between the trees.
Inputs required: Information about the economics of this activity and the effect it has on the citrus trees.
Alternatives: The farmer will either grow vegetables or not. If he does, he will have the chance of an income from the farm during the first three years of establishing the citrus trees, and of minimising costs of establishment by looking after both the citrus trees and the vegetable crops at the same time. The alternative of not growing vegetables may not give the trees the best conditions under which to grow. Our farmer decided to grow mounded vegetables.
Latest time to decide: September.
Preceding activity: 3.
Comments: Research into this is lacking; it is needed for the establishment of farms in the future.

11. GROW MOUNDED VEGETABLES
This work carries the same record of activities as for the mixed vegetable farm and is repeated annually until the end of the third year of growing citrus trees. The farmer can attend to both vegetables and citrus trees at once when irrigating, ploughing, applying fertilisers, cultivating, and even applying chemicals. The activities related to vegetable crops satisfactorily cover citrus growing needs.

12 and 13. DECIDE ON USE OF CHEMICALS AND FERTILISERS
Purpose: To decide whether or not to use chemicals and fertilisers in growing citrus trees.
Inputs required: Information on the economics of the use of chemicals and fertilisers and their effect on the growth and yield of citrus trees.
Alternatives: To use chemicals and fertilisers or not. In the first three years, having decided to grow vegetables between the citrus trees, the farmer is unavoidably applying these inputs to them also. He still has to decide about the fourth and following years. All farmers in the Valley use chemicals and fertilisers in growing citrus trees, but they differ in types and amounts. Our farmer also decided to use them. As we have said several times, research into the best method of using chemicals and fertilisers under local conditions is very limited and farmers cannot rely on it, but must turn elsewhere.
Latest time to decide: November.
Preceding activities: 3, 7 and 9.
Comments: Some farmers use leaf nutrients in addition to soil fertilisers and some use weed-killers, all of which must be taken into consideration. This decision is taken annually as long as the citrus trees are to be maintained. The decision involves the selection of the types of chemicals and fertilisers as well as the rate of application, rather than whether or not to use them. If the farmer chose to use them every year, the costs per du would be as shown in Table 4.6.

14 and 15. PURCHASE CHEMICALS AND FERTILISERS
Purpose: To obtain the chemicals and fertilisers decided upon.
Inputs required: Contact with dealers for the best deal, including service. Capital to cover annual costs is as follows:

Year	Chemicals		Fertilisers*	
	Quantity (kg)	Cost (JD)	Quantity (kg)	Cost (JD)
1	2	12	—	—
2	2	12	—	—
3	2.5	15	—	—
4	3	18	1,200	140
5 and after	3.5	21	1,200	140

* Costs of fertilisers during the first three years are included in the cost of growing mounded vegetables. If the farmer does not want to grow vegetables he needs the quantities shown in Table 4.6.

Timing: September. Time required: 3 days.
Preceding activities: 12 and 13.
Comments: Determining factors are the same as for the systems previously described.

16. APPLY CHEMICALS

Purpose: To apply chemicals selected for plant protection.

Inputs required: In addition to the chemicals, the farmer needs:

	Total cost (JD)
— Labour: 44 man-days on average	88
— Machinery to spray chemicals for 35 hours	35

These inputs change over time, as shown in Table 4.6.

Determining factors: The MOA is setting up agencies to provide treatment of citrus trees and the farmer can rent a spraying machine from them for JD 5 per hour. Private agents offer a similar service at double the price. Rates and methods of application are mainly based upon manufacturer's recommendations.

Timing: 4 days in each of March and July.

Preceding activity: 14.

17. APPLY FERTILISERS

Purpose: To treat the soil with the selected fertilisers.

Inputs required: In addition to the fertiliser, the farmer needs:

	Total cost (JD)
— Labour of 38.5 man-days at JD 2 per day	77

Detailed annual labour and chemical requirements per du are shown in Table 4.6.

Determining factors: Farmers use different quantities and rates of the same fertiliser. At present, the majority of farmers use NPK at the rate of 0.5 kg per tree during the first few years, gradually increasing the quantities applied to a maximum of 3 kg per tree. The fertiliser is added twice a year, in March and September.

Timing: March to September. Time taken: 4 days, intermittently.

Preceding activity: 16.

18. MAINTAIN TREES

Purpose: This activity is made up of a sequence of simple ones required to care for the trees, such as cultivating, pruning and general inspection.

Inputs required: In the second, third and fourth years, the farmer needs:

— Labour of 40 man-days at JD 2 per day and a total cost of JD 80 per year.

In the fifth year and each following year, he needs:

— 105 man-days at the same unit cost, making a total of JD 210 per year.

In addition to labour, the farmer needs hand tools at a total cost of JD 20 per year.

Determining factors: Trained labour is required for pruning; it is in short supply and costs are high for this service.

Timing: March to April of the second year and each following year.

Preceding activity: 9.

19. DECIDE ON IRRIGATION SYSTEM

This activity is the same as for the cucumber farm, with the exception that the farmer has been using the flood system and has not yet made up his mind to change. The number of farmers who have adopted the drop system for irrigating citrus trees are few, but their number is increasing.

20. IRRIGATE

Purpose: To supply the trees with the water they require.

Inputs required:

— Labour: 420 man-days for each of the first five years; 490 man-days in the sixth year and each following year. The total annual cost for labour is JD 980.

—Water: at an annual rate of 1,200 m^3 per du and a total of 42,000 m^3. The farmer needs to irrigate every week all year round except during heavy rainfall periods. The need for water increases after the plants are established; when they are fully grown, the farm needs a total of 42,000 m^3 at a price of 3 fils per m^3, and a total cost of JD 126. We should note that water is to be used for growing both vegetables and citrus trees in the pre-maturity period and for citrus trees only later on.

Determining factors: Farms in the southern part of the Valley need more water to irrigate citrus trees in winter than those located further north.

Timing: All the year round. Time required: 1 day per week.

Preceding activities: 9 and 19.

Comments: At present, the situation is somewhat difficult as far as irrigation is concerned. It is not easy to get permission from the JVA to put the whole farmland under citrus, and to get enough water all year round. At the moment only a quarter of any holding is permitted to be put under citrus trees.

21. DECIDE ON MARKETING OUTLET

Purpose: To select the market promising the best wholesale prices and best services.

Inputs required: Information about different marketing outlets.

Alternatives: The ACWM, Me'adi Assembly Market, and Contractors' Market. The first and second alternatives have the same merits as those previously recorded for the cucumber plastic farm. The third alternative is to sell the fruit while still on the tree on a contract basis. In this case the crop is sold according to the terms agreed upon, and the buyer takes care of the harvest. This way of selling citrus fruit is very common in the Jordan Valley, accounting for 38.5 per cent of farms according to the 1978 agricultural census. Having once signed the contract, the farmer loses the benefits of any later price increases, but at the same time he is protected from the risk of a drop in prices.

Latest time to decide: October, just before the harvest.

Preceding activity: 18.

22. HARVEST CROP

Purpose: To get the crop ready for delivery to the market.

Inputs required: The farmer has nothing to do with harvesting if he chooses to sell to a contractor. The contractor needs:

— Labour input of 245 man-days at JD 2 per day and a total cost of JD 490.

He also needs containers:

— 4,200 plastic boxes at JD 1 per box and a total cost of JD 4,200.

If the contractor chooses to sell through the ACWM, he has to pay 80 fils per box, or a total of JD 336 in transportation costs.

Timing: October. Time required: 2 weeks.

Preceding activity: 21.

23. SELL CROP
Purpose: To obtain a cash income.
Inputs required: Since the farmer has decided to sell to a contractor, he must negotiate with several contractors to secure the best price and the best possible terms for his crop. The farmer and the contractor then sign an agreement setting out the price, the payment schedule, the time schedule and any other additional terms. Once the agreement is signed, the contractor becomes responsible for harvesting and marketing and has the right to sell the crop independently.
Timing: June-October. Time taken: 3 days.
Preceding activity: 21.

24. REPAY LOANS DUE
As for the cucumber plastic farm.

Yield and Cost of Production

The human activities recorded above, together with the conditions already described, produced an annual average of 105 t of citrus fruit when the trees became established. Total annual cost, and cost of items, were detailed in Table 4.6. These costs were estimated at 1982 market prices.

Conclusions

Agricultural production in irrigated areas has increased during the period 1974-81 as a result of expansion in the areas under cultivation and development of the means of production. There is a trend towards an increase in the production of vegetables and fruit and a decrease in that of field crops. Output from irrigated farms plays a major role in agricultural exports; its share has been steadily increasing and at present amounts to 85 per cent.

In the absence of reliable technical advice for each region, the typical farmer develops his own cropping pattern and crop rotation. This means that he does not necessarily adopt the best possible ways uf utilising his own potential resources and of implementing the national agricultural objectives. The application of agricultural inputs he uses are based on his own judgement and with a virtual absence of testing under local conditions to discover the best methods. This implies non-economic usage of inputs and may result in the deterioration of agricultural resources.

Information relevant to production and marketing is not always available to the farmer to help him make the best choices and to secure the highest returns. He relies heavily upon salesmen, dealers, marketing agents and other non-official sources in this connection.

References and Other Works Consulted

Agricultural Council, 1982, *Jordan Council Resolutions, 1982*, Amman.
Agricultural Credit Corporation, 1982, *Jordan Regulations for Extending Loans*, Amman.
Arab Organisation for Agricultural Development, 1981, *Technical and Feasibility Study on Development of Zarka River Lower Catchment*, Amman.

—— 1983; *'Agricultural field survey for developing a cropping pattern for the Jordan Valley', Draft Report*, Amman.
Aretsvik, O., 1976, *The Agricultural Development of Jordan*, Pall Mall Press, London.
Department of Meteorology, *Summary of Climatological Data for the Period 1966-75*, Amman.
Department of Statistics, *Agricultural Sample Surveys in the Chors, 1974-81*, Amman.
-- *Statistical Yearbooks; 1974* and later years to 1981, Amman.
—— *Survey of Agricultural Land Irrigated by Artesian Wells; 1974-81*, Amman.
—— Agricultural Census; 1975, Amman.
—— *Results of the Agricultural Census in the Jordan Valley, 1978*, Amman.
FAO (Food and Agriculture Organisation), 1966, *Jordan Country Report*, Rome.
Jordan Cooperative Organisation, 1982, *Regulations for Extending Agricultural Loans*, Amman.
Jordan Valley Authority, 1974, *Regulations on Land Holdings, Ownership and Water Distribution*, Amman.
—— 1982, *Cropping Patterns for the Jordan Valley*, Amman.
Ministry of Agriculture, 1974, *Agricultural Atlas of Jordan*, Amman.
—— 1974, *Report on Agricultural Zoning*, Amman.
—— *Statistical Records, 1974-81*, Ammman.
—— 1982, '*A proposed cropping pattern for the Jordan Valley*', submitted to the Consultative National Council, Amman.
—— *Working Plan for 1983*, Amman.
—— 1983, 'A statement on agricultural policy', presented as a Government Paper to the Consultative National Council, Amman.
National Planning Council, *Social and Economic Development Plan, 1981-85*, Amman.
— 1983, *Land Use Capability Planning Study*, Amman.
National Resources Authority, 1980, *Water Master Plan of Jordan*, Amman.
Royal Scientific Society, 1974, *Impact of Land Consolidation*, Amman.
Steitieh, A., 1978, *A Manual for Main Vegetables Grown in the East Jordan Valley*, University of Jordan, Amman.
—— and A. H. Musa, 1980, *Vegetables Grown Under Plastic Covers and Drip Irrigation Systems in the East Jordan Valley*, University of Jordan, Amman.

5 FARM SYSTEMS IN RAIN-FED AREAS

Mahmud Duwayri

Background

The total area of the Hashemite Kingdom of Jordan is 94,800 km^2. Of this area, only 9 per cent receives more than 200 mm of rainfall and can be considered productive agricultural land (Table 5.1). The annual precipitation map is shown in Figure 5.1.

Based on the results of the agricultural census of 1975, however, the total cultivable area in Jordan, including the Ghor, was estimated to be 3.9 mn du (Table 5.2).

Table 5.1: Land Areas Classified by Rainfall, East Bank

	Average rainfall	Area (mn du)	Per cent
Arid desert	100 mm	75.0	81.1
Desert	100-200 mm	9.6	10.3
Marginal	200-300 mm	5.3	5.7
Semi-arid	300-500 mm	1.7	1.8
Semi-humid	500 mm	1.0	1.1
TOTAL		92.6	100.0

Source: NPC 1976: 172.

Table 5.2: Distribution of Holdings by District

District	Number of holdings	Area (000 du)	Per cent
Amman	7,965	981	15.7
Irbid	29,915	1,967	58.9
Balqa	5,467	229	10.8
Karak	5,896	560	11.6
Ma'an	1,548	167	3.0
Total	50,791	3,904	100.0

Source: DOS 1982: 10.

Of the total cultivable area of Jordan, approximately 93 per cent depends on natural rainfall. The remaining 7 per cent of irrigated land produces over 40 per cent of the total agricultural production, and nearly 70 per cent of the gross value of agricultural production. Irrigated agriculture provides 85 per cent of agricultural exports. In contrast, most of the products produced in the rain-fed areas of Jordan are import-replacement products. There is a large deficit in the production of cereals and livestock, and the average amount spent on wheat imports was JD 16.9 mn annually for the years 1980-82. The average amount spent on apple imports was JD 5.4 mn (Table 5.3).

Characteristics of Rain-Fed Agriculture in Jordan

Land Use: The area of cropped land varies from one year to another, depending on the amount and distribution of rainfall. The annual cropped area was 2.1 mn du in 1981 and 2.4 mn du in the better year of 1980 (Table 5.4).

Field crops occupied the largest proportion (78.9 per cent) of the cultivated areas; next were fruit trees, which occupied 14.5 per cent. It should be emphasised, however, that there is a trend towards planting more fruit trees, particularly olives. This is being encouraged by the Government through the Highland Agricultural Development Project, which is being financed by the World Food Programme (WFP) of the United Nations Food and Agriculture Organisation (FAO). The main objective of this Project is to control soil erosion through the introduction of improved land-use systems under which land can be used in accordance with its suitability, i.e. to restrict annual crops to suitable soils which have a slope of 0 to 9 per cent. Of the 798,000 ha that receive more than 200 mm of rainfall, 263,000 have a slope of less than 9 per cent. The hilly areas are severely affected by erosion and are subject to continuous deterioration and need urgent protection. The aim of the Project is to reduce erosion through a number of activities such as the construction of contour terraces, other erosion-control structures and the planting of fruit trees. Farmers are offered incentives to participate in this Project.

The dominant fruit in Jordan is the olive, followed by grapes. The dominant field crops are wheat and barley (Table 5.5).

In recent years, tobacco has been occupying more land (Table 5.6). This crop requires special care and provides higher profits than wheat or other summer crops.

Yields: Yields are generally low for all crops. This has been attributed to the low and poor distribution of rainfall. Other factors are equally or more important, however; they include poor husbandry and declining soil fertility. For example, high yields of wheat which amounted to over 5,000 kg per ha were obtained at Taibeh in Irbid in the 1983 high-rainfall season. This occurred on a demonstration plot using a high-yielding variety (Stork) in combination with early planting and the use of nitrogen fertilisers. Low yields of 1,000 kg per ha were also obtained when the variety was planted late and without fertilisation (Duwayri, unpublished data.) The national yields of wheat are presented in Table 5.7.

The data in Table 5.7 indicate that yields ranged from 16 kg per du in 1979 to 102 kg per du in 1974 and 100 kg per du in 1980. Both the seasons 1973/74 and 1979/80 were exceptionally good, whereas the 1978/79 season was extremely dry. In spite of low rainfall in the dry season, some farmers may escape complete crop failure. In years when crops fail due to drought, farmers are able to sell their unharvested crops to shepherds for animal grazing. Failure of the crop is judged to have occurred when the estimated wheat yield is of the order of 250-300 kg per ha or less.

Table 5.3: Main Agricultural Imports and Exports for Selected Commodities, 1980—82 (thousand JD)

	Imports				Exports			
Crops	1980	1981	1982	Average	1980	1981	1982	Average
Wheat	13,962	17,707	18,995	16,888	115	3	—	39
Barley	1,251	558	3,523	1,777	2.8	—	—	1
Lentils	177	279	233	230	—	429	91	173
Tomatoes	17	5	11	11	5,394	6,664	7,194	6,417
Potatoes	2,651	3,807	4,246	3,568	42	116	73	77
Olives	66	9	—	25	—	—	—	—
Grapes	110	47	21	59	231	306	327	288
Apples	4,055	4,789	7,433	5,426	15	30	29	25

Table 5.4: Land Use on the East Bank excluding the Ghor, 1980—82 (thousand dunums)

Year	Cultivated				Uncultivable	Fallow		
	Fruit trees	Veget-ables	Field crops	Sub-total		For rest	Unused	Grand total
1980	316	133	1,981	2,430	301	532	369	3,632
1981	327	158	1,650	2,135	177	808	517	3,637
1982	339	154	1,693	2,186	160	790	501	3,637
Average	327	148	1,775	2,250	213	710	462	3,635
% of total	9.0	4.1	48.8	61.9	5.9	19.5	12.7	
% of cultivated	14.5	6.6	78.9	100				

Source: DOS 1980, 1981 and 1982.

Table 5.5: Areas of Crops Cultivated under Rain-fed Conditions, 1974–82 (thousand dunums)

Crop	1974	1975	1976	1977	1978	1979	1980	1981	1982
Wheat	2,355	1,111	1,307	1,212	1,303	961	1,301	970	995
Barley	623	512	515	446	506	445	502	467	479
Lentils	216	148	229	133	143	73	86	105	108
Chick-peas	n.a.	36	16	14	13	26	29	20	20
Olives	221	251	256	338	319	324	275	291	302
Grapes	44	66	31	26	42	40	22	16	28

Source: DOS 1974-82.

Table 5.6: Area and Production of Tobacco, 1978—82

Year	Area (000 du)
1978	36.4
1979	42.5
1980	60.0
1981	80.0
1982	72.0

Source: DOS 1982.

Table 5.7: Yields of Wheat, 1974—82

Year	Yield (kg per du)
1974	102
1975	42
1976	47
1977	49
1978	40
1979	16
1980	100
1981	51.0
1982	48

Source: DOS 1974-82.

Landholdings: The Meeri land-ownership laws which govern agricultural land in Jordan give the owner freedom over his land within those prevailing laws (AOAD 1977, 1978). Land is divided among a landowner's heirs. Sometimes division is made vertically on slopes so that all recipients receive equal proportions of good bottom land and poorer stony land on the hilltops. The effect this practice has on the soil erosion problem is obvious. Furthermore, it results in small, fragmented holdings. Qasem (1979) reported that the upland agricultural area in 1975 was divided into about 45,000 holdings and 110,000 pieces. The average size of each holding was 80.4 du distributed in 2.44 pieces average, which amounted to 33 du per piece. He suggested that fragmentation worsened still further after 1975.

A study conducted by a team from the Arab Organisation for Agricultural Development (AOAD) in 1977 for Irbid and Balqa Governorates showed that the average area of holdings in each governorate was 88 and 32 du, respectively. The average ownership was 31 and 12 du respectively. This situation shows how difficult it is to cultivate these holdings efficiently unless they are grouped in a cooperative, or in some other way. In many cases, however, owners have chosen to rent or lease the land to a tenant farmer, who is more interested in exploiting resources than in conservation. This land fragmentation is thought to be one of the major constraints on agricultural development in the rain-fed lands.

Farming Practices: Field Crops

Crop Rotation: The crop rotation practised in Jordan varies from one area to another and will therefore be discussed by zone:

1. The desert zone (100-200 mm rainfall). Successful wheat or barley crop may be obtained from this zone once every five to ten years. There is no definite crop rotation, and farmers plant almost every year when the early rains fall.
2. The marginal or eastern zone (200-300 mm rainfall). A two-year crop rotation is followed: wheat/barley—fallow, wheat/barley—lentils or vetch, wheat/-barley—summer crop. The wheat—fallow rotation occupies about 50 per cent of the area in this zone. It is not a planned rotation and depends mainly on early rains. The wheat—summer crop rotation occupies about 10 per cent of this area.
3. The western plain area (300-400 mm rainfall). The rainfall pattern is reliable and therefore well-planned crop rotation is practised. There are three types in this area:
   a. Two-year crop rotation mainly of wheat—legume.
   b. Two-year crop rotation of wheat—summer vegetables.
   c. Three.year crop rotation of wheat—legume—summer vegetables. This type of rotation predominates near the city of Irbid.
4. The upland area (400 mm rainfall). A two-year rotation predominates in this zone. Wheat is either followed by legume or summer vegetables.

A survey by Arabiat *et al.* (1982), however, which included 238 farmers, reported that the primary rotation is one of wheat—fallow; 38 per cent of farmers had left their wheat fallow and 31 per cent would leave it fallow the next year. Around 20 per cent had planted vegetables, around 20 per cent included legumes in the rotation and less than 5 per cent planted wheat continuously.

Cultivation Practices

1. Seed-bed preparation: This is carried out in October or November by ploughing the land using tractors and large three-disc harrow ploughs to a depth of 20-25 cm, usually up and down the slope. This is inadequate seed-bed preparation, because these implements are not suitable for rain-fed conditions.
2. Seeding: Seeding is broadcast, although there are a few seed drills in operation. Normally the farmer divides the land into strips 10-15 m wide to help distribute evenly. He casts the seed by hand, covering one side of the strip, then comes back from the other side. The seed is covered over bŷ a multi-disc offset cultivator. Seeding depth and distribution are not uniform, and therefore large amounts of seed are required. The survey by Arabiat *et al*-(1982) indicated that 50 per cent of farmers use seed from their own stock, while 14 per cent obtain it from cooperatives and 28 per cent purchase it in the market. A very high proportion, 62 per cent of farmers, use seed treated with chemicals.
3. Date of seeding: The optimum seeding date is 15 November, but most farmers plant later, and it is not unusual for them to plant as late as February. This practice leads to poor crops and low yields.
4. Fertiliser application: Very few farmers use fertilisers, only 8 per cent use N at a very low rate and less than 1 per cent use P_2O_5 (Arabiat *et al.* (1982).
5. Weed control: The dominant method of control is hand weeding. Eight per cent of farmers use herbicides.
6. Harvesting: Combines are available for harvesting; manual harvesting,

however, is still practised when the holding is very small, when the land slopes, or when the landlord has animals and would like to use the straw for feed.

For the past 10 years the author has been involved in wheat improvement in Jordan. The 1983 season was the last one of a five-year project implemented by the Ministry of Agriculture (MAO), the Faculty of Agriculture (FAO) and the International Centre for Agricultural Research in the Dry Areas (ICARDA). Table 5.8 presents a summary of the recommendations made.

Radio and TV programmes, together with field days, made farmers aware of the technologies of this project and previous ones, like that of the Agency for International Development (AID), Oregon State University. The adoption rate of these technologies is very slow compared to other countries in the Middle East. Arabiat *et al.* (1982) commented on the slow adoption rate, pointing out that 50 per cent of wheat is used for home consumption, so that farmers are not motivated to adopt improved technologies. Furthermore, the agri-climatic conditions in Jordan, especially the variability of rainfall from year to year, introduce a high degree of uncertainty and risk to agricultural production. In general, this has negative impact on the adoption of technologies.

Lentils, vetch and chick-peas are important field crops in Jordan. They are part of a crop rotation with wheat. Lentils and vetch in particular are dealt with in the same way as wheat. New technologies have not been adopted. These crops are hand harvested and attempts to machine harvest them have been unsuccessful, despite it being the norm in the US and other developed countries. The value of both grain and straw is such that many farmers are hesitant to use machines for harvesting them. The migration of workers from rural areas has caused shortages at times of peak labour requirements, which increases the cost of hand harvesting of crops, particularly lentils. Many labourers would prefer any other manual job to pulling lentils. This has affected the growing of lentils, and unless mechanical harvesting is available this crop will be phased out.

Olive Trees

Olive trees occupy the largest area among the orchards in the rain-fed lands of Jordan and we give below an outline of farming practices for them. Olive trees are available for the tree grower from the Government and from private nurseries as rooted cuttings or grafted trees. The problem lies in getting olive trees true to type. For instance, the cultivar Nabali is well known and grown for oil production but trees sold from this cultivar by the two types of nurseries are not guaranteed. A study by al-Wir and Larsen (1981) showed that farmers are not aware of rootstocks.

Orchard Sites: A very common practice by most tree farmers is to choose infertile soils for olive trees, since they withstand poor soil conditions. Furthermore, the Fruit Tree Project provides incentives to farmers to grow trees in the uplands with an average rainfall of not less than 300 mm and a slope ranging from 9 to 25 per cent, thus saving fertile land for annual crops. However, it is highly recommended to choose fertile soil for a profitable yield.

Tree Spacing:

1. In rain-fed areas 10 × 10 m spacing is used, or 10 trees per du.
2. With supplementary irrigation, 7 × 7 m or 8 × 8 m.

Cultivation Practices: From date of planting in the orchard until trees come into bearing:

Table 5.8: 'Best Bet' Production for Cereals by Rainfall/Zone

	Zone A (> 350 mm)	Zone B (250–350 mm)	Zone C (< 250 mm)
Variety[1]	Durum wheat: DA6 & Stork	Durum wheat: Hourani, Stork & DA2 Barley: DA106	Barley: Giza 120
Seed rate (kg/ha)[2]	80	80	60
Nitrogen (kg N/ha)[3]	60 (30 at sowing & 30 at tilling if season good)	60 (30 at sowing & 30 at tilling if season good)	30 (preferably at tilling if season good)
Phosphate (kg P_2O_5/ha)[4]	30	30	40
Sowing date[5]	Before rain (Nov.)	Before rain (Nov.)	Before rain (Nov.)
Sowing method[6]	Drill	Drill	Drill
Weed control[7]	2,4-D ester, 800-1,000 g/ha	2,4-D ester, 800-1,000 g/ha	2,4-D ester, 800-1,000 g/ha

Notes: 1 Choice of variety may depend on the value placed by the farmer on the straw. The seed of some varieties will need to be increased.

2 Does not greatly affect economics. More research is needed on interaction with weed control and sowing method.

3 A high-risk input; thus split application is recommended, though experimental data are not available.

4 Individual recommendations should eventually be tied to a soil test.

5 This practice is only recommended in conjunction with deep sowing (6-8 cm) and herbicide use.

6 Drill sowing is suggested on the basis that if custom drilling becomes available, it will prove cheaper than hand sowing.

7 2,4-D is recommended for broadleaf weed control, especially for early sowing; other chemicals may be economical for grass weeds.

Source: ICARDA 1982.

1. Additional irrigation must be practised for the first three or four years, even if rainfall is 400-500 mm.
2. Proper training for trees is needed to facilitate harvesting and other practices in addition to improving the health of the tree. This includes knowledge of pruning young trees as well as fruit-bearing ones.
3. Fertilisers: The olive tree, like any other fruit tree, requires the addition of NPK fertilisers as well as other elements such as boron, which is necessary for flower development. The amount of fertiliser applied varies according to soil fertility, age of the tree and yields.
4. Most orchards are watered by rain and no supplementary irrigation is practised except in the Jordan Valley. Such practice is necessary particularly during the period when flower-bud development occurs, and especially in seasons when there is less than 400-500 mm rainfall.
5. Pesticides: A number of diseases and insects attack olive trees (roots, leaves, flowers and fruit), thus causing substantial financial loss to the tree farmer. It is necessary, therefore, for him to watch his trees very carefully and to use the proper pesticides. For example, Mustafa (1983) reported that olive psylla (*Euphyllura olivina*) caused a loss in fruit yield of 21 and 23 per cent in al-Fheas and Jerash respectively in the 1983 season.
6. Harvesting: At present, olives are harvested by hand and will be for a long time to come, since there is no chemical registered to facilitate harvesting. Hand picking increases the cost of production, particularly with the shortage of labour.

Alternate Bearing Problem (where high yields occur one year, followed by low yields the next)
This is a major problem and is due mainly to tree physiology, internal factors and to the farmer's ignorance of the proper practices, for example, not practising fruit thinning in years of high yields, not watering trees sufficiently during flower-bud development and not applying fertilisers.

Even though farmers are aware of the importance of proper cultural practices, through either radio or TV programmes, few use them. Al-Wir and Larsen (1981) made a survey of 117 growers, representing 11,264 du planted with fruit and nuts in 1979. Their findings are summarised as follows:

1. 70-80 per cent of growers were not aware of the cultivars or rootstocks they were growing.
2. 97 per cent did some annual pruning but did not seem to understand the fundamentals of tree and vine pruning.
3. Less than 3 per cent used supplementary irrigation in their orchards.
4. 37 per cent used fertilisers, mainly of manure.
5. Growers had little knowledge of the different kinds of orchard and vineyard insect, and disease.

The researchers concluded that the lack of basic understanding of fruit production fundamentals is partly due to the inadequate experience of the grower. This is reflected by relatively young trees and vine age, which averaged 10 years. Absence of extension services (rated at 0.5 on a scale of 0 to 2) also contributed to the lack of basic understanding. The growers of the Jordan Valley were found to be more advanced in their practices and had received more extension help.

Farming in the Rain-Fed Areas of Jordan

The Government has been keen to develop the rain-fed areas of Jordan. A total investment of JD 214.8 mn in the 1981-85 Five-Year Plan has been proposed. This includes investment of funds in the development of livestock and cereal industries, as well as in the forestry and soil conservation areas. The Government hopes to achieve its objectives for the improvement of the rain-fed agricultural sector by:

1. The adaption of agricultural production to agri-climatic zones on the basis of land-use planning
2. The introduction of modern technology in order to increase production
3. The improvement of rangelands to increase fodder production and the integration of rain-fed farming and livestock.

Watkins *et al.* (1983) have prepared a land-use plan for the rain-fed agricultural sector. The proposed land classification system and the crop zoning proposals based on land capacity sub-classes per agri-climatic zone are shown in Tables 5.9 and 5.10. Wheat growing is restricted to the three rainfall zones (semi-humid, semi-arid and marginal) which prevail in land classes I, II and III. Legumes are not recommended for cultivation in the marginal 200-300 mm rainfall zone. Grapes and olives are recommended for land classes IV and V, and in high rainfall areas.

Table 5.9: Proposed Land Classification System

Land	Criteria
I	Land with few limitations, soils deep (> 100 cm), non-rocky (0-10%), slopes (0-2%) with textures clay to sandy loam.
II	Soils deep to moderately deep (50-200 cm), slopes (2-7%), non-rocky with textures clay to sandy loam.
III	Soils deep to moderately deep, slopes (8-14%), non-rocky textures clay to sandy loam.
IV	Soils moderately deep, non-rocky and slopes (15-25%) of textures clay to sandy loam.
V	Soils moderately deep, with slopes (0-24%) rocky (10-25%) or soils with textures coarser than loamy sands.
VI	Soils moderately deep, with slopes (25-50%) rocky and non-rocky.
VII	All shallow soils (< 20 cm) irrespective of slope, as well as land with more than 50% surface rock exposure.
VIII	All land with slopes over 50%, irrespective of depth or amount of rockiness.

Source: Watkins *et al.* 1983.

Qasem (1979) suggested that a farming family of five members should cultivate about 200 du each year in order to obtain an income equal to the national average. He based his calculations on JD 5 being the average net productivity per du of the rain-fed areas. The number of holdings with 200 du or more was reported in 1975 to constitute about 9 per cent of the total.

Table 5.10: Crop Zoning Proposals Based on Land Capability Sub-Classes per Agri-Climatic Zone

Land class	Rainfall zones				
Class	Semi-humid (>500 mm)	Semi-humid (300-500 mm)	Marginal (200-300 mm)	Arid (<200 mm)	Area (000 du)
I	Wheat Grain legume Summer crop Horticulture	Wheat Grain legume Summer crop Horticulture	Wheat Barley Sown pasture (vetch)	Natural pasture (a)	79,381
II	Wheat Grain legume Summer crop Horticulture	Wheat Grain legume Summer crop Horticulture	Wheat Barley Sown pasture (vetch)	Natural pasture	
III	Wheat Grain legume Horticulture	Wheat Grain legume Horticulture	Wheat Barley Sown pasture (vetch)	Natural pasture	
IV	Horticulture (grapes, olives)	Horticulture (olives, grapes, figs, apricots)	Olives (pistachios) Horticulture Natural pasture	Natural pasture	4,301
V	Forestry Grazing Horticulture (grapes, olives)	Forestry Grazing Horticulture (peaches, figs, apricots)	Pasture Forestry	Natural pasture	8,868
VI	Forest Grazing	Forest Grazing	Pasture Forestry	Natural pasture	
VII	Forest pasture (forage rein-forcement)	Forest pasture (forage rein-forcement)	Natural pasture	Natural pasture	
VIII	Forest	Forest	Natural pasture	Natural pasture	
Area (000 du)	989	1,700	5,301	84,560	92,550

Note: [1] Natural pasture includes species reinforcing, moisture conservation practices, and where possible reseeding pastures into arable land to be retired.

Source: Watkins *et al.* 1983.

The survey conducted by Arabiat *et al.* (1982) showed that farming was diversified and that farmers planted an average of 105 du of wheat in an average of 2.1 land parcels. They also planted an average of 43 du of barley, plus a variety of other crops such as legumes, vegetables, tobacco, and so on. Land was primarily owned, although share-cropping was also prevalent. Sixty-nine per cent of the parcels analysed were owned and 23.5 per cent share-cropped. From the data reported by al-Wir and Larsen (1981), we can conclude that a farmer was planting 96 du or 1,810 trees and vines.

The list of 26 farmers who received loans from the Agricultural Credit Corporation (ACC) to use for the Highland Agricultural Development Project indicates that the average area of land was 92.6 du, ranging from 10 to 756 du. This background makes it difficult to generalise about the average size of the farm in rain-fed agriculture.

A characteristic rain-fed farm utilising the resources available in Jordan would be comprised of land in classes I, II and III for wheat and legumes, and in classes IV and V for olive trees.

In this chapter we shall be dealing with a farmer who has 100 du of land in classes I, II and IV in a semi-arid rainfall zone. He is using his class I land for growing wheat and planting his orchard in class IV. He has 70 du in classes I and II, where he is planting 40 du of wheat, 15 du of lentils and 15 du of other summer or annual crops. The orchard comprises 30 du.

Environments and Activities Related to a Rain-Fed Farm System

Off-Farm Environments

Physical

Access: The land parcels are located 2-3 km from the village, which is near Irbid. There is an asphalted road with bus and taxi services to Irbid, in addition to telephone communications.

Seeds and Seedlings:

1. Wheat: The farmer can use seeds from his previous year's stocks. Many farmers either send their seeds to one of the stationary cleaners which are owned by the Government in different agricultural centres. There are at least five of them in Irbid Governorate. The farmer may, however, choose to clean his seeds at home. This is normally done by women by hand, using sieves. The farmer can also buy seeds from a neighbour or from the Jordan Cooperative Organisation (JCO) which sells clean seeds true to variety. The JCO sells seeds at a subsidised price of JD 100 per t. The most common varieties are traditional ones like Hourani and F8. These are preferred because they involve the least risk; neither agriculture officials nor farmers are very willing to make seeds of high-yielding varieties available, or to plant them, because of the risk factor involved. Mercuric compounds are commonly used in seed treatment for the control of fungal diseases. Now non-mercuric compounds are available, and farmers are encouraged to use them.
2. Lentils: The farmer obtains seeds from his own sources, i.e. his previous year's stocks or his neighbour's, and the JCO sells limited amounts.
3. Tree seedlings: These are available from private and government nurseries. However, the Government sells them at subsidised prices mainly to the participants of the Highland Agricultural Development Project. (This is the Fruit Tree Project sponsored by the WFP.)

existence of a wide variety of farm machinery and equipment—such as stone pickers, fruit harvesters, orchard sprayers and mechanical pruners—that could save them labour and money. Furthermore, they reported that private custom service operators are neither well trained nor skilled. The Jordanian Government has realised the importance of and need for farm machinery and it has supported the JCO in establishing custom service operations for farm machinery and equipment. The JCO now has three stations, in Irbid, Madaba and Karak. These three stations can serve a maximum area of 150,000 du per annum. The JCO aims

Availability of Inputs: The JCO and agri-business firms provide chemicals for seed treatment, fertilisers and herbicides.

Machinery: Most farmers obtain their machinery through private custom services. However, a study conducted by Snobar and Arabiat (1983) showed the services provided by such custom operators to be limited to land preparation, threshing and cereal harvesting. Large numbers of farmers were unaware of the to introduce new sets of farm equipment, demonstrating proper use and timing; and to encourage the private sector rather than to replace it.

Institutional

Ministry of Agriculture—Agriculture Directorate: The Directorate is staffed by agricultural engineers whose function is to provide assistance and extension help to farmers. They also have links with researchers of the FOA at the University of Jordan who keep them informed of research activities in the district and occasionally help in coordinating field days.

Agricultural Research Stations and Seed Increase Farms: There are two of these, one called the Ramtha Agriculture Station and the other the Marrow Agriculture Station. Their main purpose is to propagate certified seeds, but they also carry out research into cereals and legumes for the different research organisations and programmes of Jordan. Field days are held in these Stations and farmers are invited to attend.

Jordan Cooperative Organisation: The JCO has a higher degree of contact with farmers than any other organisation in Jordan. In 1982, of the 39,311 members of different societies, 16,062 belonged to agricultural societies, of which there are 169. Although the major function of the JCO is to extend seasonal or medium-term loans to members, it also provides agricultural supplies for farmers regardless of membership. Cooperative societies also assist farmers by hiring out cooperative-owned machinery and by the handling or processing of agricultural produce.

Yarmouk University: A Faculty of Agriculture will start up in 1985 but research programmes on cereals improvement and fruit-tree production may start in 1984.

Credit Institutions: The three major sources of financial credit are the commercial banks, the ACC and the JCO. Commercial banks finance agri-business and machinery firms. The ACC has given $US 2.5 mn in loans to farmers in the highlands, most of which has been received by project beneficiaries (Project 2108 and Expansion) for carrying out land improvement. Farmers in areas receiving over 300 mm of rainfall are granted these loans at an interest rate of 7 per cent (USAID 1983). In 1982, 26 farmers received JD 100,380 in loans (ACC, unpublished data). The JCO also provided loans in the Highland Agricultural Development Project area between 1975 and 1981 totalling JD 600,000 to some 400 farmers with holdings of between 2 and 3 ha, at an interest rate of 6 per cent. The International Fund for Agricultural Development (IFAD) has provided a loan of $US 10.6 mn to the JCO. This loan is being given to small farmers to be utilised, among other things, for orchard development, including olives, stone fruit and grapes.

Marketing:

1. Wheat: There are a number of government interventions in wheat growing. These include domestic wheat price support, control over imports and exports, distribution of wheat at subsidised prices, and price control of flour and bread. The Government has provided price support to producers of domestic wheat since 1974. The price of wheat is fixed each season at harvest time. A committee from the Ministry of Supply (MOS) the MOA and the JCO is formed to buy wheat at the fixed prices, which are higher than those on the world market. This has established a minimum price which has encouraged farmers to grow wheat. The price set in 1979 was JD 90 per t and for 1983 it was JD 100. However, not all domestic wheat is sold to the Government. The amount purchased in 1983 totalled 30,000 t, which is about 30 per cent of production. There are also private channels; farmers sell their wheat to consumers or wholesalers who sell it to consumers, mainly living in city suburbs or in rural areas. After buying it from the wholesalers, consumers have to carry it to small mills located in their neighbourhoods, to be milled at their own expense, before they use it for baking bread.

   All imports of wheat are controlled by the Government through the MOS, which distributes wheat to flour mills at highly subsidised prices. The price of flour sold by millers to bakers, and of bread sold by bakers to consumers, is regulated by the Government. Currently the price of bread made from imported wheat or flour is JD 0.075 per kg. In the past, the only way for local wheat to be used for bread was for people to make their own bread, or to take their wheat to custom millers and custom bakers. This year the MOS is selling a new kind of bread: local bread made from local wheat, milled and baked by the Ministry and certain bakeries. The official price is JD 0.110 per kg. This is a very good incentive for farmers to grow wheat.
2. Lentils: Lentil prices are also set by the Government. Farmers can sell to the Government, to wholesalers or to other consumers. The MOS fixes the retail price to final consumers.
3. Olives and Olive Oil: By the middle of October, farmers start harvesting their olive crop. Farmers may sell olives wholesale or send them to a factory for pressing. There are factories situated close to the areas of production. When pressed, every 100 kg of olives yields an average of 20-30 kg of oil. The pressing factory charges the farmer in kind. The majority of farmers sell their oil to wholesalers, who then sell it to retailers. The selling unit at the wholesale level is the tin can, and at the retail level it is usually sold by kilogramme.

Informational

Research: Research aimed at finding the best practices for crops grown in the rainfed sector, particularly for cereal crops, started in the late 1950s. The MOA was then testing varieties in the Irbid nursery as well as in other nurseries in Jordan. In 1967 the MOA started a programme in collaboration with Oregon State University which was sponsored by AID to increase wheat production through better practices which conserve soil moisture, control weeds and make better use of fertilisers. In 1974 the FOA launched a research programme and in 1978 it collaborated with the MOA and ICARDA in a cooperative winter cereal research programme, which was partially sponsored by the Ford Foundation (see Table 5.8 above). Many farmers know about this programme because part of the research and demonstration was conducted on their lands. Research on food legumes was neglected until 1979, when a research programme was started at the FOA aimed at finding better varieties and best practices. It also emphasised ways to use mechanical harvesting of lentils, providing support for the development of the prototype of a mechanical lentil harvester designed by Snobar (Snobar, 1981). The project is partially financed by the International Development Research Centre (IDRC). Research efforts on fruit trees have not been as vigorous as for annual crops, and although some work on fertilisation of olive trees and pest control research projects has been started, no research results are available yet. There has also been some research emphasis on the use of growth regulators.

Extension: The MOA in Amman has a Department of Agricultural Research and Extension which writes bulletins and distributes them to farmers through Agricultural Directorates in the different governorates. The Agricultural Directorates hold frequent field days and invite farmers to experimental stations. There are two experimental stations for annual crops, one in Ramtha and the other in Marrow, Irbid Governorate. Farmers are also invited to field days to review, with researchers, the results of experiments. Extension agents also hold meetings on pruning and disease control of fruit trees. Farmers seem to be aware of the new technologies in cereals, and possibly in legumes and tree planting. But it would appear that they put risk minimisation before profit maximisation since the adoption rate of new technologies is very slow.

On-Farm Environments

Physical

Location: The characteristic farm is located in one of the villages in the Irbid Governorate. (See Figure 5.1.) The farmer has three parcels of land, one of 30 du located in the highlands where the soil is moderately deep, non-rocky and has a slope of 15-25 per cent, the texture being clay to loam. Average rainfall is 450 mm. The other two parcels are classified as class I and II land and have deep soil (approximately 200 cm) and a low slope. Average rainfall is 350-400 mm for both. One of these parcels (40 du) is planted with wheat. Half the other is planted with lentils and the other half will be planted with summer vegetables. The farmer is following a two-year crop rotation pattern.

Climate: The winters are cold but not too severe. The lowest temperature is close to Irbid's, which was -5.0°C between 1966 and 1975. The mean daily temperature is 10°C in January and 25°C in August. Average annual rainfall for the years 1955-82 in Irbid was 460.2 mm. The rainy season starts in October with the most precipitation from December to March, January having the highest (104 mm).

Housing: The average farming family is estimated at 6-8 persons, with an adult-worker capability of 2.1 full-time persons. It is estimated that more than 80 per cent of farmers are over the age of 35 years. Arabiat *et al.* (1982) reported in their survey that the average age was 53 years. The farmer's elder sons have jobs in Irbid or Amman, or in neighbouring countries. The farmer is living in a three-room concrete house built recently with his sons' help. Electricity and piped water are available in the village, which is connected to Irbid by an asphalted road. The farmer lives within 2 to 3 kms of his holdings.

Buildings and Constructions: Farmers do not have permanent buildings on land for annual crops. Some may have a small hut in an orchard. This room is used for storing hand tools and in the summer is used as a dwelling for the family or guards. Also in the orchard there may be a well dug out of the rocks for collecting

Figure 5.1: Land capability map for the areas surrounding the city of Irbid as an example of the rainfed characteristic farm.

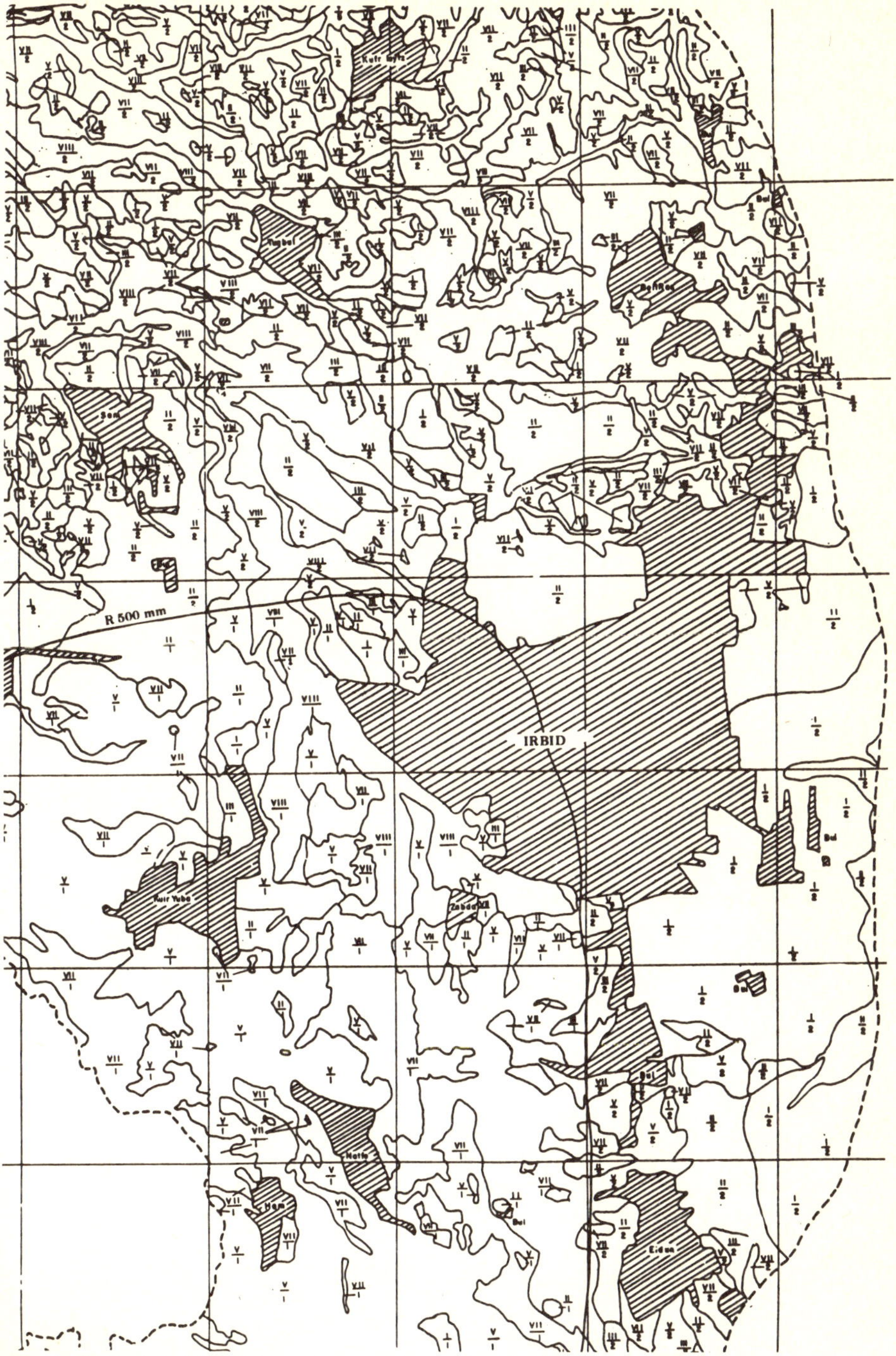

(Scale 1:50,000) I-VIII: Land Cap. Class (See Table 9)
1-4: Rainfall.
Rainfall 1: more than 500, 2: 300-500, 3: 200-300, 4: Less than 200 mm.

and storing water. In some cases there is a reservoir made of concrete or aluminium. This is used for watering young trees or for supplementary irrigation in the dry season.

Outputs and Accounts: For class I land planted with wheat, the yield is estimated at 160 kg per du if the farmer uses modern techniques. It is estimated to be 33 per cent less, i.e. 110 kg per du, if he uses traditional methods. The comparative gross margin for traditional versus modern techniques is shown in Table 5.11. The gross margins in JD per du are 4.92 for traditional techniques as against 9.34 for modern techniques. The gross margin for growing lentils at improved techniques is JD 4.4. (See Table 5.12.) The gross margin for olive growing is estimated at JD 15.8 per du. (See Table 5.13.) Therefore the rain-fed farmer's net income in an average year will be 40 × 4.92 = JD 196.8 from wheat farming if he uses traditional

Table 5.11: Comparative Gross Margins: Traditional versus Modern Techniques for Wheat Grown on Class I Land Receiving 400mm Rainfall

Income	(JD)
Traditional techniques	
Wheat, 110 kg per du at JD 100 per t	11.00
Total income	11.00
Production expenses:	
2 machine ploughings at JD 0.8 each	1.60
Chemical seed treatment at JD 25 per t	0.30
Seed cost (12 kg per du at JD 0.11 per kg)	1.32
Hand sow (0.05 man-day per du at JD 4[1] per man-day)	0.20
Disc harrow to cover seed	0.50
Harvesting machinery hire	1.50
Bags at JD 0.3 per bag	0.33
Sewing & transport at JD 0.3 per bag	*0.33*
TOTAL PRODUCTION EXPENSES	6.08
GROSS MARGIN per du	4.92
Modern techniques	
Wheat, 160 kg per du at JD 100 per t	*16.00*
Total income	16.00
Production expenses:	
2 cultivations at JD 0.4 each[2]	0.80
Chemical seed treatment at JD 25 per t	0.25
Seed cost (10 kg per du at JD 0.11 per kg)	1.10
Fertiliser, 10 kg mecaphos at JD 0.105 per kg	1.05
Seeding machinery hire	0.50
Chemical weed control	0.50
Harvesting machinery hire	1.50
Bags at JD 0.3 per bag	0.48
Sewing & transport at JD 0.3 per bag	*0.48*
TOTAL PRODUCTION EXPENSES	6.66
GROSS MARGIN per du	9.34

Notes: [1] Assumes an opportunity cost to the farmer's labour.
[2] Shallow cultivations at reduced cost.
Source: Adapted from Watkins *et al.* 1983.

farming methods, which is more than likely. The income will be 40 × 9.34 = JD 373.6 if he uses modern technology. He will also receive JD 66 from lentils and JD 474 from his orchard. His income will be JD 913.6 if modern wheat farming is utilised, or JD 736.8 if he keeps to the traditional practices.

Informational

The farmer can read and write, although his wife is illiterate. The younger sons and daughters go to school, while the oldest ones are out at work elsewhere. The farmer has a radio and TV. There are daily radio agricultural programmes which are normally broadcast in the morning. These programmes concentrate on the seasonal activity; for example, emphasis is given to cereal planting and the different practices needed during October and November. Early planting, use of chisel ploughs and grain drills, use of improved and treated seeds and use of fertilisers and herbicides are stressed.

TV agricultural programmes are shown weekly and also emphasises the seasonal activity. Extension agents pay visits to farmers, particularly horticultural extension workers, who mainly come to inspect the activities of farmers involved in tree-planting projects. They examine terracing, tree planting and other activities which are tied to the provision of material assistance or to the supervision of loans.

Table 5.12 Gross Margin for Lentils Grown on Class II Land Receiving 400 mm Rainfall

Income	(JD)	Labour (man-days)
90 kg lentils at JD 180 per t	16.2	
90 kg tibin at JD 60 per t	5.4	
TOTAL INCOME	21.6	
Production expenses		
2 cultivations at JD 0.4 each	0.8	0.04
Seed cost at JD 190 per t	1.3	
Fertiliser — Triple Super at JD 105 per t	0.6	
Seeding machinery hire	0.5	0.02
Hand Weeding	4.0	1.00
Picking by hand	4.0	1.20
Collecting lentils	1.7	0.67
Transporting lentils & tibin[1]	2.1	0.20
Threshing by contractor	0.8	0.11
Bags — lentils & tibin[2]	1.1	
Sewing of bags	0.1	
(Labour included under transport)		
TOTAL PRODUCTION EXPENSES	17.0	
GROSS MARGIN (JD per du)	4.4	

Notes: [1] Labout included under transport.
[2] Based on JD 0.3 per kg bag lentils; JD 0.40 per 90 kg bag straw.

Source: Adapted from Watkins *et al.* 1983.

Table 5.13: Gross Margin for Olive Production in an Average Year for an Olive Orchard Grown on Class IV Land Receiving 400 mm Rainfall

Income	Average Year	Labour (man-days)
Olives at JD 325 per t[1]	48.8	
TOTAL INCOME	48.8	
PRODUCTION EXPENSES		
1 cultivation at JD 0.5	0.5	0.03
Fertiliser — Mecaphos at JD 105 per t	2.1	
Fertiliser application	1.1	0.20
Insecticide & application[2]	4.2	0.30
Pruning	2.0	0.50
Hand weeding	1.6	0.40
Harvesting[3]	12.0	4.60
Bags at JD 0.3 each	0.5	
Transport & storage[4]	0.9	0.12
Cans for oil[5]	1.1	
Pressing[6]	4.9	
Replacement cost[7]	2.1	0.16
TOTAL PRODUCTION EXPENSES	33.0	
GROSS MARGIN (JD per du)	15.8	
TOTAL LABOUR (man-days per du)		6.31
LESS OPPORTUNITY COST[8]	15.0	
REAL MARGIN	0.8	

Notes: 1 Based on olive oil price of JD 1,300 per t and oil yield of 25%.
2 1.5 1 per du or JD 2 per l and 0.3 man-days at JD 4 per man-day.
3 Based on 50 kg per man-day.
4 Based on JD 6 per t.
5 JD 0.7 per 25 kg oil.
6 No labour assumed, as pressing is an industrial rather than agricultural activity.
7 JD 150 per du every 70 years.
8 Opportunity cost at 10 per cent of capital required to establish an orchard.

Source: Adapted from Watkins *et al.* 1983.

Activity Analysis

The activities and decisions involved in the operations of the selected characteristic farm which produces wheat, lentils and olives will now be described. The information is based on direct observation of current practices. Flow charts are included to summarise the sequence of operations and their relations with each other. The operations involved are presented separately for each of the three products.

Analysis of a Wheat Farm System

The activities involved in the production of wheat under rain-fed conditions in Jordan are set out in Figure 5.2.

Physical Activities

1. ASSESS ASSETS
Duration: September-November.
Comments: The farmer must know what wheat sources he has for annual cropping. This mainly includes seed sources.

2. REMOVE CROP RESIDUE
Timing: 2-3 days in October or November.
Environmental requirements: Rains have started so tomatoes will not produce any longer. The watermelon crop can be grazed or could have been removed a long time earlier.
Preceding activity: Previous season must be completed.

Inputs required	*Type*	*Quantity*	*Price*	*Comment*
Labour	Unskilled	—	Can be paid in tomatoes	This could be done by family members

3. PREPARE SEED-BEDS
Duration: July-November.

Environmental requirements: Dry soil, one ploughing before rain in October and the other after rain. Soil should not be ploughed when it is wet.
Preceding activity: 2. The land must have been cleaned of the crop residue such as tomatoes, tobacco or watermelon plants.

Inputs required	*Type*	*Quantity*	*Price*
Plough mounted on tractor	Disc or mouldboard	2	JD 0.8 per time or JD 1.6 per du

Comments: Normally first ploughing is done in June-July if after lentils and the other after the rain in November.

4. PURCHASE SEEDS
Duration: October and November.
Preceding activity: Decision on variety to be used.

Inputs required	*Type*	*Quantity*	*Price*
Purchase seeds	—	12 kg per du	JD 0.11 per kg

5. TREAT SEEDS
Duration: One day during October and November.
Preceding activity: 4.

Figure 5.2: Rain Fed Wheat Farm System

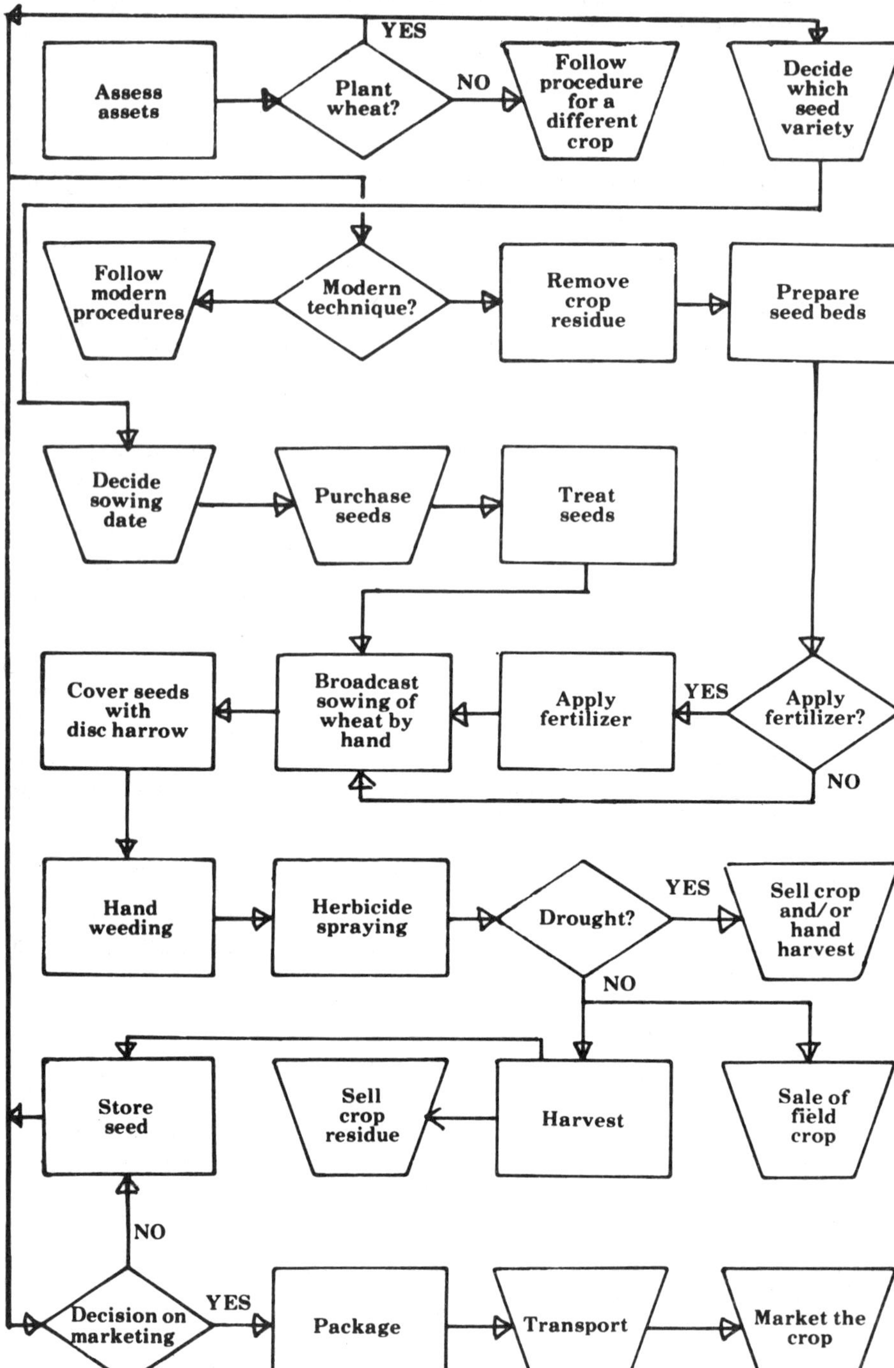

Inputs required	*Type*	*Quantity*	*Price*	*Comment*
Fungicide	Preferably non-mercuric	—	JD 0.3 per du	Includes cost and chemical

Comments: Seed treatment for smut, to which the local varieties are susceptible.

6. SOW (BROADCAST) WHEAT BY HAND
Duration: One man can sow 20 du per day.
Environment requirements: The soil should have been ploughed and weeded. It should not be muddy. The weather should not be rainy to allow time for covering.
Preceding activities: 2, 3, 4 and 5.

Inputs required	*Type*	*Quantity*	*Price*
Labourers	Skilled	0.05 man-days per du	JD 4 per du

Comments: Only skilled labourers can do a good job. Drills are preferred, but they are not available.

7. COVER SEEDS WITH DISC HARROW
Duration: One day to cover 100 du.
Environmental requirements: The soil should not be muddy and it must not be raining.
Preceding activity: 6.

Inputs required	*Type*	*Quantity*	*Price*
Machine	Disc harrow	—	JD 0.5 per du

Comments: Drills are preferred, but they are not available.

8. APPLY FERTILISERS
Duration: One man can broadcast 20 du in one day.
Environmental requirements: The soil should have been ploughed and weeded. It should not be muddy. The weather should not be rainy.
Preceding activity: 3.

Inputs required	*Type*	*Quantity*	*Price*
Nitrogen fertiliser	Ammonium sulphate	12 kg per du	JD 0.7 per du
Labour	skilled		JD 0.2 per du
Total			JD 0.9 per du

Comments: Few farmers apply fertilisers.

9. HAND-WEED
Duration: February-April.
Environmental requirements: Weed incidence is high; there should be no rain or mud.
Preceding activity: The crop must be standing in the field.

Inputs required	*Type*	*Quantity*	*Price*
Labour	Pulling weeds weeds	—	No price

Comments: This activity is done by the farmer's family and other farmers in the region. They pull weeds as a source of animal feed.

10. SPRAY HERBICIDE
Duration: February-April. One day, depending on the stage of growth of the crop.

Environmental requirements: The sky should be clear, the soil not muddy and the crop at the right stage of growth.
Preceding activity: The crop must be standing in the field.

Inputs required	*Type*	*Quantity*	*Price*
Sprayer	Knapsack or machine	—	JD 0.5 per du, including chemicals

Comments: Chemicals to control broadleaf weeds are the only ones available. The decision to spray depends on the density of weeds and on the availability of inputs.

11-13. HARVEST, PACKAGE, AND TRANSPORT CROP
Duration: late June/July.
Environmental requirments: Mature crop and warm weather.
Preceding activities: All previous activities.

Inputs required	*Type*	*Quantity*	*Price*
Machine	Combine	—	JD 1.5 per du
Labour	Packaging and sewing	—	JD 0.3 per du
Transport	Packing and sewing	—	JD 0.3 per du

14. MARKET CROP
Duration: All year, but mostly during July/August.
Preceding activities: 11-13.
Comments: The crop is sold to the Government, to the wholesalers, or to neighbours. Farmers may store some of the wheat for their own consumption and for seed.

15. SELL WHEAT FIELD CROP
Duration: Decision is taken in April.
Environmental requirements: Drought conditions.
Comments: The farmer is selling the crop at a rate of JD 1-5 per du. The price depends on the crop growth.

16. SELL CROP RESIDUE
Duration: June-July.
Environmental requirements: Dry, hot weather. The season was below average and therefore there is a great demand for straw.
Preceding activity: 11.
Comments: The farmer may sell the crop residue in the field at JD 1 per du.

Decision-Making Activities

1. PLANT WHEAT
Taken by: Farmer, during September-November.
Possible outcome (decisions):
— Prepare seed-bed
— Purchase seeds and inputs
Information
Desirable: a. Expected weather conditions

b. Prices of wheat at harvest
c. Availability of machinery

Available: a. Previous crops were lentils and summer vegetables.
b. Wheat prices during previous season were high.

Comments: The farmer is using a two-year crop rotation and the average rainfall is > 300 mm, so annual cropping prevails in this area.

2. DECIDE WHICH SEED VARIETY

Taken by: Farmer.

Possible outcome (decisions):

— Decide to purchase the seeds from the cooperative
— Decide to seed own stocks

Information

Desirable: a. Ability to sell own crop for seed purposes
b. Minimum price for wheat at next harvesting season

Available: a. Superiority of the cooperative-treated seeds
b. Reliable reputation of seeds from cooperatives

Comments: Farmers may know or have seen the superiority of some varieties, but these are not available to them, mainly because agriculture officials like to minimise the risks involved in introducing new varieties.

3. DECIDE SOWING DATE

Taken by: Farmer.

Possible outcome (decisions):

— Obtain clean, treated seeds
— Prepare the seed-bed

Information

Desirable: a. The start of the rainy season
b. Weather conditions of next season
c. Neighbours' willingness to plant
d. Availability of sprayers for herbicides

Available: a. Early planting increases yields.
b. Weeds prevail in early planted fields.
c. Competition for machinery is high in mid-season (late November/December).

Comments: Early planting, i.e. around 15 November, is recommended. A critical factor in increasing wheat yield is the date of seeding (Duwayri 1979, and Duwayri 1980). Semi-dwarf varieties will respond better to early planting but this has to be combined with optimum fertilisation and weed-control practices.

4. DECIDE ON TRADITIONAL OR MODERN TECHNIQUES

Taken by: Farmer in October/November.

Possible outcome (decisions):

— Look for soil moisture conservation machinery
— Use grain drill
— Plant early
— Use herbicide

Information

Desirable: *a.* Availability of machinery
b. Input and output prices
c. Weather bulletins to encourage the use of inputs like fertilisers

Available: *a.* Data on superiority of using fertilisers and grain drill *b. Lower cost using chisels and sweeps*

Comments: The JCO is trying to introduce the necessary machinery recommended for soil moisture conservation and energy saving.

5. DECIDE ON ACTION IF DROUGHT PREVAILS DURING SEASON
   Taken by: Farmer in mid-April.
   Possible outcome (decisions):
   — Sell crop for animal grazing
   — Sell crop to others
   — Keep crop and harvest, probably by hand, since combine use will be limited in dry seasons
   Information
   Desirable: a. Price offered per du
   b. Action taken by neighbouring farmers
   Available: a. Observation of crop conditions in the field
   b. Previous weather conditions at this time of the year
   c. Previous experience: difficulty of finding labourers to hand harvest, or availability of a combine during a drought
   Comments: Farmers are faced with such decisions once every three to five years.

6. DECIDE ON MARKETING OUTLET
   Taken by: Farmer in June/July.
   Possible outcome (decisions):
   — Decide to store own consumption needs plus seeds
   — Decide to sell at harvest to the Government or others
   — Decide to store
   Information
   Desirable: a. Prices in the coming months and during winter
   Available: a. Prices set by the Government
   b. Prices offered by buyers
   c. Storage will probably involve losses and insect damage.
   Comments: Some farmers like to store their wheat and sell when they need subsistence money.

Analysis of a Lentil Farm System

The activities involved in the production of lentils under rain-fed conditions in Jordan are set out in Figure 5.3.

Physical Activities

1. ASSESS ASSETS
   Duration: July-October.
   Comments: The farmer is preparing a cropping plan for next season.
2. PREPARE SEED-BEDS
   Duration: Normally during July-November. Two ploughings are done.
   Environmental requirements: Land should be free of crops. Normally lentils follow wheat in a two-year crop rotation.
   Preceding activity: Wheat must have been harvested.

Figure 5.3: Activities Related to a Lentil Farm System

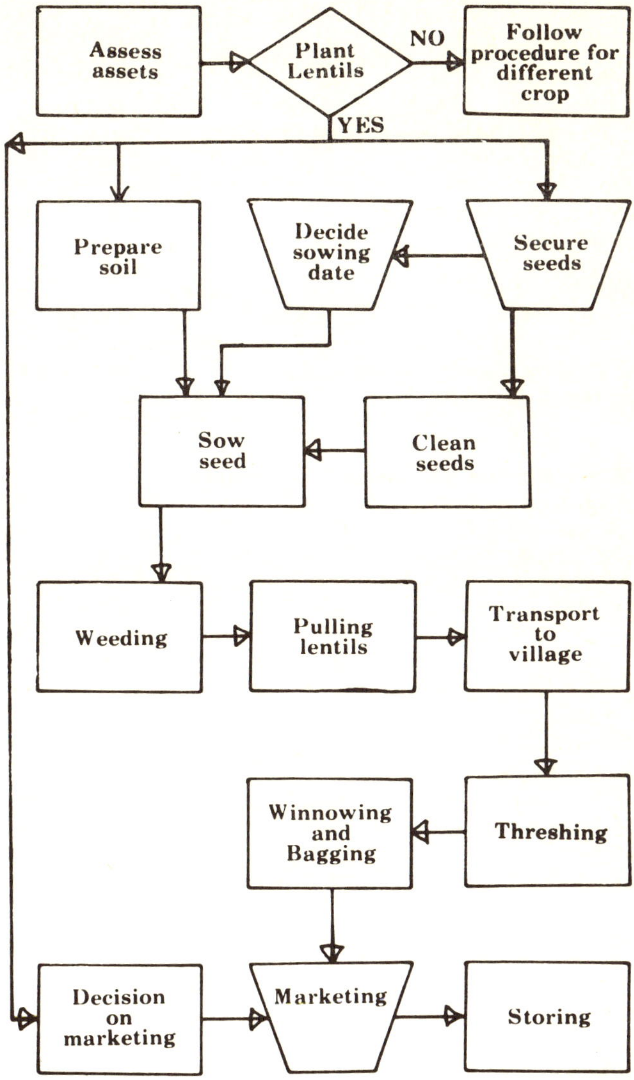

Inputs required	*Type*	*Quantity*	*Price*	*Comment*
Tractor	Disc or mouldboard plough	2	JD 0.8 per du	JD 1.6 per du

3. SECURE SEEDS

Duration: July-November.

Inputs required	*Type*	*Quantity*	*Price*
Seeds	Local	8 kg/du	JD 1.3 per du

Comments: Farmers obtain seeds from their farm market sources.

4. CLEAN SEEDS

Duration: October-November.
Preceding activity: 3.

Inputs required	*Type*	*Comment*
Labour	Women	Done by women on the farm; takes 0.25 man-hour per du

Comments: This process is normally done at home since it involves only small amounts.

5. SOW SEEDS

Duration: November and late December (normally).
Environmental requirements: Land should have been prepared, and the soil should not be wet.
Preceding activity: 2.

Inputs required	*Type*	*Quantity*	*Price*
Labour	Skilled	A labourer can plant 20 du/day	JD 0.2 per du

Comments: Skilled labourers should do this job in order to ensure even planting.

6. WEED

Duration: February-April.
Environmental requirements:The weather should permit working outside. Weeding is done by hand since selective herbicides for weeds in lentils are not in common use.
Preceding activities: Those which allow for crop establishment.

Inputs required	*Type*	*Comment*
Labourers	Family and neighbours	Done by family members

Comments: Mainly undertaken to collect animal fodder. Time required depends on density of weeds.

7. PULL LENTILS

Duration: Two weeks in May.
Environemental requirements: Lentils are liable to shatter and be lost when pulled if this activity is delayed. It is preferable to pull them in the morning to reduce losses which may amount to 20 per cent of yield.

Inputs required	*Type*	*Quantity*	*Price*	*Comment*
Labour	Men	—	JD 4 per du	Sometimes family members help.

Comments: Hand pulling of lentils is considered one of the hardest jobs on the farm as well as being the most expensive. This difficulty is causing a decrease in the area under lentils in Jordan.

8. TRANSPORT CROP TO VILLAGE
Duration: This is done in May, one to two weeks after harvesting.
Environmental requirements: Moist environment preferred, particularly in the morning.
Preceding activities: 7.

Inputs required	*Price*
Trailers or trucks	JD 3.8 per du (including cost of labour)

Comments: Lentils are transported and then placed in heaps in the village. Trucks used for transport are custom-rented.

9. THRESH CROP
Duration: May.
Environmental requirements: Dry weather.
Preceding activities: 7 and 8.

Inputs required	*Price*
Stationery thresher	JD 0.8 per du

Comments: Some farmers still use animals to do the threshing.

10. WINNOW AND BAG CROP
Duration: End of May/Early June.
Environmental requirements: Light winds to help separate tibin from seeds.
Preceding activities: 9.

Inputs required	*Price*	*Comment*
Skilled labour	JD 1.5	2-3 hours per du

Comments: This includes the bagging.

11. MARKET CROP
Duration: June, and continuing in winter.
Preceding activities: 10.

Inputs required	*Comment*
Store	Old room in the house for tibin

Comments: The farmer may sell the whole of his crop and tibin, or part of it, or may store mainly the tibin to get a better price for it in the winter especially in dry seasons. Grain is stored in bags. Insects may damage stored grain.

Decision-Making Activities

1. PLANT LENTILS
Taken by: Farmer.
Possible outcome (decisions):
- Buy seeds
- Look for labourers to hand-pull in the future
Information
Desirable: Government intention to buy lentils
Available: Lentil prices are profitable this year; lentil tibin is in demand.

2. DECIDE SOWING DATE
Taken by: Farmer.
Possible outcome (decisions):
— Weeding is required if crop is planted early.
— Tractors are available for early planting.
Information
Available: Early planting will give higher yields.
Family members will pull weeds.

3. DECIDE ON MARKETING OUTLET
Taken by: Farmer.
Possible outcome (decisions):
- *Storage of seeds may bring higher profits.*
- *Storage of tibin will bring higher prices, especially if the season was good and a poorer season is expected.*
Information
Desirable: Minimum prices are guaranteed.
Available: Prices set by the Government are encouraging.

Analysis of an Olive Farm System

The activities involved in the production of olives under rain-fed conditions in Jordan are set out in Figure 5.4.

Physical Activities

1. PRUNE TREES
Duration: November-December.
Environmental requirements: Suitable conditions for the work to proceed.
Preceding activities: Harvesting the previous crop.

Inputs required	*Price*	*Comment*
Labour	JD 2.0 per du	Labour = 0.5 man-day

2. APPLY FERTILISERS
Duration: September-October for manure.
December-January for chemical fertilisers.
Environmental requirements: Soil should be neither muddy nor wet.
Preceding activities: 1.

Inputs required	*Price*
Fertilisers and application	JD 3.2 per du

3. CULTIVATE IN WINTER
Duration: December to allow for rain to penetrate and mix in fertilisers. Also at end of spring, to reduce competition from weeds.
Environmental requirements: Soil must be neither muddy nor wet.
Preceding activities:
— Harvesting of the crop
— Pruning of the orchard in the autumn

Inputs required	*Type*	*Price*
Tractor	Disc plough	JD 0.5 per du

Figure 5.4: Activities Related to an Olive Farm System

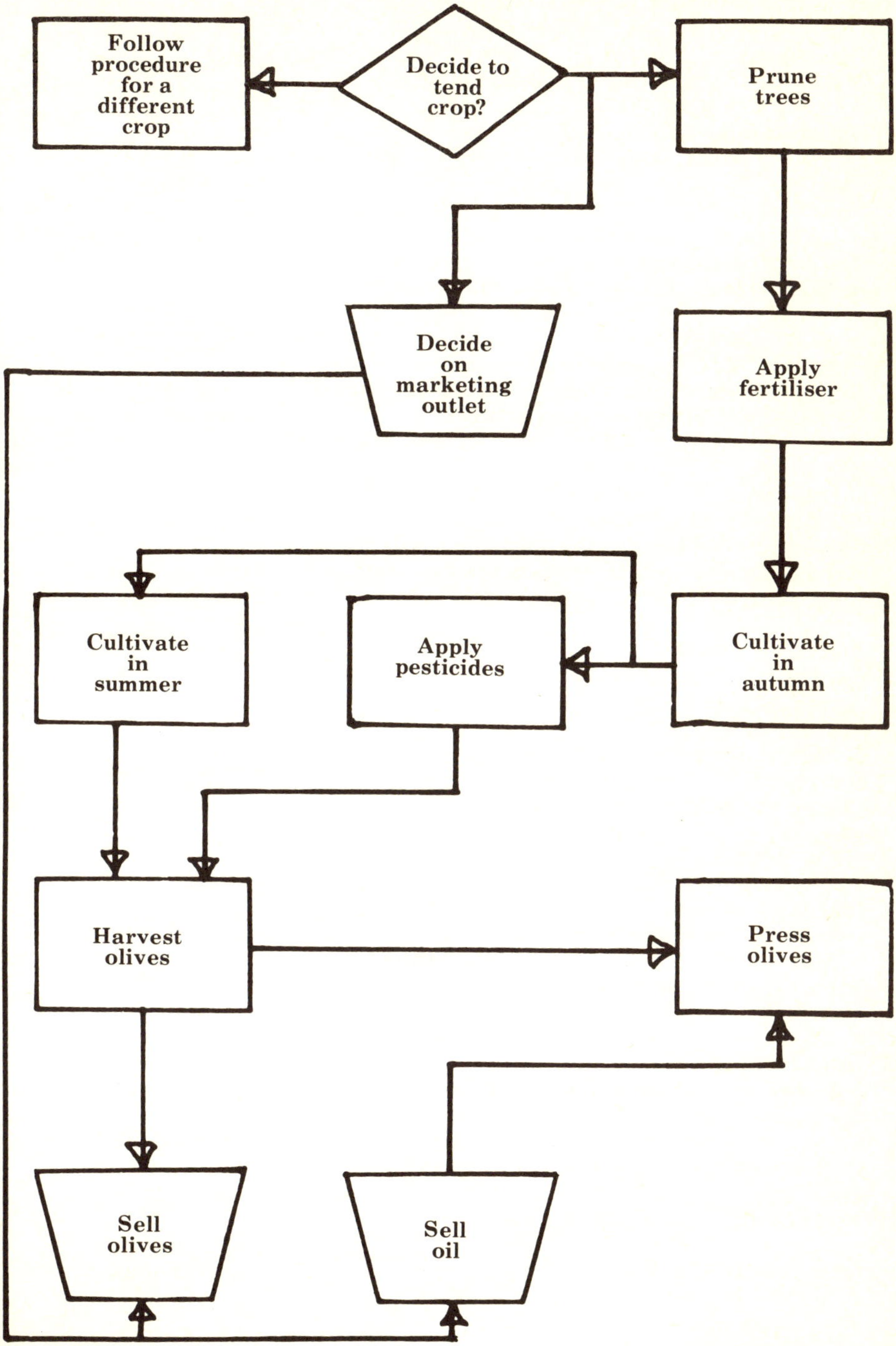

4. APPLY PESTICIDES
Duration: June/July spraying for insects and diseases.
Environmental requirements: Sunny days.
Preceding activities: Fertilisation in the winter.

Inputs required	*Price*
Pesticides and application	JD 4.2 per du

5. CULTIVATE IN SUMMER
Duration: June-July.
Preceding activities: 4.

Inputs required	*Type*	*Price*
Cultivation	Disc plough	JD 0.5 per du

Comments: Sometimes animals are used when the tractor cannot be used on rocks and slopes. Costs therefore may reach JD 2.

6. HARVEST OLIVES
Duration: Mid-October to mid-December.
Environmental requirements: The soil should not be muddy and weather conditions should favour harvesting.
Preceding activities: All previous physical activities.

Inputs required	*Price*
Labour	JD 12 per du

Comments: The family may help which will reduce cost and increase family income. Labour costs depend on the age, yield and position of the tree: the range is between JD 5 and 20 per du. Harvest losses are less than 5 per cent.

7. MARKET OLIVES
Duration: October-December.
Preceding activities: 6.

Inputs required	*Price*
Transport	JD 1 per du

Comments: Sold for table olives or for pressing.

8. PRESS OLIVES
Duration: October-December.
Preceding activities: 6.
Comments: There are traditional and modern pressing factories. In the new factories, the percentage extracted and the qualities are higher. Normally farmers pay the price of pressing in kind. (For each 10 tin cans, the factory usually takes one).

9. MARKET OIL
Duration: November, and all the year round.
Preceding activities: 8.
Comments: The farmer can sell his oil on the retail market in the village for the most part or on the wholesale market.

Decision-Making Activities

1. TEND CROP CAREFULLY
Taken by: Farmer.

Possible outcome (decisions):
— Apply fertilisers
— Pest control
— Cultivation
Information
Available: All these cultural practices will lead to higher yield and better quality.

2. DECIDE ON MARKETING OUTLET
Taken by: Farmer.
Possible outcome (decisions):
— Market for table olive
— Sell olive oil at pressing time
— Store
Information
Desirable: Quantities to be harvested this year.
Available: Amounts of olive oil stored in past years will help the farmer to decide on alternatives.

3. DECIDE ON USE OF MODERN OR TRADITIONAL PRESSING FACTORIES
Taken by: Farmer.
Possible outcome (decisions):
— Fix date for pressing
— Transport for longer distance
Information
Desirable: Prices of the oil pressed by the modern method will be higher.
Available: a. Quality of oil from new factories is better.
b. Extraction percentage is higher.

References

AOAD (Arab Organisation for Agricultural Development) 1977, *Economic and Technical Feasibility Study for the Development of Rain-fed Agriculture in the Amman Governorate of the Hashemite Kingdom of Jordan*, Khartoum.

— 1978, *Technical and Economic Feasibility Study for the Development of Rain-fed Agriculture in the Balqa and Irbid Governorates of the Hashemite Kingdom of Jordan*, Khartoum.

Arabiat, S., D. Nygaurd and K. Somel, 1982, *Issues of Improving Wheat Production in Jordan. Results from a Survey*, University of Jordan and ICARDA (International Centre for Agricultural Research in the Dry Areas), Amman.

DOS (Department of Statistics), *Agricultural Sample Survey, 1974-82: Agricultural Statistical Yearbooks and Agricultural Sample Survey*, Amman.

Duwayri, M., 'Effect of sowing date on yield of wheat under rain-fed conditions in Jordan', *Dirasat*, No. 6, 1979, pp. 99-108.

— 1980, 'Effect of seeding date and rate on yield and yield components of Stork wheat grown under rain-fed conditions in Jordan, Proceedings of Third International Wheat Conference, Madrid, 22 May-3 June 1980, pp. 407-13.

Mustafa, T.M., 1983, 'Factors affecting the distribution of olive *Euphyllura olivina* Costa on olive trees, Report submitted to Seminar on Uses of Olive Oil, Amman, 17-20 October 1983.

NPC (National Planning Council), 1976, *Five-Year Plan For Social and Economic Development, 1976-1980*, Amman.

Qasem, S., 1979, *Agricultural Sector in Jordan, Status and Changed (1974-1978) and Future Outlook (1980-1987),* University of Jordan, Amman.

Snobar, B., 'Prototype of a mechanical lentil harvester', *Dirasat Agricultural Studies: Special Issue,* No.8, 1981, pp. 79-80.

—— and S. Arabiat, 'Mechanisation of Agriculture and Socio-Economic Development in Jordan', *Dirasat,* No. 10: in press.

Steitieh, A.M., 'Marketing costs and margins of major food commodities in Amman', Dirasat Agricultural Studies: Special Issue, No.8, 1981. pp. 89-114.

USAID (United States Agency for International Development), 1983, Development of the Highland Agricultural Regions project, Jordan 2108, Report submitted to Ministry of Agriculture, Amman.

Watkins, D.J., I.T. Grierson and D.R. Harvey, 1983, *Land Use Capability: Planning Study Jordan,* Jordanian National Planning Council and Australian Development Assistance Bureau, Sagric International Pty. Ltd, South Australia.

Al-Wir, A. and F. Larsen, 'An evaluation of fruit grower practices and needs in Jordan', *Dirasat Agricultural Studies: Special Issue,* No.8, 1981, pp. 9-15.

6 POULTRY AND LIVESTOCK FARM SYSTEMS

Adib H. Duayfi

Poultry Sector: General Background

The poultry industry in Jordan has developed rapidly since the beginning of the 1960s, with the establishment of commercial poultry farms which had hitherto relied on the importation of day-old chicks and feedstuffs. The main factors which encouraged this rapid development are:

1. the demand by Jordanian consumers for broiler meat and eggs as a source of animal protein in their diets
2. an increase in the price of red meat produced locally
3. the role of governmental and semi-governmental bodies in the poultry development encapsulated in services extended to this sector, to be discussed later in this chapter

Broiler* Farms and their Geographic Distribution

The total number of broiler farms in 1981 amounted to 798. Most broiler farms are concentrated in the Amman and Irbid Governorates, where they account for 72.8 per cent of all farms. The remaining broiler farms are located in other governorates. Table 6.1 shows the number of broiler farms, their size and geographic distribution. Here we see that 43 per cent of broiler farms are small, with fewer than 5,000 birds each, and 41 per cent have a capacity of 10-20,000 birds.

Table 6.1: Broiler Farms: Number, Size and Distribution by Governorate

Farm size (birds per cycle)	Amman	Irbid	Balqa	Karak	Ma'an	Total	(%)
Less than 1,000	—	28	1	—	—	29	3.6
1,001-5,000	77	183	39	18	1	318	40.0
5,001-10,000	115	93	69	38	10	325	40.7
10,001-20,000	34	30	29	9	—	102	12.8
20,001-50,000	13	6	3	—	—	22	2.7
More than 50,000	—	2	—	—	—	2	0.2
TOTAL	239	342	141	65	11	798	

Source: MOA, ACC and JCO 1981.

Layer Farms† and their Geographic Distribution

The total number of layer farms in 1981 amounted to 194, of which 80 per cent were located in Amman and Irbid Governorates, the remainder in other

* Broiler is the term used for intensively reared chickens intended for consumption as meat.
† Layer farms refer to poultry units specialising in egg production.

governorates. Table 6.2 shows the size of layer farms, their number and distribution. Layer farms are generally larger than broiler farms: 49 per cent of layer farms have a capacity of 5-20,000 hens and 18 per cent have a capacity of more than 20,000 hens. Although the Ministry of Agriculture (MOA) controls the licensing of both broiler and layer farms, farmers turn broiler farms into layer farms and vice versa according to the prevailing market conditions. This causes confusion when classifying broiler and layer farms according to production.

Production

Broiler production in Jordan is estimated by taking a count of broilers on commercial poultry farms once a year and multiplying the number by five, on the assumption that all farms have five cycles of broilers per year. Table 6.3 shows production estimates for the period 1974-81. Fluctuations from one year to another are evident. The reasons for such fluctuations are the change-over from one type of production to another because of changes in inputs, disease hazards and management practices. Local broiler production meets almost 85 per cent of the demand for broiler meat.

Table 6.2: Layer Farms: Number, Size and Distribution by Governorate

Farm size (birds per farm)	Amman	Irbid	Balqa	Karak	Ma'an	Total	(%)
Less than 1,000	5	1	—	—	6	—	3.1
1,001-5,000	34	13	13	—	—	59	30.4
5,001-10,000	30	10	11	3	—	54	27.9
10,001-20,000	27	5	5	4	—	41	21.1
20,001-50,000	20	3	1	2	—	26	13.4
More than 50,000	6	1	—	—	1	8	4.1
TOTAL	122	33	29	9	1	194	

Source: MOA, ACC and JCO 1981.

Egg production is estimated by adding up eggs produced from layers, from local hens and from culled hatching eggs. Table 6.3 shows the number of eggs produced for the period 1974-81. We see that egg production has increased very rapidly over the last few years both to meet the demand at home and to export the surplus to neighbouring Arab countries.

Production Systems

Poultry houses in Jordan are of the open type, where ventilation is controlled through windows located on both sides. Broilers and layers are reared on the floor litter. The farmer has groups of different ages on the farm. Broilers are raised for a period of 50-55 days, when they reach a live body weight of 1.50-1.75 kg. Layers are kept for a period of 18 months, when they are sold as second-quality meat.

Table 6.3: Production of Poultry Meat and Eggs, 1974-81

Year	Broiler Meat (t)	Eggs (mn)
1974	17,250	51
1975	19,100	86
1976	22,500	161
1977	27,000	214
1978	23,609	265
1979	31,700	285
1980	30,825	349
1981	29,600	350

Sources: Duayfi and Qadi 1980.
Animal Production and Health Department 1980 and 81.

Production Inputs

Financial Resources

There are two main sources of loans to poultry farms. First, government and semi-government institutions provide short-, medium- and long-term loans, which are used to build farms and purchase equipment, chicks, feed, and so on. The second source consists of agricultural companies involved in chick production and feedmills, and providing middleman services; these creditors usually provide loans to poultry farmers on a short-term basis.

Day-old Chicks

All day-old chicks raised on broiler farms are locally produced by the 23 hatcheries located in Jordan. The total capacity of these hatcheries is about 4.5 mn hatching eggs per set, or the production of 4.5 mn day-old chicks per month. Most of the hatching eggs for broilers are produced in Jordan or imported from Europe.

Poultry Feed

There are five licensed feedmills producing poultry feed with a capacity of 120,000 t per year. In addition, there are small grinder-mixer units which can produce 30,000 t of feed per year. Most small poultry farms purchase their feed from these two sources. Large poultry farms prepare their feed on the farm. Poultry feed ingredients—mainly corn, soya bean meal and concentrate pre-mixes—are imported.

Vaccines and Medicines

There are two factories producing a number of veterinary medicines. The Animal Health Institute, a veterinary laboratory in the MOA, also produces a number of poultry vaccines. In addition, commercial companies import veterinary medicines and sell them to poultry farmers.

Poultry Equipment

Poultry equipment includes feeders, waterers, brooders, laying nests, and so on. Most of these are imported and available in local markets.

Labour and Technical Expertise

Poultry production is a specialised career which requires sound technical experience. The technical expertise of the labour force has a direct impact on the cost of production and the resulting profit. Most labourers on poultry farms have gained their experience on the job and they lack the proper technical know-how. The total number of labourers working on poultry farms is about 2,500. Most of them have an educational level lower than high school, some are graduates from agricultural school, and a very small number of farm managers have a university degree. Some poultry farms also employ labourers from abroad, who similarly lack technical experience.

There are a number of services contributing to the growth of technical experience on poultry farms; these include the Agricultural Research and Extension Service, the Veterinary Service, which diagnoses poultry diseases, and agricultural companies.

Poultry Slaughter-Houses

There are two poultry slaughter-houses in Jordan. The first is located in Amman and has a capacity of 3,000 birds per hour. The quality of broiler meat produced at this slaughter-house is poor, due to a lack of processing channels and poor management. The second slaughter-house is located in Irbid, with a capacity of 300 birds per hour. This also lacks some processing channels and is not working at present. In addition, there are about 200 retail shops which sell live broilers and have manual slaughter processing units.

Analysis of a Broiler Farm System

On-Farm Environments

Physical

Area: The broiler farm is built on an area of 25 du or more. According to regulations, it should be 1 km away from other poultry farms.

Buildings and Equipment: The typical broiler farm has the following buildings; two broiler houses each of 50×10 m^2 with a 10,000-bird capacity. Broiler houses are built from available materials, such as gravel, sand and cement. Their height ranges from 2.75 m to 3.25 m, the roof being made either from asbestos or zinc sheets. There is also a small house for the workers.

Poultry equipment includes:

- 5 brooders run either on gas or gasoline
- 65 baby-chick feeders
- 35 small waterers
- 120 broiler feeders
- 30 automatic waterers, each 2 m long

Power Source: There is no mains power supply on broiler farms. Gas and gasoline are used to provide lighting and heating.

Broiler Numbers and Breeds: Our typical farm has a 10,000 broiler capacity. The birds are raised in two age groups. The chicks raised in Jordan are hatched in hatcheries. Chick strains are Hubbard, Hybro, Hypco, Ross and Tetra. No real grading of baby chicks is carried out by the hatcheries, so that the quality of baby chicks the farmer receives is not standard. In one batch the farmer receives different sizes of chicks and some abnormal ones. The differing quality of baby chicks results in an inconsistent growth rate and in high production costs.

Number of Labourers: There is one unskilled labourer living and working on the farm.

Duration of the Growing Period: Broilers are raised on the farm for a period of 50-55 days, when the live body weight of the bird reaches 1.5-1.75 kg. Then one week is allowed for cleaning and disinfecting the poultry houses and equipment. There are usually 5.5 cycles of production per year.

Financial Assets and Liabilities:

			(JD)
Assets:	—	land	2,500
	—	buildings	24,500
	—	equipment	1,000
Liabilities:	—	long-term loans	12,250
	—	short-term loans	5,800

Inputs and Yields:

Inputs:	—	baby chicks	10,000
	—	feed	35 t
	—	water	120 m^3
	—	vaccines and medicines	
	—	gas	150 cylinders
	—	gasoline	200 l
	—	labour	1 worker for 2 months
	—	litter	250 sacks
Yields:	—	broilers	9,500 of 1.5 kg live body weight
	—	litter	100 m^3

Institutional
Most of the broiler farms are owned by farmers who hire labourers to carry out the daily work. The poultry farmer visits the farm every day and makes all the necessary off-farm communications. Broiler farms are exempted from income tax.

Informational
Most of the workers on the poultry farm are skilled labourers who have gained their experience on the job. They lack technical knowledge and receive their daily instructions from the farm owner. Most of the workers received a primary or preparatory education and a few had an agricultural education. Large broiler farms have agricultural engineers as farm managers.

Off-Farm Environments
Physical
Broiler farms are located outside cities and villages at a distance of 500 m from the municipality borders. To limit the spread of disease, broiler farms should be not less than 1 km apart. Most broiler farms have access to roads for transporting inputs to the farm and outputs to the market.

Institutional
The institutions concerned with broiler farms are:
The MOA: This has the following main functions:
- issuing of laws, regulations and licences for poultry farms
- provision of extension service through different communications media
- provision of free veterinary services to farms
- production of some poultry vaccines

The Ministry of Supply (MOS): The Ministry is involved in the control of prices of broiler meat and feed. It is also concerned in making feed available on the market.
The Agricultural Credit Corporation (ACC) and the Jordan Cooperative Organisation (JCO): These two bodies provide long-, medium- and short-term loans to farmers.
The University of Jordan: The University's contribution to the poultry sector lies in teaching poultry science, in training students in practical poultry production,

and in carrying out research in the field of poultry feeding.
Agricultural Companies: These supply the farmer with baby chicks, feed and medicines and also provide him with technical bulletins and leaflets.

Informational
The poultry farmer has access to information on input prices of items whose prices are controlled by the MOS: for example, broiler wholesale and retail prices and feed prices. He also has access to information on the market prices for baby chicks, medicine and poultry equipment from commercial companies. He hears weather forecasts on the radio and TV.

Further information, such as research results and advice, is channelled from research workers to the producers through the extension service. Research in the field of poultry feeding and management is very limited, and no research priorities are set.

Activity analysis

Physical Activities
The activities related to a broiler farm system are set out in Figure 6.1.

1. ASSESS ASSETS

Inputs required	*Type*	*Quantity*	*Timing*	*Price*
Labour	Farm manager	1	8 hours for 3 days	—

Timing: Normally after the previous cycle is finished and broilers and litter are sold, and also during the growing period.
Comments: This activity requires the farmer to:
- estimate capital in hand
- estimate returns from sales of product
- estimate cost of production
- estimate loans
- estimate expenditure
- estimate profit

2. DECIDE TO GO INTO BROILER FARMING

Inputs required	*Type*	*Quantity*	*Timing*	*Price*
Labour	Farm manager	1	6 hours	—

Alternatives: There are many alternatives for the farmer to select from: broiler production, egg production, cattle, sheep or other agricultural projects. Our farmer selected broiler production.
Determining factors:
- loans can be obtained easily.
- The life cycle of broilers is short, so the farm can become capitalised in a short time.
- There is less risk than in layer farms.
- There is a good local market for broilers.
- The farmer's neighbour has started a broiler farm which seems to be profitable.

Figure 6.1: Activities Related to a Broiler Farm System

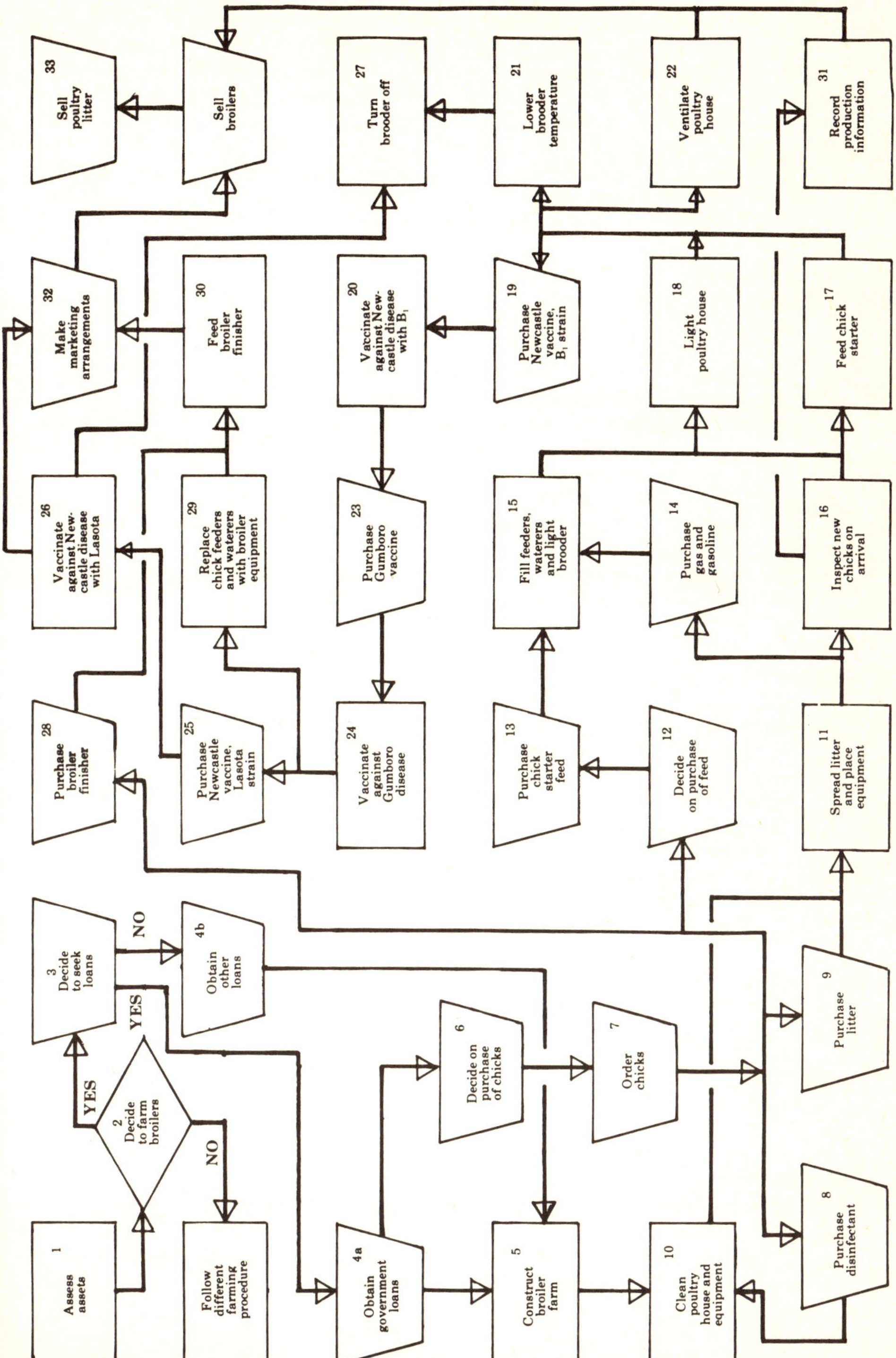

3. DECIDE TO SEEK LOANS

Inputs required	*Type*	*Quantity*	*Timing*	*Price*
Labour	Farm manager	1	6 hours	—

Alternatives: There are two alternatives: either to obtain loans from government and semi-government bodies—in which case the farmer has to apply for a loan and he should have a mortgage—or to obtain credit from commercial poultry companies which provide him with baby chicks, feed, vaccines, and so on, on a loan basis.
Determining factors: For short-term loans needed for the purchase of chicks, feed, vaccines, etc., the farmer prefers to go to commercial poultry companies in order to minimise time and red tape.

4. OBTAIN LOANS

Inputs required	*Type*	*Quantity*	*Timing*	*Price (JD)*
Labour	Farm manager	1	6 hours	—
Transportation	Car	Car seat	1 hour	0.5

Timing: Usually before the start of each growing season.
Comments: The farmer goes to the agricultural companies he deals with and discusses the possibility of obtaining chicks, feed and vaccines on a loan basis. He has to pay back the loan after marketing the broilers, which is also done through commercial poultry companies or middlemen.

For building the farm and purchasing equipment the farmer goes to the ACC.

Total capital required for building: JD 24,500.
Total capital required for equipment: JD 1,000.

5. CONSTRUCT BROILER FARM

Inputs required	*Type*	*Quantity*	*Timing*	*Price*
Capital	Cash	—	—	—
Labour	Farm manager	1	3 hours daily for 2 months	—

Timing: After obtaining the loan.
Comments: The farmer usually looks for a contractor to undertake the construction of the farm buildings; he may purchase the building materials and look for a mason to build it. He then outfits the farm with poultry equipment, namely feeders, waterers and brooders.

Total capital required for the buildings: JD 24,500.
Total cost of the equipment: JD 1,000.

6. DECIDE ON PURCHASE OF CHICKS

Inputs required	*Type*	*Quantity*	*Timing*	*Price*
Labour	Farm manager	1	2 hours	—

Alternatives: There are a number of broiler strains on the market from which the farmer can choose. These include Hubbard, Hybro, Hypco, Ross and Tetra.

Determining factors: There are a number of factors that affect the choice of a certain strain:

- the productive performance of the strain
- the farmer's satisfaction with the performance of the strain
- the ease of dealing with the commercial poultry companies
- the time at which the chicks will be available

7. ORDER CHICKS

Inputs required	*Type*	*Quantity*	*Timing*	*Price (JD)*
Labour	Farm	1	2 hours	—
Transportation	manager			
Transportation	Car	Car seat	1 hour	0.5

Environmental factors: 10,000 chicks of Hubbard strain are needed at JD 150 per thousand. Sometimes the farmer obtains fewer baby chicks than he orders. For example, during the summer he does not usually get the total number of baby chicks he orders because there is a higher demand than in winter.
Timing: Usually before the start of the new growing cycle, and throughout the year.
Determining factors:
— obtaining a loan
— decision on the chick strain

8. PURCHASE DISINFECTANT

Inputs required	*Type*	*Quantity*	*Timing*	*Price (JD)*
Labour	Unskilled	1	3 hours	—
Transportation	Small car	1	1 hour	2

Environmental factors: There are a number of disinfectants available on the market which have similar potencies. The price per gallon is JD 4.5 and 10 gallons are needed.
Timing: Usually after the selling of the broilers and before the new cycle starts.

9. *Purchase litter*

Inputs required	*Type*	*Quantity*	*Timing*	*Price (JD)*
Labour	Unskilled	1	4 hours	—
Transportation	Truck	1	2 hours	10

Environmental factors: The kind of litter used is wood shavings and it is usually purchased from carpenters by the sack. The price is JD 0.9 per sack and 250 sacks are needed.
Timing: After the selling of the broilers and before the new cycle starts.

10. CLEAN AND DISINFECT POULTRY HOUSE AND EQUIPMENT

Inputs required	*Type*	*Quantity*	*Timing*	*Price (JD)*
Water	—	2 m^3	-	*25 per* m_3
Disinfectant	Vapcofor	10 gallons	-	*4.5 per gallon*
Labour	Skilled	1 labourer	*6 hours*	-

Timing: Before the new cycle starts.
Preceding activities:
— selling of previously grown broilers
— selling of used litter

11. SPREAD LITTER AND PLACE EQUIPMENT

Inputs required	*Type*	*Quantity*	*Timing*	*Price (JD)*
Litter	Wood shavings	250 sacks	—	0.9 per sack
Feeders	Feeders, 1 m	65	—	0.5 per feeder
Waterers	Plastic waterers	35	—	0.75 per waterer

Brooders	Gas brooders	5	—	50 per brooder
Labour	Skilled	1	6 hours	—

Timing: Before the arrival of the baby chicks.
Determining factor:
— Clean and disinfect the poultry houses and equipment.
Comments: Poultry equipment depreciates at 20 per cent per year. It is used at the rate of 5.5 cycles per year. The cost we give in the price column is the purchasing price of the equipment.

12. DECIDE ON PURCHASE OF FEED

Inputs required	*Type*	*Quantity*	*Timing*	*Price*
Labour	Farm manager	1	3 hours	—

Alternatives: There are two alternatives: the first is to purchase ready-made feed at a price of JD 120 per t. The quantity of feed needed for the whole growing period is 35 t. The second alternative is to purchase feed ingredients and prepare and mix them on the farm. The estimated cost of the feed ingredients is JD 102 per t. However, the farmer needs the following items in order to prepare the feed on the farm:
— a grinder mixer costing JD 1,500
— a generator costing JD 2,500
— a room for the generator at a cost of JD 100
Our typical farmer purchased ready-made feed.
Determining factors:
— Capital is needed to prepare feed on the farm.
— The quality of the prepared product is better.
— Extra work is involved when feed is prepared on the farm.

13. PURCHASE CHICK STARTER FEED

Inputs required	*Type*	*Quantity*	*Timing*	*Price*
Labour	Skilled	1	3 hours	—

Environmental factors: The quantity of feed starter needed is 15 t. It is purchased all together or purchased by instalment. Agricultural companies usually transport the feed to the poultry farms by trucks. The price is JD 120 per t, including delivery.
Timing: Before the arrival of chicks on the farm.

14. PURCHASE GAS AND GASOLINE

Inputs required	*Type*	*Quantity*	*Timing*	*Price*
Labour	Unskilled	1	1 hour	—

Environmental factors: The farmer needs a total of 150 gas cylinders with a 12·5 l capacity. Gas is transported to the farm by the gas dealer at a price of JD 1.65 per cylinder. Gasoline is purchased from fuel stations at JD 1.2 for 20 l. During the whole growing period, 200 l of gasoline are needed.
Timing: Gas is purchased during the first three weeks of the growing period. Gasoline is purchased throughout the growing period.

15. FILL FEEDERS, WATERERS AND LIGHT BROODER

Inputs required	*Type*	*Quantity*	*Timing*	*Price*
Feeders	—	65	—	—
Waterers	—	35	—	—
Brooders	—	5	—	—
Labour	Skilled	1	1 hour daily throughout the growing period	—

Environmental factors: The feeders are filled to the top with feed. This causes some loss when the chicks eat, due to scattering. This loss is estimated at 5 per cent. However, feed loss could be avoided by filling the feeders to one third of their capacity. In this case more, frequent filling of the feeders would be necessary.

Brooders are lit 12 hours before the arrival of the chicks to ensure that they are working properly. The temperature should be 95°F at 30 cm from the ground.

Timing: Daily until marketing of the broilers; brooders are lit daily until 3 weeks after the start of the growing period, when they are extinguished.

16. INSPECT NEW CHICKS ON ARRIVAL

Inputs required	*Type*	*Quantity*	*Timing*	*Price*
Labour	Skilled	1 labourer	3 hours	—
Baby chicks	Broiler	10,000	—	—

Environmental factors: Chicks are usually transported to the farm in boxes holding 100. The farmer inspects the delivery for dead and abnormal chicks. At the same time he puts normal chicks around the brooder. The farmer watches them to see that they are comfortable and start to eat and drink, and that they are not crowded under the brooder or away from it, as such behaviour indicates a heating problem which should be corrected.

Timing: On arrival of the chicks; observing the chicks is done intensively during the first 3 days on the farm.

17. FEED CHICK STARTER

Inputs required	*Type*	*Quantity*	*Timing*	*Price (JD)*
Feeders	—	65	—	—
Chick starter	Starter	15 t	—	120 per t
Labour	Skilled	1 labourer	2 hours daily for 35 days	—

Environmental factors: The farmers feed chick starter to chicks from 1 day old to 35 days. The estimated amount of feed consumed by a chick throughout the 35 days is 1.5 kg.

Timing: The period is 35 days.

Preceding activities:

- purchase of chick starter feed
- filling of feeders
- arrival of chicks on the farm

18. LIGHT POULTRY HOUSE

Inputs required	*Type*	*Quantity*	*Timing*	*Price (JD)*
Lamps	—	10	—	—
Gasoline	—	200 l	—	1.2 per 20 l
Labour	Skilled	1 labourer	0.5 hours daily	—

Environmental factors: Poultry houses are lit at night to allow the chicks to eat and to avoid crowding of chicks on one side of the house and consequent mortality to some of them.

Timing: Daily throughout the growing period.

19. PURCHASE NEWCASTLE VACCINE, B_1 STRAIN

Inputs required	*Type*	*Quantity*	*Timing*	*Price*
Labour	Skilled	1 labourer	3 hours	—

Environmental factors: The farmer purchases 10,000 doses of Newcastle

B^1 strain for vaccination against Newcastle Disease. The price of the vaccine is JD 3.5.
Timing: The day before vaccination.

20. VACCINATE AGAINST NEWCASTLE DISEASE WITH B_1

Inputs required	*Type*	*Quantity*	*Timing*	*Price (JD)*
Newcastle vaccine	B_1	10,000 doses	—	3.5
Labour	Skilled	1 labourer	1 hour	—

Environmental factors: The baby chicks are vaccinated against Newcastle Disease with B_1 strain. Failure to do so may subject them to an outbreak of the disease which could cause complete loss of the flock. This vaccine is usually given in water.
Timing: When the chicks are 7 days old.

21. LOWER BROODER TEMPERATURE

Inputs required	*Type*	*Quantity*	*Timing*	*Price*
Brooders	—	5	—	—
Labour	Skilled	1 labourer	15 minutes per week for 3 weeks	—

Environmental factors The temperature around the brooder at 30 cm above the ground should be 95°F. It should be lowered by 5°F a week. Failure to do so may decrease food consumption.
Timing: Once a week up to 3 weeks, when the brooder is extinguished.

22. VENTILATE POULTRY HOUSE

Inputs required	*Type*	*Quantity*	*Timing*	*Price*
Labour	Skilled	1 labourer	15 minutes daily	—

Environmental factors: Ventilation in poultry houses is very important. However, farmers close the windows because they are afraid of cold weather. This causes a build-up of ammonia in the house and may affect the health of the chicks. Proper ventilation is vital.
Timing: During the whole growing period.

23. PURCHASE GUMBORO VACCINE

Inputs required	*Type*	*Quantity*	*Timing*	*Price*
Labour	Skilled	1 labourer	3 hours	—

Environmental factors: The farmer purchases 10,000 doses of Gumboro vaccine. The price of the vaccine is JD 3.
Timing: 1 day before vaccination.

24. VACCINATE AGAINST GUMBORO DISEASE

Inputs required	*Type*	*Quantity*	*Timing*	*Price*
Gumboro vaccine	—	10,000 doses	—	—
Labour	Skilled	1 labourer	1 hour	—

Environmental factors: The chicks are vaccinated against Gumboro Disease. The vaccine is given in water. Failure to do so may subject the chicks to an outbreak of the disease.
Timing: Usually when the chicks are 12 days old.

25. PURCHASE NEWCASTLE VACCINE, LASOTA STRAIN

Inputs required	*Type*	*Quantity*	*Timing*	*Price*
Labour	Skilled	1 labourer	3 hours	—

Environmental factors: The farmer purchases 10,000 doses of Newcastle vaccine, Lasota strain. The price of the vaccine is JD 3.5.
Timing: 1 day before vaccination.

26. VACCINATE AGAINST NEWCASTLE DISEASE WITH LASOTA

Inputs required	*Type*	*Quantity*	*Timing*	*Price (JD)*
Newcastle vaccine	Lasota	10,000	—	3.5
Labour	Skilled	1 labourer	1 hour	—

Environmental factors: The birds are vaccinated against Newcastle Disease with Lasota strain in water.
Timing: When the birds are 21 days old.

27. EXTINGUISH BROODER

Inputs required	*Type*	*Quantity*	*Timing*	*Price*
Brooders	Gas	5	—	—
Labour	Skilled	1 labourer	10 minutes	—

Environmental factors: The brooders are extinguished after 3 weeks from the start of raising the chicks.
Timing: 3 weeks after starting to raise the chicks.

28. PURCHASE BROILER FINISHER

Inputs required	*Type*	*Quantity*	*Timing*	*Price*
Labour	Skilled	1 labourer	3 hours	—

Environmental factors: 20 t of broiler finisher is needed for the farm. The price is JD 120 per t and includes delivery to the farm. The quantity required is purchased either all at one time or by instalments.
Timing: 30 days before the start of the cycle.

29. REPLACE CHICK FEEDERS AND WATERERS WITH BROILER EQUIPMENT

Inputs required	*Type*	*Quantity*	*Timing*	*Price (JD)*
Broiler feeders	2 m long	120	—	2 per feeder
Broiler waterers	Automatic	30	—	15 per waterer
Labour	Skilled	1 labourer	2 hours	—

Environmental factors: Broiler feeders and waterers should be put into the poultry houses.
Timing: 2 weeks from the beginning of the growing cycle. Chick feeders and waterers should be removed when the birds are accustomed to the larger feeders and waterers.

30. FEED BROILER FINISHER

Inputs required	*Type*	*Quantity*	*Timing*	*Price (JD)*
Feeders	—	120	—	—
Broiler finisher		20 t	—	120 per t
Labour	Skilled	1 labourer	2 hours for 20 days	—

Environmental factor: The farmer feeds broiler finisher from 35 days to marketing of the birds. The estimated amount of feed consumed per bird up to marketing is 2 kg.
Timing: From 35 days up to marketing time.

31. RECORD PRODUCTION INFORMATION

Inputs required	*Type*	*Quantity*	*Timing*	*Price*
Stationery	—	—	—	—
Labour	Skilled	1 labourer	15 minutes daily	—

Environmental factors: The farmer keeps very simple records in which he notes the number of chicks purchased, feed purchased and mortality. The recorded information does not show the productive performance of the flock. Detailed information, such as daily feed intake, weekly feed intake, body weight and feed conversion, could be of help to farmers in improving management and consequently increasing profit.
Timing: During the whole growing period.

32. MAKE MARKETING ARRANGEMENTS

Inputs required	*Type*	*Quantity*	*Timing*	*Price*
Labour	Farm manager	1 labourer	12 hours	—
Broilers	—	9,500 birds	—	—

Environmental factors: When the broiler reaches an average body weight of 1.5 kg the farm manager arranges to market his product through:

- commercial poultry companies
- middlemen
- direct sales to retailer

A mortality rate of 5 per cent among the flock is considered normal during the rearing period. The farm-gate price is JD 0.54 per kg live weight. It is fixed by the MOS and revised from time to time.
Duration: Marketing the whole flock usually takes 3-4 days.

33. SELL POULTRY LITTER

Inputs required	*Type*	*Quantity*	*Timing*	*Price*
Poultry litter	—	100 m^3	—	—
Labour	Farm manager	1 labourer	3 hours	—

Environmental factors: After selling the broilers, the farm manager arranges the sale of poultry litter. The average farm-gate price of poultry litter is JD 4 per m^3. It ranges from JD 2 to JD 6 per m^3.
Timing: After selling all the broilers.

Analysis of a Layer Farm System

On-Farm Environments

Physical

Area: The area of a typical layer farm is 25 du or more. The farm should be 1 km away from the municipal border.

Buildings and Equipment: The typical layer farm consists of the following buildings:

- 2 poultry houses, each 100 × 10 m^2
- feed and equipment storage, 50 m^2
- labourers' houses, 50 m^2
- power house, 20 m^2

Poultry equipment includes:

- 5 brooders — at JD 50
- 65 baby-chick feeders — at JD 0.5

— 35 small waterers	at JD 0.75
— 2 automatic feeders	at JD 1,500
— 40 automatic waterers	at JD 15
— egg nests No. 200, each with 10 holes	at JD 7
— grinder mixer, with a capacity of 0.5 t per hour	at JD 1,400

Power Source: The source of power on the farm is a generator which provides power for the automatic feeders and the grinder mixer and light. The generator required has a 25 kw capacity and costs JD 2,500.

Layer Numbers and Breeds: The typical farm has a capacity of 10,000 layers. The hens are raised in two age groups in order to ensure uniform production of eggs. The chick strains are Babcock, Hyline, Shaver and Tetra. Chicks are sold as female and 1 day old.

Number of Labourers: The total number of skilled labourers working on the farm is three. Two labourers work in the poultry houses and the third is in charge of the feedmill.

Duration of the Production Period: The length of the production period from 1 day old to the end of egg production ranges from 72 to 74 weeks. This consists of two distinct periods. The period from 1 day old up to 20 weeks is the growing period, and the period from 20 weeks up to the end of egg production is the egg production period.

Financial Assets and Liabilities:

		(JD)
Assets:	— land	2,500
	— buildings	45,000
	— equipment	6,709
	— power house	2,500
Liabilities:	— long-term loans	30,000
	— short-term loans	42,142

Inputs and Yields:

Inputs:	— female baby chicks	10,000
	— feed	450 t
	— water	
	— fuel and gas	
	— litter	
	— vaccine, medicines and disinfectants	
	— labour	3 workers for 74 weeks
	— 30 egg trays	75,000 trays
Yields	— eggs for the table	2,280,000
	— hens at the end of production	8,300
	— poultry litter	400 m^3

Institutional

Most layer farms are owned by farmers who hire labourers to do the daily work. The farmers visit the farm daily and make all off-farm communications. Layer farms are exempted from income tax.

Informational

The labourers who work on layer farms are skilled workers who have gained their experience by working on poultry farms. They have had either primary or preparatory education. Few of them have had agricultural education. Large layer farms have agricultural engineers as farm managers.

Off-Farm Environments

Physical

Layer farms are located beyond city boundaries at not less than 500 m outside the municipality borders. The farms are 1 km apart by law, in order to limit the spread of poultry disease. They are close to roads for transportation of farm inputs and products.

Institutional

The institutions concerned with layer farms are:

- the MOA
- the MOS
- the ACC and the JCO
- the University of Jordan
- agricultural companies

Informational

Poultry farmers have access to market prices of poultry meat and eggs. They also have access to different inputs for poultry production from poultry companies. The weather forecast is communicated through radio and TV.

Activity Analysis

Physical Activities

The activities related to a layer farm system are set out in Figure 6.2.

1. ASSESS ASSETS

Inputs required	*Type*	*Quantity*	*Timing*	*Price*
Labour	Farm manager	1 labourer	3 days	—

Timing: Normally undertaken before the start of work and after each growing cycle.

Comments: This activity requires the farmer to:

- estimate capital in hand
- estimate capital needed
- estimate cost of production
- estimate loans
- estimate expected profit

2. DECIDE TO GO INTO LAYER PRODUCTION

Inputs required	*Type*	*Quantity*	*Timing*	*Price*
Labour	Farm manager	1 labourer	6 hours	—

Alternatives: There are many alternatives. The farmer has to decide whether to go into egg production, broiler production, cattle or sheep production, or any other agricultural project. Our farmer selected layer production because it yields a good and stable profit. Furthermore, there is a good market for eggs in Jordan, as well as an outside market in neighbouring Arab countries. Egg production is more profitable than broiler production; it is also, however, more risky because of the relatively longer period of keeping the layers on the farm.

Figure 6.2: Activities Related to a Layer Farm System

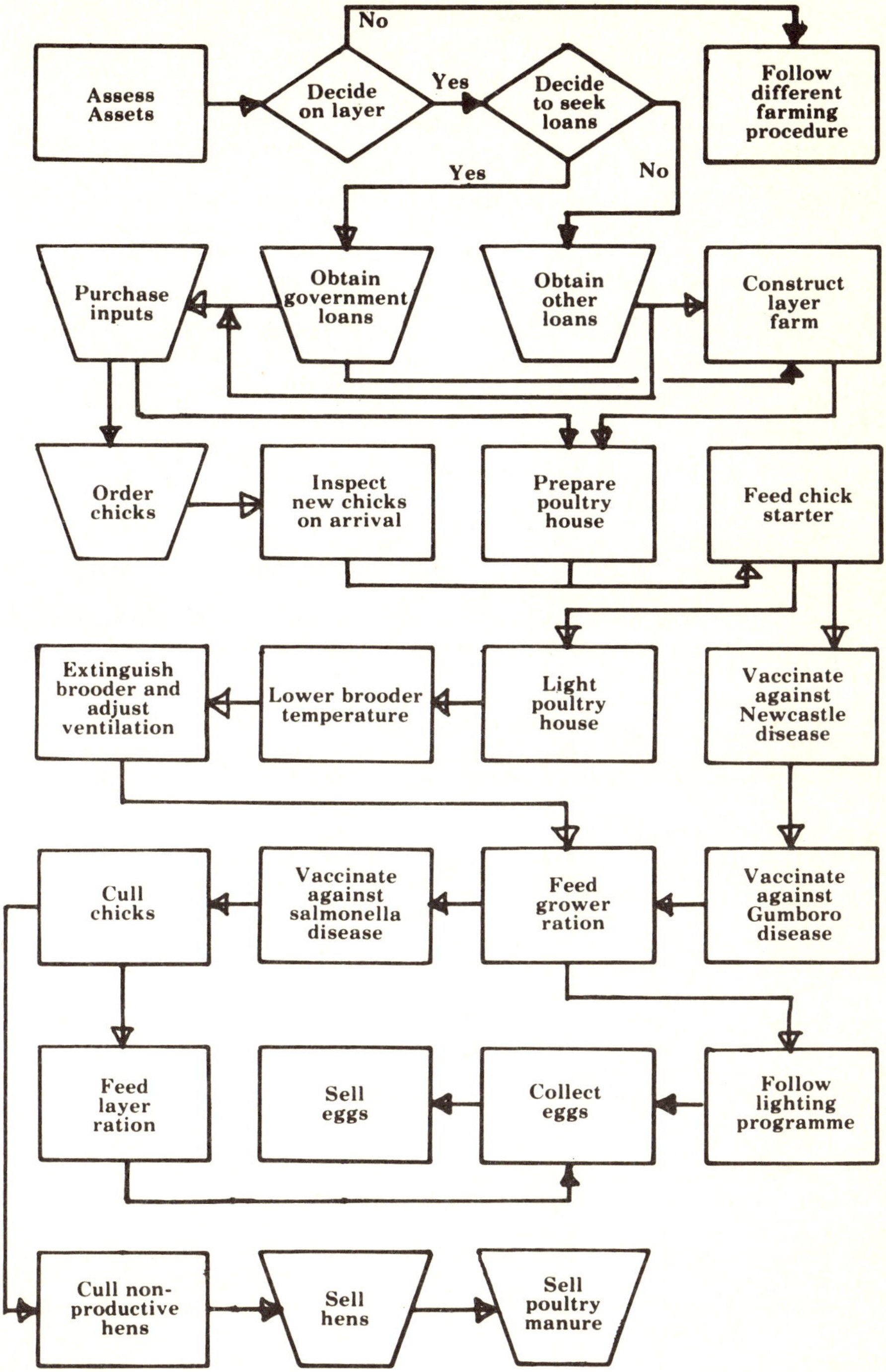

Determining factors:
- Loans can be obtained easily
- Feed is available on the market.
- The farmer's neighbour has already started a layer project and seems to be making a profit.

3. DECIDE TO SEEK LOANS

Inputs required	*Type*	*Quantity*	*Timing*	*Price*
Labour	Farm manager	1 labourer	2 hours	—

Alternatives: There are two alternatives. The first is to obtain loans from government and semi-government bodies to finance building the farm. The second (if the farmer has the funds to cover the cost of building but needs capital for purchasing inputs) is to obtain financing from commercial poultry companies.

Determining factors: For long-term loans the farmer prefers to borrow from government bodies because interest on capital is low. For short-term loans he prefers to go to commercial companies in order to cut down the time involved.

4. OBTAIN LOANS

Inputs required	*Type*	*Quantity*	*Timing*	*Price (JD)*
Labour	Farm manager	1 labourer	6 hours	—
Transportation	Car	Car seat	1 hour	0.5

Timing: After the farmer has decided to go into egg production and at the beginning of each growing cycle.

Comments: The farmer goes to the ACC to obtain a loan for building the poultry houses. He fills in the application form and completes the procedure. For short-term loans needed for the purchase of baby chicks, feed, vaccines, and so on, he goes to the commercial poultry companies.

Total capital required for buildings: JD 45,000.

Total capital required for equipment, including feeders, waterers, brooders, egg nests, grinder mixer and generator: JD 9,209.

5. CONSTRUCT LAYER FARM

Inputs required	*Type*	*Quantity*	*Timing*	*Price*
Capital	Cash	JD 54,209	—	—
Labour	Farm manager	1 labourer	3 hours daily for 2 months	—

Timing: After obtaining the loans.

Comments: The farmer usually approaches several building contractors and makes an agreement with one of them to build the farm. He pays the contractor either on a turnkey basis or by square metre. On completion, the farmer has to equip the farm with the necessary feeders, waterers, brooders, egg nests, a grinder mixer, and a generator.

6. PURCHASE INPUTS

Inputs required	*Type*	*Quantity*	*Timing*	*Price (JD)*
Labour	Farm manager	1 labourer	8 hours for 3 days	—
Transportation	Car	Car seat	1 hour	2

Environmental factors: It is not advisable to order chicks in December or January because of cold weather and the high cost of heating, and in order to avoid subjecting the chicks to respiratory diseases which are prevalent at

that time. The feed includes grains, meals and concentrates. Grains and soya-bean oil meal can be purchased on a seasonal basis. Concentrates should be purchased by instalment in order to avoid loss of certain vitamins.
Timing: Before the new cycle of production starts and when there is a shortage of a certain input on the farm.
Comments: The quantities and prices of different inputs are listed below:

Type	*Quantity*	*Unit price* (JD)	*Total price* (JD)
Female chicks	10,000	0.3	3,000
Feed	450 t	—	39,142
Litter	1,000 sacks	0.9	900
Vaccines, disinfectants & medicines	—	—	600
Gas & fuel	—	—	730

7. PREPARE POULTRY HOUSE

Inputs required	*Type*	*Quantity*	*Timing*	*Unit price* (JD)
Disinfectants	—	10 gallons	—	4.5 per gallon
Water	—	2 m³	—	0.25 per m³
Litter	Wood shavings	1,000 sacks	—	0.9 per sack
Feeders	Feeders, 1 m	65	—	0.5 per feeder
Waterers	Plastic	35	—	0.75 per waterer
Brooders	Gas brooders	5	—	0.50 per brooder
Feed	—	100 kg	—	95 per t
Labour	Skilled	1 labourer	6 hours for 2 days	—

Comments: The labourer should clean the poultry house thoroughly, using high pressurised water with disinfectants, and then allow it to dry. He then spreads the litter, places the feeders and waterers half-full and lights the brooder 24 hours before the arrival of the chicks.

Poultry equipment depreciates at a rate of 20 per cent per year.
Timing: Before the arrival of chicks on the farm.

8. INSPECT NEW CHICKS ON ARRIVAL

Inputs required	*Type*	*Quantity*	*Timing*	*Price*
Labour	Skilled	1 labourer	3 hours	—
Baby chicks	Female	10,000	—	—

Comments: The farmer checks and counts the number of chicks. He checks for dead and abnormal chicks, while putting the normal female chicks around the brooder. He watches them closely to see that they are comfortable and start to eat and drink.
Timing: On the arrival of baby chicks to the farm.

9. FEED CHICK STARTER

Inputs required	*Type*	*Quantity*	*Timing*	*Price* (JD)
Feeders	—	65	—	—
Chick starter	—	14 t	—	95 per t
Labour	Skilled	1 labourer	2 hours for 6 weeks	—

Comments: The farmer prepares chick starter on the farm by grinding and mixing feed ingredients. The estimated cost of 1 t of chick starter is JD 95. However, if he purchases ready-made chick starter he has to pay JD 120 per t. Feeders should be filled to half-full in order to avoid losses due to scattering while chicks are eating. A saving of 5 per cent is thus attained.
Timing: Daily for 1-day up to 6-weeks old chicks.

10. LIGHT POULTRY HOUSE

Inputs required	*Type*	*Quantity*	*Timing*	*Price*
Power house	—	—	—	—
Labour	Skilled	1 labourer	15 minutes daily	—

Comments: The farmer should follow the recommended lighting programme suggested by the breeder.
Timing: At night, for 6 weeks after the arrival of chicks.

11. VACCINATE AGAINST NEWCASTLE DISEASE

required	*Type*	*Quantity*	*Timing*	*Price* (JD)	*Comments*
Newcastle vaccine	B_1	100,000 doses	—	3.5	at 7 days
Newcastle vaccine	Lasota	10,000 doses	—	3.5	at 21 days
Newcastle vaccine	Lasota	10,000 doses	—	3.5	at 70 days
Newcastle vaccine	K	10,000 doses	—	3.5	at 17 weeks
Labour	Skilled	3 labourers	16 hours	—	—

Environmental factors: None.
Timing: This vaccination programme for Newcastle Disease is continued throughout the life of the chicken. Vaccination with B_1 and Lasota in water is carried out at the age of 7, 21 and 70 days. Vaccination with K strain is done at 17 weeks by injection of each hen in the muscle.
Comments: The production of eggs is more complex than that of broilers because hens are kept on the farm for 18 months, while broilers are kept for 55 days. Moreover, the management of layers is more difficult than that of broilers.

12. LOWER BROODER TEMPERATURE

Inputs required	*Type*	*Quantity*	*Timing*	*Price*
Brooders	—	5	—	—
Labour	Skilled	1 labourer	0.5 hours	—

Comments: The temperature around the brooder at 30 cm above ground should be 95°F when the chicks arrive. It is lowered by 5°F each week up to three weeks, when the brooder is turned off.
Timing: At 7, 14 and 21 days.

13. VACCINATE AGAINST GUMBORO DISEASE

Inputs required	*Type*	*Quantity*	*Timing*	*Price* (JD)
Gumboro vaccine	—	10,000	—	3.5
Labour	Skilled	1 labourer	1 hour	—

Timing: At the age of 12 days. The vaccination is given in water.

14. EXTINGUISH BROODER AND ADJUST VENTILATION

Inputs required	*Type*	*Quantity*	*Timing*	*Price*
Brooder	—	5	—	—
Labour	Skilled	1 labourer	15 minutes daily	—

Environmental factors: The brooders are turned off 3 weeks after lighting. However, if the weather is cold and the chicks are clustering around, the brooders should be kept on for another week.
Timing: Three weeks after lighting the brooders.

15. FEED GROWER RATION

Inputs required	*Type*	*Quantity*	*Timing*	*Price* (JD)
Automatic feeders	—	—	—	3,000
Grower feed	—	60 t	—	85 per t
Labour	Skilled	1 labourer	1 hour daily from 6-20 weeks of age of chicks	—

Comments: The grower feed is prepared on the farm. The estimated cost of 1 t of feed is JD 85. The market price of ready-made grower feed is JD 100 per t.
Timing: Daily from 6 to 20 weeks.

16. CULL CHICKS

Inputs required	*Type*	*Quantity*	*Timing*	*Price*
Labour	Skilled	1 labourer	8 hours	—
Chicks	Abnormal	Variable	—	—

Environmental factors: Culling continues throughout the production cycle. However, mass culling for abnormality or stunted growth is carried out at 8 weeks and at 20 weeks. Culled chicks can be sold as broilers, but at a lower price.
Timing: (Mass culling) at the age of 8 weeks.

17. VACCINATE AGAINST SALMONELLA DISEASE

Inputs required	*Type*	*Quantity*	*Timing*	*Price* (JD)
Vaccine	—	20,000 doses	—	7
Labour	Skilled	1 labourer	2 hours	—

Environmental factors: None.
Timing: At 6 and 15 weeks. The vaccination is given in water.

18. FEED LAYER RATION

Inputs required	*Type*	*Quantity*	*Timing*	*Price* (JD)
Feeders	Automatic	—	—	3,000
Layer feed	—	376 t	—	87 per t
Labour	Skilled	1 labourer	2 hours daily	—

Timing: From 20 weeks of the pullet up to the end of egg production at 72 weeks.

19. FOLLOW RECOMMENDED LIGHTING PROGRAMME

Inputs required	*Type*	*Quantity*	*Timing*	*Price* (JD)
Generator	—	1	—	2,500
Labour	Skilled	1 labourer	15 minutes daily	—

Comments: The farmer should follow the breeder's recommended lighting programme for growing chicks and layers for optimum egg production. Failure to do so may decrease egg production by 10 to 15 per cent.
Timing: From 6 weeks up to the end of production at 72 weeks.

20. COLLECT EGGS

Inputs required	*Type*	*Quantity*	*Timing*	Price (JD)
Eggs	—	—	—	—
Egg baskets	—	2	—	2
Labour	Skilled	1 labourer	2 hours daily	—

Comments: Collection of eggs from egg nests is carried out three to four times daily. If done more than three times a day this results in clean eggs and fewer breakages, and eliminates the eating of eggs among layers.
Timing: From the start of egg production at about 20 weeks up to its end and the selling of the flock.

21. SELL EGGS

Inputs required	*Type*	*Quantity*	*Timing*	*Price* (JD)
Eggs	—	2,280,000	—	—
Egg trays	Carton	76,000	—	0.02 per tray
Labour	Farm manager	1 labourer	2 hours daily	—

Comments: The number of eggs given is the total egg production per cycle. Egg production usually starts when the birds are 20-22 weeks old, increasing weekly until it reaches its peak after about 8 weeks of production, and then decreasing slowly over the rest of the laying period. Eggs are sold in trays of 30 eggs according to weight. The classified categories of weight and their prices are set by the MOS and revised from time to time.

Weight of tray (g)	*Price of 360 eggs (12 trays) (JD)*
1,200 - 1,350	*6.0*
1,351 - 1,500	*6.5*
1,501 - 1,650	*9.2*
1,651 - 1,800	*9.8*
1,801 - 1,950	*10.4*
1,951 and more	*11.0*

An additional JD 1 per 12 trays goes to the different marketing channels.
Timing: Daily or every other day, the eggs being graded by eye and put in egg trays. The middleman or egg marketing society collects the eggs from the farm.

22. CULL NON-PRODUCTIVE HENS

Inputs required	*Type*	*Quantity*	*Timing*	*Price*
Hens	Abnormal	12%	—	—
Labour	Skilled	1 labourer	8 hours	—

Comments: Culling is a continuous process during the laying period. The estimated culling rate is 1 per cent each month during this time. Culled hens can be sold as broilers, but at a lower price of JD 0.3 per hen.
Timing: Daily as required throughout the laying period.

23. SELL HENS AT END OF PRODUCTION

Inputs required	*Type*	*Quantity*	*Timing*	*Price* (JD)
Hens at the end of production	—	8,300	—	—
Labour	Farm manager	1 labourer	6 hours for 3 days	—
Transportation	Pick-up	1	For 3 days	30

Comments: Layer farms have difficulties in marketing hens at the end of production because of the poor quality of the meat and the lack of poultry-processing plants. Hens are usually sold live to consumers at a price of JD 0.45 per bird.
Timing: At the end of production.

24. SELL POULTRY MANURE

Inputs required	*Type*	*Quantity*	*Timing*	*Price* (JD)
Poultry manure	—	400 m^3	—	—
Labour	Farm manager	1 labourer	2 hours	—
Transportation	Car	Car seat	—	0.5

Comments: Poultry manure is sold from the farm at a price of JD 7 per t. It is collected from the poultry house by the buyer and there is a demand for it for vegetables and fruit growing.
Timing: At the end of the production cycle.

Livestock Sector: General Background

The different kinds of livestock raised in Jordan include cattle, sheep, goats and camels. The amount of livestock fluctuates from one year to another, depending upon the condition of rangelands and consequent prices of feed, and the outbreak of epidemic diseases.

Table 6.4 Number of Livestock, 1974-81 (thousands)

Year	Cattle	Sheep	Goats	Camels
1974	36	850	502	18
1975	36	792	399	16
1976	37	772	454	18
1977	37	565	382	14
1978	38	807	379	18
1979	37	856	382	10
1980	37	743	381	11
1981	35	852	446	12

Source: Animal Production and Health Department 1974-81.

Cattle
The two breeds of cattle in Jordan are the Baladi and the Friesian. Baladi cattle constitute about 75 per cent of the total number. The Baladi cow is small in size and stocky in appearance. The colour is solid black, or red, or a combination of both. Baladi cattle are extremely hardy and have great powers of endurance even with little feed and poor management, and are resistant to prevailing diseases. The productive characteristics of a herd of Baladi cattle have been studied at Deir Alla Agricultural Experiment Station and found to be inferior to those of imported cattle.

The second breed is the Friesian, which was introduced to Jordan early in 1950 and has adapted to the different climatic conditions in the country. Table 6.5 shows the distribution of cattle by governorate.

Table 6.5 Number of Livestock by Governorate, 1981

Governorate	Cattle		Sheep	Goats
	Baladi	Friesian		
Amman	6,055	5,565	202,550	52,300
Balqa	5,073	2,275	117,800	60,900
Irbid	13,698	2,478	231,650	130,650
Karak	1,770	204	182,950	65,840
Ma'an	278	40	117,500	134,965
TOTAL	26,874	10,562	852,450	445,655

Source: Animal Production and Health Department 1981.

Sheep
The breed of sheep raised in Jordan is the Awassi, which is also raised in Syria, Lebanon and Iraq. It is characterised by its fat tail and semi-coarse wool. Its colour is white with a red-brown head. The ram weighs 60-75 kg live and the ewe 45-60 kg. The sheep are mostly owned by bedouin tribes who travel from one place to another according to available pasture and crop residues. The ranging season in Jordan is from late November until April. During this time sheep are kept on the range during the day and fed in the evening on small quantities of straw. At other times of the year, they receive supplementary feeds of grain mixtures. Table 6.5 shows the distribution of sheep in Jordan by governorate.

Goats
There are two breeds of goat in Jordan: the Baladi, which constitutes 96 per cent of the goat population, and the Shami, or Damascus goat. The Baladi goat is considered to be a dual-purpose animal for milk and meat. It can withstand the tough environmental conditions of Jordan, but it is a threat to forest trees. The estimated milk yield per goat per day is 0.5 kg over a lactation period of four to five months. Baladi goats are owned, like sheep, by Bedouin tribes travelling from one place to another looking for pasture. Shami goats are kept indoors mainly for the production of milk. Surplus males are kept for meat production. Shami goats have a high twinning rate.

The estimated milk yield per goat per day is 2.5-3.0 kg over a lactation period of five months. Table 6.5 shows the distribution of goats in Jordan by governorate.

Red Meat Production
Red meat production comes from four main sources: sheep, goats, cattle and camels. Table 6.6 shows the estimated production of red meat for the period 1974-81. Local production meets about 30 per cent of demand in Jordan. Production fluctuates from one year to another. The shortage of red meat is met by the importation of live animals and fresh, chilled and frozen meat.

Milk Production
Milk produced in Jordan comes from three main sources: cattle, sheep and goats. Table 6.7 shows the estimated milk production for the period 1974-81. Local production meets about 30 per cent of demand. Milk and dairy products are imported to meet the shortfall. The estimated milk yield of the Baladi cow is 500 kg per lactation; the estimated yield of the Friesian cow is 3,000 kg per lactation.

Table 6.6 Red Meat Production, 1974-81 (tonnes)

Year	Sheep	Goats	Cattle	Camels	Total
1974	3,604	1,676	1,512	840	7,632
1975	3,513	1,907	1,554	945	7,919
1976	2,571	1,604	1,554	735	6,464
1977	3,672	1,592	1,596	945	7,805
1978	3,895	1,604	1,554	525	7,578
1979	3,381	1,600	1,550	575	7,106
1980	3,825	1,980	1,554	755	8,114
1981	4,050	2,025	1,525	945	8,545

Source: Animal Production and Health Department 1981: 34.

Table 6.7 Milk Production, 1974-81 (tonnes)

Year	Cattle	Sheep	Goats	Total
1974	20,056	12,664	9,569	42,289
1975	17,606	12,368	10,896	40,920
1976	15,034	10,405	9,220	34,609
1977	16,805	13,715	9,090	39,610
1978	21,179	13,704	9,176	44,059
1979	17,407	12,159	9,142	38,705
1980	21,256	13,600	11,112	45,968
1981	16,112	14,400	11,458	41,970

Source: Animal Production and Health Department 1981: 34.

Analysis of a Dairy Farm System

On-Farm Environments

Physical

Area: A characteristic dairy farm is built on an area of 10 du or more. Dairy farms should be 300 m away from each other in order to reduce the likelihood of disease.

Buildings and Equipment: The typical dairy farm comprises the following buildings:

- barn, with an area of 10.5 × 9.5 m^2 for 10 milking cows plus dependents
- feed store, 4 × 6 m^2
- milking parlour, 4 × 4 m^2
- house for the owner

Dairy equipment includes utensils, milk cans and waterers.

Power Source: Power is obtained off the mains.

Dairy Numbers and Breed: The cattle breed on the dairy farm is Friesian, black and white, numbering:

- 10 milking cows
- 2 heifers from 1-2 years
- 4 female calves under 1 year old

All male calves are sold after being weaned at two months, or even earlier.

Number of Workers: The owner of the farm and his family usually provide the necessary labour.

Financial Assets and Liabilities:

		(JD)
Assets:	— land	2,000
	— buildings	8,100
	— cattle	5,000
Liabilities:	— long-term loans	4,050
	— short-term loans	4,550

Inputs and Yields:

Inputs:	— animals	10 heifers in calf
	— feed	80 t tibin* + 80 t concentrate
	— water	300 m³ per year
	— medicines, vaccines, semen	
	— labour	1 man plus his family
Yields (per year):	— milk	30 t
	— 4 male calves	
	— 2 culled cows	
	— manure	20 t

Informational
Most of the owners of dairy farms are farmers who have gained their experience by working on dairy farms and have had a primary education. Few have had an agricultural education.

Off-Farm Environments
Physical
Most dairy farms are located beyond the municipal border. They have access to good roads.

Institutional
The institutions concerned with dairy farms are:
- the MOA
- the MOS
- the ACC
- the University of Jordan
- agricultural companies and dairy plants

Informational
Dairy farmers have information on input prices set by the MOS, and on market prices. Information on improving the performance of dairy farms and other new innovations is communicated to dairy farmers through extension agents and agricultural companies.

Activity Analysis

Physical Activities
The activities related to a dairy farm system are set out in Figure 6.3.

1. ASSESS ASSETS

Inputs required	*Type*	*Quantity*	*Timing*	*Price*
Labour	Farm manager	1 labourer	3 days	—

*Wheat straw.

Timing: Normally before the start of dairying.
Comments: This activity requires the farmer to:
- estimate capital in hand
- estimate capital needed
- estimate cost of production
- estimate loans
- estimate expected profit

2. DECIDE TO GO INTO DAIRY PRODUCTION

Inputs required	*Type*	*Quantity*	*Timing*	*Price*
Labour	Farm manager	1 labourer	6 hours	—

Alternatives: There are many alternatives for the farmer to choose from. He can decide to go into poultry, sheep or dairy production, or some other agricultural project.
Determining factors:
- There is a high demand for milk and dairy products, and consequently a good local market.
- Jordan imports large quantities of powdered milk and dairy products.
- Loans can easily be obtained.
- The MOA encourages dairy farms.
- The farmer's neighbour has started a dairy farm.

3. DECIDE TO SEEK LOANS

Inputs required	*Type*	*Quantity*	*Timing*	*Price*
Labour	Farm manager	1 labourer	6 hours	—

Alternatives: There are two alternatives: The first is to obtain loans from the ACC to build the farm and to purchase cattle and feed. The second is to have funds available but take out a loan for the purchase of some inputs; the farmer can get this loan from the agricultural companies.
Determining factors: For long-term loans the farmer prefers to go to government agencies because of their low interest rates. For short-term loans he prefers to go to the agricultural companies.

4. OBTAIN LOANS

Inputs required	*Type*	*Quantity*	*Timing*	*Price* (JD)
Labour	Farm manager	1 labourer	6 hours	—
Transportation	Car	Car seat	—	0.5

Environmental factors: None.
Timing: After the farmer has decided to obtain loans and decided on the sources.
Comments: The farmer goes to the ACC to obtain a loan for building the farm and for purchasing cattle.
Total capital required for building the farm: JD 8,100.
Total capital required for cattle: JD 5,000.

5. CONSTRUCT DAIRY FARM

Inputs required	*Type*	*Quantity*	*Timing*	*Price*
Capital	Cash	JD 8,100	—	—
Labour	Farm manager	1 labourer	3 hours daily for 3 months	—

Timing: After obtaining the loans.
Comments: The farmer usually approaches several building contractors and makes an agreement with one of them to build the farm according to his

specifications. He will pay the contractor according to the terms of their agreement.

6. PURCHASE INPUTS: cattle, feed, semen, medicines and disinfectants

Inputs required	*Type*	*Quantity*	*Timing*	*Price* (JD)
Labour	Farm manager	1 labourer	8 hours for 1 week	—
Transportation	Car	Car seat	—	2.0

Timing: At the start of dairying and when a certain input is not available on the farm.

Comments: 1. The quantities and prices of different inputs are listed below:

Type	*Quantity*	*Unit price* (JD)	*Total Price* (JD)
Heifers in calf	10	500	5,000
Feed (concentrate)	40 t per year	65	2,600
Feed (tibin)	30 t per year	65 *	1,950
Semen	10 doses per year	1	10
Medicine and disinfectants	—	—	25
Utensils	—	—	50
Labour	1 man	80 per month	960 per year
Water	300 m^3 per year	0.25	75

* Tibin is usually purchased in August. Straw bought out of season costs JD 75-80 per t.

2. Cattle are purchased from cattle merchants who deliver the cattle to the farm.
3. Feed and other inputs are purchased from the market.

7. FEED CATTLE

Inputs required	*Type*	*Quantity*	*Timing*	*Price* (JD)
Feed	Tibin	30 t	—	65 per t
Feed	Concentrate	40	—	65 per t
Labour	Skilled	1 labourer	2 hours daily	—

Comments: The farmer gives each milking cow 6 kg of tibin and 8 kg of concentrate per day. The quality of tibin as hay is very poor. If the farmer feeds alfalfa hay instead of tibin, the milk yield increases from 3 to 4 t per year. However, alfalfa hay is very scarce and its price ranges from JD 80 to JD 100 per t. Farmers look for cheap sources of feed in the market.

Timing: Daily. Feed is offered to cattle 3 times a day.

8. CARE FOR COWS AT CALVING TIME

Inputs required	*Type*	*Quantity*	*Timing*	*Price*
Labour	Skilled	1 labourer	3 hours per cow per year	—

Comments: The farmer should keep a careful watch on cows expecting calves and see that they are not nervous, and have plenty of food and water. He should check cows daily for signs of approaching labour. Help during labour is available when needed.

Timing: During the 2 weeks before delivery.

9. CARE FOR NEWBORN CALVES

Inputs required	*Type*	*Quantity*	*Timing*	*Price*
Calves	Newborn	—	—	—
Labour	Skilled	1 labourer	2 hours per calf	—

Comments: The farmer dries the newborn calf with a piece of cloth, removes any mucous from its nose and mouth and ties the umbilical cord. Then he watches the calf stand up and start nursing.
Timing: After the delivery of the calves.

10. FEED CALVES UP TO WEANING

Inputs required	*Type*	*Quantity*	*Timing*	*Price* (JD)
Calves	—		—	—
Colostrum		Nursing from the cow for 3 days	—	—
Milk	Cow's milk	300 kg per calf	—	0.14 per kg
Milk utensils	Bucket	1	—	0.5
Feed	Concentrate	*Ad lib*	—	—
Labour	Skilled	1 labourer	1 hour daily for 2 months	—

Comments: The calf receives colostrum by nursing from its mother for the first 3 days. It is then fed cow's milk 3 times a day. After 3 weeks it is fed concentrates *ad lib.* Powdered calf's milk is available on the market and is fed to calves on a very limited scale.
Timing: From 1 day old up to weaning, at 2 months.

11. MILK COWS

Input required	*Type*	*Quantity*	*Timing*	*Price* (JD)
Cows	Milking	10	—	—
Milk utensils	—	—	—	50
Labour	Skilled	1 labourer	3 hours daily	—

Comments: Cows are milked twice a day by hand. Milk is then filtered into cans either through a cloth or a milk funnel.
Timing: Daily.

12. SELL MILK

Inputs required	*Type*	*Quantity*	*Timing*	*Price*
Milk	—	200 kg	—	—
Transportation	—	Pick-up	—	—
Labour	Unskilled	1 labourer	2 hours daily	—

Comments: The farmer sells his milk to dairies or directly to the consumer. The selling price to dairies is 140 fils per kg; to the consumer it is 240 fils per kg. However, direct marketing to consumers calls for additional transportation and labour.
Timing: Daily.

13. WEAN CALVES

Input required	*Type*	*Quantity*	*Timing*	*Price*
Calves	2 months old	—	—	—
Labour	Skilled	1 labourer	15 minutes	—

Comments: The calves are weaned and fed concentrates and hay.
Timing: When calves reach 2 months of age and are in good health.

14. SELL MALE CALVES

Inputs required	*Type*	*Quantity*	*Timing*	*Price* (JD)
Male calves	2 months old	—	—	—
Labour	Skilled	1 labourer	3 hours	—
Transportation	Pick-up	1	1 hour	3

Figure 6.3: Activities Related to a Dairy Farm System

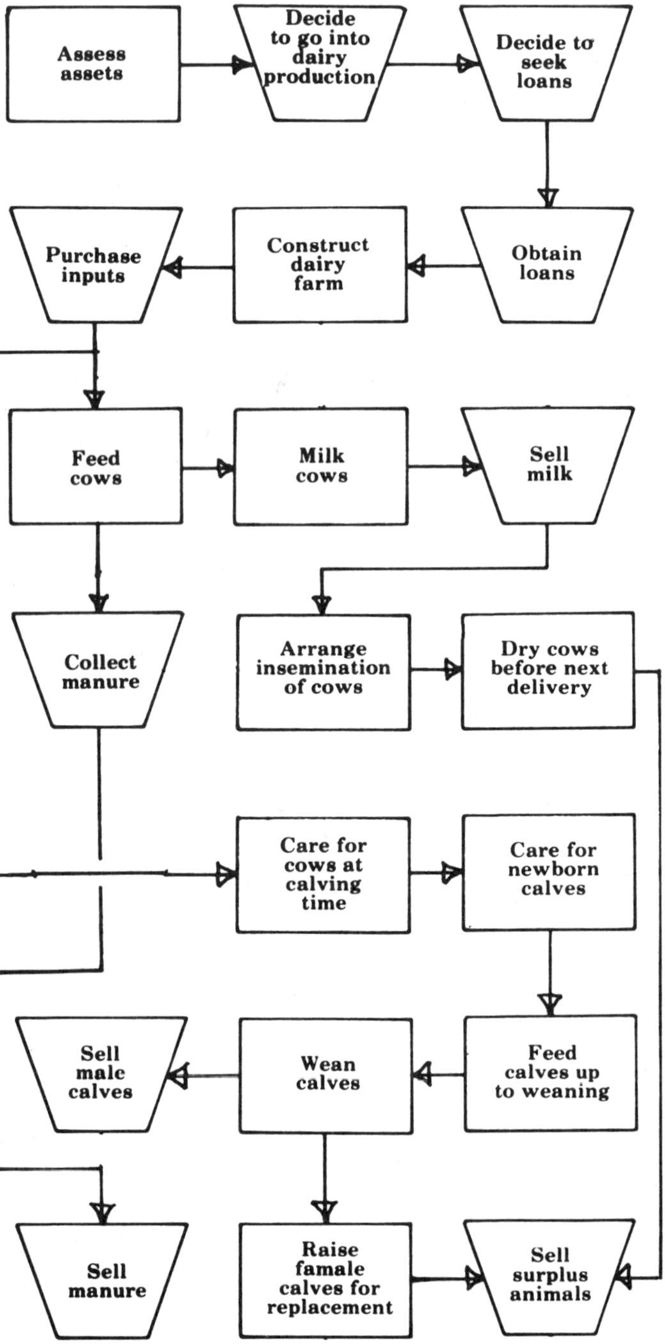

Comments: The farmer sells his male calves after weaning at an average body weight of 75 kg, at a price of JD 1 per kg live weight. However, he can keep his calves until they reach 300 kg body weight. By then a calf could have consumed an additional 1,500 kg of feed. The price of feed is 65 fils per kg.
Timing: When male calves reach 2 months of age and are weaned.

15. RAISE FEMALE CALVES FOR REPLACEMENT

Inputs required	*Type*	*Quantity*	*Timing*	Price (JD)
Calves	Female	4	—	—
Labour	Skilled	1 labourer	0.5 hour daily	—
Feed	Straw	1 t	--	65
Feed	Concentrate	1.5 t	—	97.5

Comments: The quantities of feed needed are calculated in Activity No. 7 (Feed cattle).
Timing: From weaning until 2 years old.

16. ARRANGE INSEMINATION OF COWS

Inputs required	*Type*	*Quantity*	*Timing*	*Price* (JD)
Cows	On heat	—	—	—
Semen	—	1 dose per cow	—	0.1
Labour	—	1 labourer	2 hours per cow	—

Environmental factors: When cows come on heat the farmer calls in the AI technician. If the farmer uses natural servicing he takes his cow to his farmer neighbour who has a bull for stud purposes.
Timing: When the cow comes on heat.

17. DRY COWS BEFORE THEIR NEXT DELIVERY

Inputs required	*Type*	*Quantity*	*Timing*	*Price*
Cows	—	—	—	—
Labour	Skilled	1 labourer	5 minutes per cow	—

Timing: 2 months before the expected new delivery.

18. *MARKET SURPLUS ANIMALS*

Inputs required	*Type*	*Quantity*	*Timing*	*Price*
Cows	Culled	2	—	—
Heifers	In calf	2	—	—
Labour	Skilled	1 labourer	3 hours	—

Comments: Culled cows are sold for meat at a price of JD 350. Heifers in calf are sold at JD 500.
Timing: When heifers are 5-6 months pregnant.

19. SELL MANURE

Inputs required	*Type*	*Quantity*	*Timing*	*Price*
Manure	—	20 t	—	—
Labour	Skilled	1 labourer	3 hours	—

Comments: Cattle manure is sold at JD 5 per t. Manure is collected by the purchaser. The price fluctuates according to supply and demand.
Timing: Any time during the year.

Analysis of a Sheep Farm System

On-Farm Environments

Physical

Area: A flock of sheep is kept in a pen with an area of 1-2 du or more.

Buildings and Equipment: The typical sheep farm comprises the following buildings:

- — barn, with an area of 200 m^2
- — pen, with an area of 400 m^2
- — food store 4 × 6 m^2
- — house for the shepherd

Sheep farming equipment includes feeders, waterers and milk utensils.

Power Source: There is no power source on the sheep farm. Light at night is provided by gasoline lamps.

Sheep Numbers and Breed: The breed of sheep is the Awassi. The flock is composed of:

- — 100 ewes
- — 38 yearling females
- — 40 female lambs
- — 4 rams

(All male lambs are sold before weaning or directly after weaning.)

Number of Workers: One shepherd.

Financial Assets and Liabilities:

		(JD)
Assets:	— land	1,000
	— buildings	7,000
	— sheep	5,260
Liabilities:	— long-term loans	6,130

Inputs and Yields:

Inputs:	— animals	104
	— supplementary feed	
	— water	
	— medicines and vaccines	
	— labour	1 shepherd
Yield *(per year)*:	— milk	
	— 40 female lambs	
	— 40 male lambs	
	— wool	
	— culled animals	

Informational

Most sheep owners are bedouins or tribal chieftains who gained their experience from their fathers and grandfathers. Some of them have had a primary education and a few have had agricultural school education.

Off-Farm Environments

Physical

Sheep farming depends on natural ranges as the main source of feed. There are three types of range in Jordan:

1. The desert ranges in the east and north with a total precipitation of less than 150 mm. The area is covered with annual plants and is subject to severe overgrazing.
2. The sandstone ranges in the south with a total precipitation of 25-100 mm. The area is covered with occasional annual plants, and is exposed to severe

and irregular overgrazing.
3. The steppe ranges with a total precipitation of 150-250 mm. The area has a good grázing potential.

The deterioration of ranges is due to overgrazing, and the cultivation of rangeland for the production of cereal crops. Under present range conditions, the estimated carrying capacity is 50-75 du per sheep. A short- and a long-term improvement programme is essential.

Sheep are owned either by bedouin tribes who roam according to availability of water and pastures, or by village dwellers who take the sheep to range in the morning and return them to the village in the evening for supplementary feeding of tibin and barley. Water on the range is provided from available cisterns.

Institutional
The institutions concerned with sheep farming are:
- the MOA
- the MOS
- the ACC and the JCO
- agricultural companies and sheep merchants

Informational
Sheep farmers have access to information on inputs and market prices. Information on improving production is communicated to them through agricultural agents. However, such communication is rare.

Activity Analysis

Physical Activities
The activities related to a sheep farm system are set out in Figure 6.4.

1. ASSESS ASSETS

Inputs required	*Type*	*Quantity*	*Timing*	*Price*
Labour	Sheep owner	1 labourer	8 hours	—

Timing: Before building up a flock of sheep.

2. DECIDE TO GO INTO SHEEP PRODUCTION

Inputs required	*Type*	*Quantity*	*Timing*	*Price*
Labour	Sheep owner	1 labourer	6 hours	—

Alternatives: Sheep owners in the villages have a few alternatives to choose from: they can decide to go into either sheep or cattle production. But the bedouin keep sheep and goats as their only choice because of the nature of their tribal life-style.

Determining factors:
- There is a high demand for dairy products.
- There is a high demand for red meat.
- The risk of failure is low.
- Loans can be obtained easily.
- The farmer's father and grandfather have already kept sheep.

3. DECIDE TO SEEK LOANS

Inputs required	*Type*	*Quantity*	*Timing*	*Price*
Labour	Sheep owner	1 labourer	6 hours	—

Alternatives: There are two alternatives. The first is to obtain a loan from the ACC to build sheep pens, and buy sheep and feed; the second is to obtain a

Figure 6.4: Activities Related to a Sheep Farm System

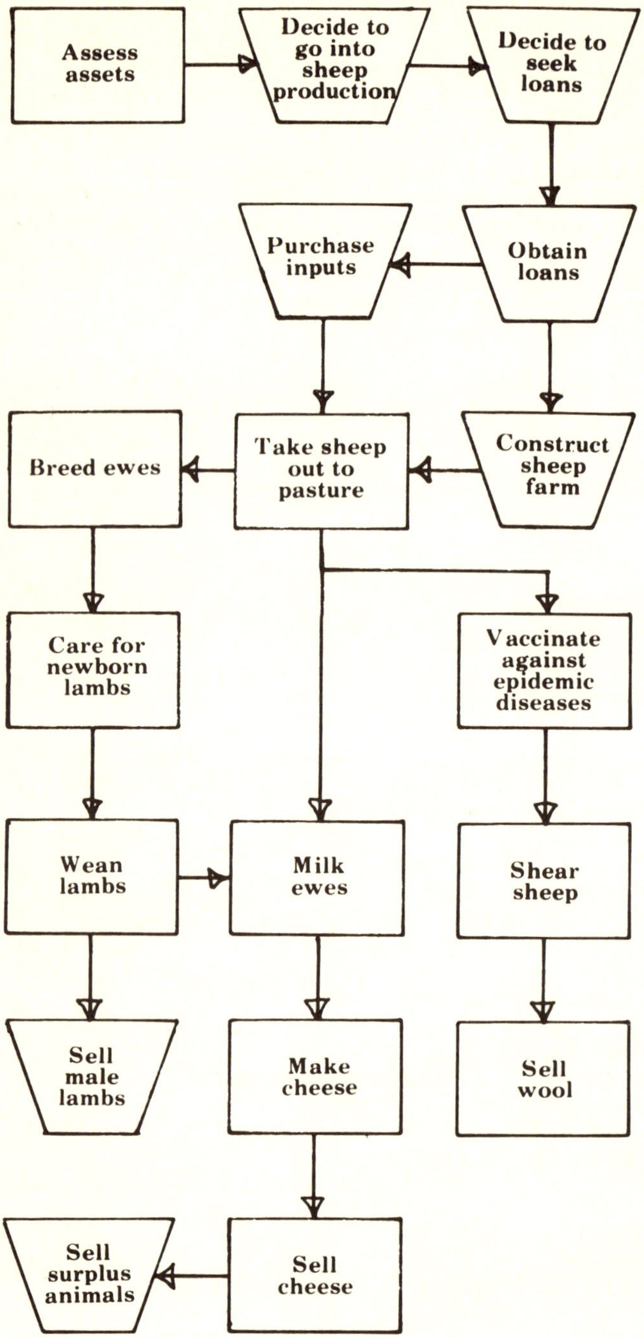

loan from the ACC and/or JCO for the purchase of sheep and feed only.
Determining factors: For long-term loans the sheep owner goes to the ACC. For short-term loans he approaches the JCO.

4. OBTAIN LOANS

Inputs required	*Type*	*Quantity*	*Timing*	*Price* (JD)
Labour	Sheep owner	1 labourer	6 hours	—
Transportation	Car	Car seat	—	0.5

Timing: After the farmer has decided to seek loans.
Comments: The farmer goes to the ACC to obtain a loan for building the sheep farm and purchase of sheep.
Total capital required for buildings: JD 7,000.
Total capital required for purchasing sheep: JD 5,260.

5. CONSTRUCT SHEEP FARM

Inputs required	*Type*	*Quantity*	*Timing*	*Price*
Capital	Cash	JD 7,000		—
Labour	Sheep owner	1 labourer	2 hours daily for 2 months	—

Timing: After obtaining the loan.
Comments: The farmer chooses a contractor to build the sheep farm according to his specifications.

6. PURCHASE INPUTS

Inputs required	*Type*	*Quantity*	*Timing*	*Price* (JD)
Labour	Sheep owner	1 labourer	1 week	—
Transportation	Car	Car seat	—	3

Timing: At the start of the project and when a certain input is needed.
Comments: 1. The quantities and prices of different inputs are as follows:

Type	*Quantity*	*Unit price* (JD)	*Total price* (JD)
Ewes	100	50	5,000
Rams	4	65	260
Feed	14 t	65	910
Medicines and vaccines	—	—	22
Water	100 per m^3	0.25	25
Labour	1	90 per month	1,080 per year
Milk utensils	—	—	25

2. Sheep are purchased from other sheep merchants and are delivered to the farm by truck.

7. TAKE SHEEP OUT TO PASTURE

Inputs required	*Type*	*Quantity*	*Timing*	*Price*
Labour	Shepherd	1	10 hours daily	—

Comments: Sheep are taken to pasture early in the morning and are brought back in the evening. On returning home, they receive a supplementary feed of 0.5 kg per head per day for 8 months of the year and when grazing is insufficient. On rainy and snowy days the sheep are kept in barns.
Timing: Daily all year round and when pasturage is good.

8. VACCINATE AGAINST EPIDEMIC DISEASES

Inputs required	*Type*	*Quantity*	*Timing*	*Price*
Sheep	—	104	—	—
Vaccine	—	As required	—	—
Labour	Skilled	2 labourers	8 hours	—

Comments: The sheep owner arranges with the veterinarian to vaccinate his sheep against epidemic diseases. Vaccination is given free of charge and veterinary services are also provided free all over the country.
Timing: According to the vaccination programme recommended by veterinarians.

9. BREED EWES

Inputs required	*Type*	*Quantity*	*Timing*	*Price*
Ewes	—	100	—	—
Rams	—	4	—	—
Labour	—	1 labourer	0.5 hours daily for 2 months	—

Comments: Sheep are seasonal breeders. They do not come on heat in hot weather. Flushing the sheep 2 weeks before the breeding season improves the conception rate, as does separating the ewes from the rams at night at this time.
Timing: August and September. During the breeding season rams run with ewes.

10. CARE FOR NEWBORN LAMBS

Inputs required	*Type*	*Quantity*	*Timing*	*Price*
Lambs	Newborn	—	—	—
Labour	Skilled	1 labourer	1 hour per lamb	—

Comments: The shepherd dries the newborn lamb with a piece of cloth, removes any mucous from its mouth and nose, ties the umbilical cord and notes that the lamb is nursing.
Timing: After the lambing of ewes.

11. WEAN LAMBS

Inputs required	*Type*	*Quantity*	*Timing*	*Price*
Lambs	2 months old	80	—	—
Labour	Skilled	1 labourer	15 minutes	—

Timing: At the age of about 2 months.

12. SELL MALE LAMBS

Inputs required	*Type*	*Quantity*	*Timing*	*Price* (JD)
Male lambs	2 months old	40	—	—
Labour	Skilled	1 labourer	8 hours	—
Transportation	Pick-up	1	—	3

Comments: The sheep owner sells his male lambs after weaning and at a body weight of 18-20 kg at a price of JD 1.2 per kg live weight. Prices fluctuate from season to season according to supply and demand.
Timing: When the male lambs are weaned.

13. MILK EWES

Inputs required	*Type*	*Quantity*	*Timing*	*Price* (JD)
Ewes	—	80	—	—
Milk utensils	—	—	—	25
Labour	Skilled	4 labourers	4 hours for 5 months	—

Comments: Ewes are milked twice a day. Many factors affect the yield: breed, feeding, age at first calving, calving intervals, number of milkings, calving season, herd management and others.
Timing: Daily from March to the end of July.

14. MAKE CHEESE

Inputs required	*Type*	*Quantity*	*Timing*	*Price* (JD)
Milk	—	6,400 kg	—	—
Milk utensils	—	—	—	25
Cheese cloth	—	—	—	1
Rennet	—	—	—	10
Salt	—	—	—	8
Labour	Skilled	2 labourers	3 hours for 5 months	—

Comments: The quantities of different inputs listed above are the total quantities for the whole milking season.
Timing: Daily from weaning until the drying up of ewes.

15. SELL CHEESE

Inputs required	*Type*	*Quantity*	*Timing*	*Price* (JD)
Cheese	White	1,400 kg	—	—
Transportation	Pick-up	1	—	10
Labour	Sheep owner	1 labourer	2 hours daily for 5 months	—

Environmental factors: White cheese is marketed either to merchants and retail shops or directly to consumers. The average price of cheese is JD 1.1 per kg. It fluctuates between JD 1 and JD 1.4 per kg.
Timing: During the milking season.

16. SHEAR SHEEP

Inputs required	*Type*	*Quantity*	*Timing*	*Price* (JD)
Sheep	—	100	—	—
Shears	—	1	—	1
Labour	Skilled	1 labourer	50 hours	—

Timing: May.

17. SELL WOOL

Inputs required	*Type*	*Quantity*	*Timing*	*Price* (JD)
Wool	Oily	125 kg	—	—
Labour	Sheep owner	1 labourer	4 hours	—
Transportation	Pick-up	Seat	—	0.5

Comments: The average selling price of oily wool is JD 0.75 per kg. It ranges between JD 0.5 and JD 1 per kg, depending on quality.
Timing: After sheep shearing.

18. SELL SURPLUS ANIMALS

Inputs required	*Type*	*Quantity*	*Timing*	*Price* (JD)
Ewes	Culled	15		—

Yearlings	Females	18	—	—
Rams	—	1	—	—
Labour	Sheep owner	1 labourer	6 hours	—
Transportation	Pick-up	1	2 hours	10

Comments: Culled ewes and rams are sold for meat at a price of JD 1 per kg. Yearling females are sold for raising at a price of JD 45 per head.
Timing: At culling time in summer.

References and Other Works Consulted

Animal Production and Health Department, *Annual Reports, 1974-81*, Ministry of Agriculture, Amman (in Arabic).

Dahoud, M.N., R. Kura'an and A. Monna', 1983, 'Poultry industry in Jordan and its problems', Amman (mimeo; in Arabic).

Duayfi, A. 1982, 'Experience of Arab Countries (Jordan) in the improvement of livestock production', Proceedings of the Fifth Technical Conference of Arab Agricultural Engineers, Kuwait (in Arabic).

— *and A. F. Qadi, 1980, 'Animal wealth in Jordan', Agricultural Seminar,* Amman (mimeo; in Arabic).

Gun Rural Management Pty. Ltd., 1978, *The Livestock Sector: Sector and Project Identification,* Report submitted to National Planning Council, Amman.

MOA (Ministry of Agriculture), ACC (Agricultural Credit Corporation) and JCO (Jordan Cooperative Organisation), 1981, *A Survey of Poultry Farms*, Amman.

NPC (National Planning Council) 1980, *Three-Year Plan for Social and Economic Development, 1973–75,* Amman.

—— 1975, *Five-Year Plan For Social and Economic Development, 1976–80,* Amman.

—— 1972, *Five-Year Plan For Social and Economic Development, 1981–85,* Amman.

—— 1983, 'Poultry complex master-plan', Amman (mimeo).

PART III

SUPPORT SYSTEMS

7 WATER SUPPLY FOR THE AGRICULTURAL SECTOR

Fahd Salih Natur

Introduction

In this chapter, the physical, institutional, economic, financial and informational environments of the water resources system of Jordan will be described. The activities, decisions and policy instruments of the water resources systems will also be discussed. Activity analysis and general verbal descriptive modelling will be attempted. The model is meant to describe the system as it works at present without optimisation attempts. Aggregation is made over the major system components and details are used only as the interaction between the water resources system and the crop production system becomes more direct.

The importance of the water resources system cannot be over-emphasised. An adequate supply of water is essential for the well-being of any society. With population increase and improved incomes, demands for water for irrigation and for municipal and industrial (M&I) purposes increase rapidly. Water resources development involves heavy investment and usually irreversible decisions (or decisions whose reversal is very costly). Problems of non-uniform supply and demand in space and time and problems of pollution and return flow make the study of the water resources system very important.

In an arid country like Jordan, where 91.5 per cent of the land area receives less than 200 mm of rainfall annually (Arar 1978), the water resources system becomes even more important for agricultural development. This is especially so because of the increasing demand for water for other competing uses, such as M&I. Water resources development plays a major role as a catalyst for agricultural development (and hence for economic development). Choice of the appropriate agricultural and rural strategy is a very important part of the development process. Identifying policy options for the water resources system, evaluating their impact on the agricultural sector and choosing the optimum water policy are all central to the agricultural sector.

The water resources system has interdependencies not only with the agricultural system, but also with many other systems such as:

1. The meteorological/hydrological system
2. The energy system
3. The political system
4. The planning system
5. The ecological system
6. The economic system
7. The social system
8. The legal system
9. The organisational system
10. The public health system
11. The technology system

With such a variety of interdependencies, the methods of systems analysis become very important tools for the study of water resources systems. Simulation is one powerful method for such studies. Yet the powers and limitations of systems analysis should be made clear. Operations research and systems analysis can only assist the decision-making process in water resources; they do not replace it.

The water system will be treated as a system of human economic activities. The system model will be governed by criteria of simplicity, workability and credibility. The system is described as it works at present, not as it should work. Water resources development projects, however, usually have a long time horizon (political preparation to maturation 30 years, and amortisation 50 years). Technology changes with time. Elements of uncertainty are always present. Because of this, simulation models of the water resources system should not be thought of as static models, but rather as dynamic models with frameworks changing over time.

Because the water resources system had a great number of components, activities, decisions and decision-makers, aggregation will be used. Large aggregate decision units will be used in decisions farther displaced from direct contact with the farmer. More detail will be used in describing the system at the level of the farmer-system interaction. Thus design and construction at the project level will be aggregated, although each consists of a long chain of activities and decisions, while water distribution to the farmer will be discussed in more detail.

The water resources system will be treated as a separate system, with inputs to and interactions with the farm systems. The on-farm irrigation system is not only a sub-system of the agricultural system, but also a sub-system of the water resources system.

Objectives of the Water Resources System

The general objective for developing the water resources system is to obtain optimum social welfare. In the absence of a satisfactory measurement for social welfare, the objective is translated into economic goals:

1. Economic growth, economic efficiency and maximising the national income
2. Regional development and advancement
3. Improved income distribution among regions, sections and individuals in society
4. Environmental conservation and enhancement
5. Obtaining satisfactory levels of employment

These goals are further details into goals of the development plans as:

1. To provide more water for irrigation and M&I
2. To increase the irrigated area to produce more food and fibre
3. To use water more efficiently
4. To protect water resources from deterioration and pollution

The Environment of the Water Resources System

The Physical Environment

The Climate: The climate of Jordan ranges from Mediterranean to desert. Two main seasons are recognised: summer, from May to October, and winter, from November to April. Summer is hot and dry, with a mean temperature of 28-32°C (National Water Master Plan (NWMP) 1977). Winter is mild and rainy in some parts. Monthly and regional differences in temperature in both seasons are significant

Rainfall: Rainfall is a very important element in the physical environment of the water resources system, because it it the basis of the water supply sub-system. Jordan is characterised by low rainfall with the larger percentage of the area being classified as arid or semi-arid. Rainfall occurs during the months from October to April, with the maximum rainfall occurring in December or January.

The total average annual rainfall on catchments draining into Jordon is 8,065 mcm, of which 6,000 mcm occur within Jordon (NWMP 1977). The monthly distribution of rainfall is given in Table 7.1.

Table 7.1: Average Monthly Rainfall, all Stations, 1954-62 (millimetres)

Oct.	Nov.	Dec.	Jan.	Feb.	Mar.	Apr.	May	June	July	Aug.	Sept.	Total
4.0	39.8	76.4	82.8	60.6	46.8	16.1	6.0	—	—	—	—	332.5

Source: FAO 1966.

The regional variation of rainfall is given in Table 7.2.

Table 7.2: Average Annual Rainfall (millimetres)

Region	Average annual rainfall
Northern Jordan Valley	300
Southern Jordan Valley	150
Southern Ghors	50
Northern Highlands	200–500
Southern Highlands	175
Southern Mountains	250
Desert	50–80

Source: NWMP 1977.

Areas of the different zones of rainfall are given in Table 7.3.

Table 7.3: Average Rainfall and Areas Covered

Zone	Rainfall (mm)	Area (km^2)	% of total area	Total rainfall (mcm)
Desert	100	73,178	79.1	2,522
Arid	100-200	11,392	12.4	1,630
Sub-total		84,570	91.5	4,152
Marginal for rain-fed agriculture	200-300	3,948	4.3	987
Semi-arid	300-500	3,041	3.2	1,190
Semi-humid	500	989	1.0	566
GRAND TOTAL		92,548	100.0	6,895

Source: Arar 1978.

Figure 7.1 shows the isohyetal map of the mean annual rainfall. The density of rainfall stations is considered adequate for assessment of annual and regional averages, but inadequate for assessment of the storm run-off pattern (NWMP 1977).

Water Resources Supply

Surface Water Supply

Except in certain relatively well-studied area like the Jordan Valley, the data base is limited and sometimes inconsistent. Sixty per cent of the catchment area is ungauged (NWMP 1977). Run-off measurements are insufficient. In most regions, annual average flows only are available, and seasonal variations are not recorded.

Total water resources in an average rainfall year are estimated in Table 7.4 below.

Table 7.4: Water Resources

	(mcm per year)	(%)
Rain volume	8,065	100
Rain volume related to Jordanian territory	6,000	75
Total annual stream flow	880	
Yarmouk River at Adaniya	400	
Total annual recharge	580	
Available annual ground water	220	
Stored ground water	12,000	

Source: NWMP 1977.

Figure 7.1: Mean Annual Rainfall, 1938—67 (millimetres)

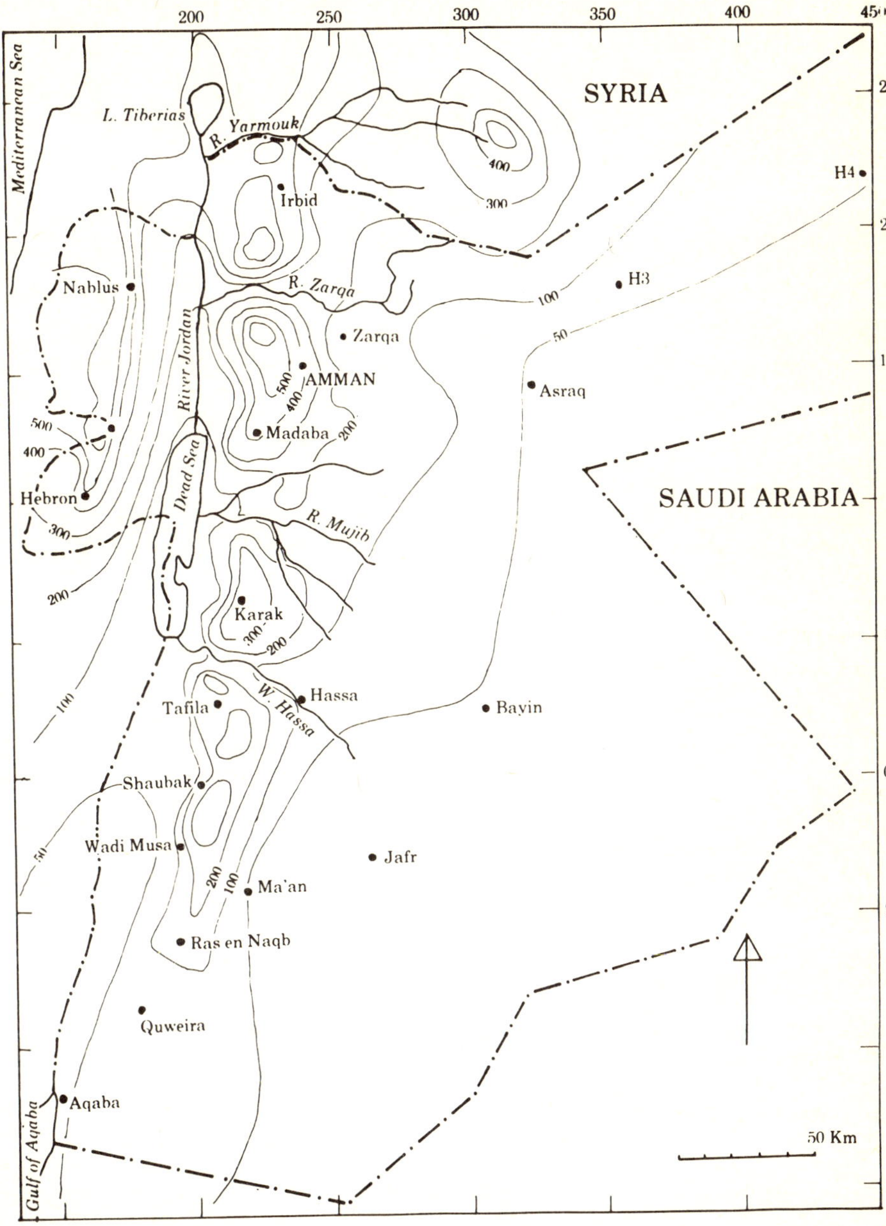

Source: Rashid 1980.

Table 7.5: Discharge of Rivers and Wadis (million cubic metres per year)

River or wadi	Average	Stream flow Minimum	Maximum
Yarmouk River	438	240	870
Zarqa River	85	29	148
Wadi al-Arab	35	14	42
Wadi Ziglab	13	7	16
Wadi Jurm	13	9	15
Wadi Yabis	6	1	11
Wadi Kufranja	12	2	25
Wadi Rajib	4	1	8
Wadi Shueib	11	4	20
Wadi Kafrein	12	3	26
Wadi Hisban	6	2	9
Total Jordan Valley	635	312	1,190
Zarqa Ma'in	20	15	25
Wadi Mujib	114	—	—
Wadi Karak	5	2.5	7.5
Wadi Hassa	30	2	127
Wadi Isal	1	1	2
Wadi Feinan	6	3	—
Wadi Khsheiba	10	7	—
Wadi Fifa	4	3	5
Wadi Araba	21	14	00
TOTAL	846		

Source: NPC 1975.

Table 7.5. gives the surface water supplies by river and wadi.

Table 7.6 give a more comprehensive estimate of surface run-off according to hydrological basins.

Surface water resources of the Jordan Valley are relatively the most studied resources, and since most of the irrigation potential lies in the Jordan Valley and the Southern Ghors, the monthly stream flows available for these two areas are given in Tables 7.7 and 7.8.

Ground Water Supply

Ground water replenishment in East Jordan is estimated at 580 mcm per year, of which 380 mcm per year emerge as dry season surface run-off. This leaves 200 mcm per year as net available water resources (NWMP 1977).

Four main aquifer systems were recognised:

1. The Disi group, which is the deepest and oldest of the four. This is a sandstone aquifer system that is exploited in the Wadi Yutum—Qaa' al-Disi—Mudawara region.
2. The Kurnub-Zarqa group above the Disi group, which is also a sandstone aquifer system. These two groups are considered as one basal system having an immense storage volume. They discharge as springs and baseflow in the escarpment of the rift valley.

Table 7.6: Assessment of Surface Water Resources (potential streamflow in million cubic metres per year)

	Area	Rainfall	Run-off coefficient		Estimated streamflow			Remarks
		(mcm)	storm (%)	total (%)	storm run-off	base flow	total	
Dry	Eastern Jordan	3,186	3.8	12.8	123	284	407	1976–80 Five-Year
Average	Valley area	4,140	6.1	14.7	250	357	607	Plan, p. 226:635
Wet		5,092	8.3	17.0	422	443	865	mcm
Dry	Dead Sea basin	938	1.8	15.5	17	128	145	1976–80 Five-Year
Average		1,406	3.6	13.6	50	141	191	Plan: 194 mcm
Wet		1,874	5.4	13.8	101	157	258	
Dry	Wadi Araba basin	253	0.4	8.7	1.2	20.9	22.1	1976–80 Five-Year
Average		451	2.1	6.9	9.4	21.6	31.0	Plan: 21 mcm
Wet		648	4.1	7.5	26.4	22.3	48 7	
Dry	Desert basins	1,381	0.3	1.7	4.7	18.6	23.3	
Average		2,068	1.4	2.4	29.2	20.1	49.3	
Wet		2,755	2.8	3.6	78.3	21.6	99.9	
Dry	TOTAL	5,758	2.5	10.3	145.9	451.5	597	
Average		8,065	4.2	10.9	338.6	539.7	878[1]	
Wet		10,370	6.1	12.3	627.7	643.9	1,272	

Note: [1] Without Jordan River flow for Lake Tiberias and West Bank.
Source: NWMP 1977.

Table 7.7: Surface Water Resources of the Jordan Valley (million cubic metres)

Data	Annual Average	Monthly Average Jan.	Feb.	Mar.	Apr.	May	June	July	Aug.	Sept.	Oct.	Nov.	Dec.
Yarmouk River at Adasiya	408.6	63.1	68.0	58.7	27.9	22.6	19.6	20.4	20.7	20.7	22.8	25.4	38.7
Wadi al-Arab	28.8	2.8	2.6	2.6	2.3	2.2	2.1	2.1	2.2	2.2	2.3	2.5	2.9
Wadi Ziglab	9.5	1.0	0.8	0.8	0.7	0.7	0.7	0.7	0.7	0.7	0.8	0.8	1.0
Wadi Jurm	11.2	1.0	0.9	0.9	0.9	0.9	0.9	0.9	1.0	0.9	1.0	0.9	1.0
Wadi Yabis	3.3	0.4	0.5	0.5	0.3	0.2	0.2	0.2	0.2	0.2	0.2	0.2	0.3
Wadi Kufranja	6.1	0.8	0.9	0.9	0.6	0.4	0.3	0.3	0.3	0.3	0.3	0.4	0.5
Wadi Rajib	7.1	0.8	0.9	1.0	0.6	0.5	0.5	0.4	0.4	0.4	0.5	0.5	0.6
Zarqa River at Deir Alla	67.3	11.5	11.5	9.1	6.0	4.0	2.8	2.6	2.6	2.6	3.3	4.1	7.1
Wadi Shueib	7.9	1.0	1.0	1.0	0.7	0.6	0.5	0.5	0.5	0.4	0.4	0.6	0.8
Wadi Kafrein	14.3	1.9	2.1	2.4	1.8	1.3	1.0	0.8	0.6	0.5	0.4	0.5	1.1
Wadi Hisban	5.0	0.5	0.6	0.6	0.5	0.4	0.4	0.4	0.3	0.3	0.3	0.3	0.4

Source: Harza and Dar al-Handasah 1978.

Table 7.8: Estimated 50% Reliable (Median) Monthly Baseflows[1] (litres per second)

Month	Wadi Mujib confluence	Ain Maghara & Wadi Ibn Hammad	Ain Sikkin	Wadi Karak upstream	Wadi Dhraa'	Wadi Isal	Wadi Numeira	Wadi Hassa at Ghor Safi	Wadi Fifa	Wadi Khanzira
Oct.	1,100	335	65	170	47	33	30	790	110	50
Nov.	1,100	315	70	178	49	33	32	850	115	53
Dec.	1,100	330	70	178	49	35	38	860	115	53
Jan.	1,100	365	70	187	52	35	44	900	120	55
Feb.	1,100	350	70	178	49	30	35	900	115	53
Mar.	1,100	330	77	178	49	30	35	860	115	53
Apr.	1,100	320	75	170	47	30	35	825	110	50
May	1,100	320	75	170	47	30	35	810	110	50
June	1,100	320	75	160	45	30	30	750	105	47
July	1,100	320	70	150	42	30	30	750	100	45
Aug.	1,100	300	65	150	42	30	30	750	100	45
Sept.	1,100	320	65	150	42	30	30	730	100	45
Year[2]	1,100	330	70	170	47	31	34	810	110	50
(mcm/yr)	34.7	10.4	2.2	5.4	1.5	0.98	1.1	25.2	3.5	1.6

Notes: [1] These statistical estimates are computed from actual records. They imply that 50 per cent of the time, the indicated flow will be equalled or exceeded with a 50 per cent probability.
[2] Average flow for the year in 1 per second.
Source: Binnie and Jouzy, 1979.

Table 7.9: Ground Water Resources of East Jordan (million cubic metres per year)

Code no.	Ground water basin	Total recharge	Discharge Total	Spring	Wells	Baseflow	Underflow	Evapo trans-piration
1	Yarmouk basin	221.2	222.9		3.5	218.0	2.0	
2	Northern escarpment to Jordan Valley	123.5	123.2		2.7	91.1	29.6	
3	Jordan Valley floor	91.2	82.1	15.4	8.1	56.0	2.0	16.0
4	Zarqa River basin	90.4	105.9	37.8	66.4	41.8	1.0	5.0
5	Central escarpment to Dead Sea	157.2	157.5	79.6	8.0	137.5	7.0	5.0
7	Escarpment to Wadi Araba	35.7	34.7		0.2	14.7	1.0	19.0
8	Red Sea basin	9.1	9.7		1.7	0.6	1.0	6.4
9	Jafr basin	67.8	71.4	2.4	5.3	2.1	61.0	3.0
10	Azraq basin	67.8	89.2	15.0	4.7	15.1	58.0	11.5
11	Sirhan basin	39.5	39.5				35.5	4.0
12	Wadi Hammad	15.0	15.5		0.5		5.0	10.0
(a)	Total East Jordan	753.1	767.6		101.1	392.9	203.1	79.9
	With Yarmouk	937.1	951.6		101.1	576.9		79.9
(b)	Without internal underflow	583.5	598.0		101.1	392.9	33.5	79.9
	With Yarmouk	767.5	782.0		101.1	576.9	33.5	79.9

Source: NWMP 1977.

Table 7.10: Present and Projected Ground Water Resources, 1975—2000

			Resources 1975		1985		2000	
Code no.	Ground water basin	Natural recharge (mcm/yr)	Available resources (mcm/yr)	Available storage (mcm)	Available resources (mcm/yr)	Available storage (mcm)	Available resources (mcm/yr)	Available storage (mcm)
1	Yarmouk basin	35.5	15.0	600.0	15.0	590.0	15.0	560.0
2	Northern Escarpment to Jordan Valley	118.1	18.0	0.0	18.0	0.0	18.0	0.0
3	Jordan Valley floor	4.1	12.0	3,550.0	12.0	3,650.0	12.0	3,800.0
4	Zarqa River basin	65.2	56.0	1,880.0	50.0	1,660.0	50.0	1,530.0
5	Central Escarpment to Dead Sea	92.5	55.5	2,900.0	55.5	2,800.0	55.5	2,610.0
6	West Bank	—	—	—	—	—	—	—
7	Escarpment to Wadi Araba	33.3	6.0	1,000.0	6.0	1,000.0	6.0	1,000.0
8	Red Sea basin	8.2	7.0	450.0	7.0	400.0	5.5	400.0
9	Jafr basin	66.0	45.0	4,300.0	45.0	4,000.0	45.0	3,700.0
10	Azraq basin	34.0	42.0	1,680.0	42.0	1,680.0	42.0	1,680.0
11	Sirhan basin	10.5	25.0	50.0	25.0	50.0	25.0	50.0
12	Wadi Hammad	15.0	5.0	0.0	5.0	0.0	5.0	0.0

Source: NWMP 1977.

3. The carbonate rock aquifer system, which is the most important aquifer system. The Amman—Wadi Sir aquifer is part of this system.
4. The shallow aquifer system which includes the basalt aquifer of Azraq—Wadi Dhuleil and the valley fill shallow aquifers of the Jordan Valley, the Southern Ghors and Wadi Araba.

The ground water system on the East Bank of Jordan is divided into 12 ground water basins, as shown in Tables 7.9 and 7.10. These tables show the estimates of ground water available in 1975, and projections for 1985 and 2000. They also show the amounts of stored ground water available for mining, and the amounts extracted from wells in 1975. Figure 7.2 shows the location of the ground water resources.

Return Flow and Re-Use of Water

Return flows occur from excess irrigation water to the ground water, and from M&I waste water to the ground water and surface streams. Recycling of water for successive uses will become very important in Jordan because the available water supply is short and limited, and the demands for the different uses are growing rapidly. Growing demands for M&I water can only be met at the expense of irrigation water, and only the recycling of M&I waste water to irrigation can keep the damage to irrigated agriculture to a minimum.

Intentional recycling of treated M&I waste water for municipal re-use is not acceptable to society. A plan to supply Amman with this kind of water from King Talal Reservoir was abandoned because the Reservoir receives, along with flood flows, the treated waste water from Amman, Zarqa and Ruseifa. Unintentional recycling of M&I waste water to M&I use does, however, occur in the Amman—Zarqa area. The Vappen Byggnas Byran (VBB) (1977) estimated that infiltration from cesspools to the aquifer in this area was 9 mcm per year in 1975, and that 6 mcm per year were recycled by pumping from ground water in the area to serve M & I and agricultural uses.

The VBB (1977) estimated the expected effluent from Amman and the area that can be irrigated by these effluents to be as given in Table 7.11.

Table 7.11: Expected Effluent from Amman, 1980-2005

Year	Effluent (mcm/yr)	Potential irrigated area (du)
1980	9	5,000
1990	24	14,000
1995	35	19,000
2000	46	26,000
2005	56	31,000

Source: VBB 1977.

Figure 7.2: Groundwater Resources of Jordan

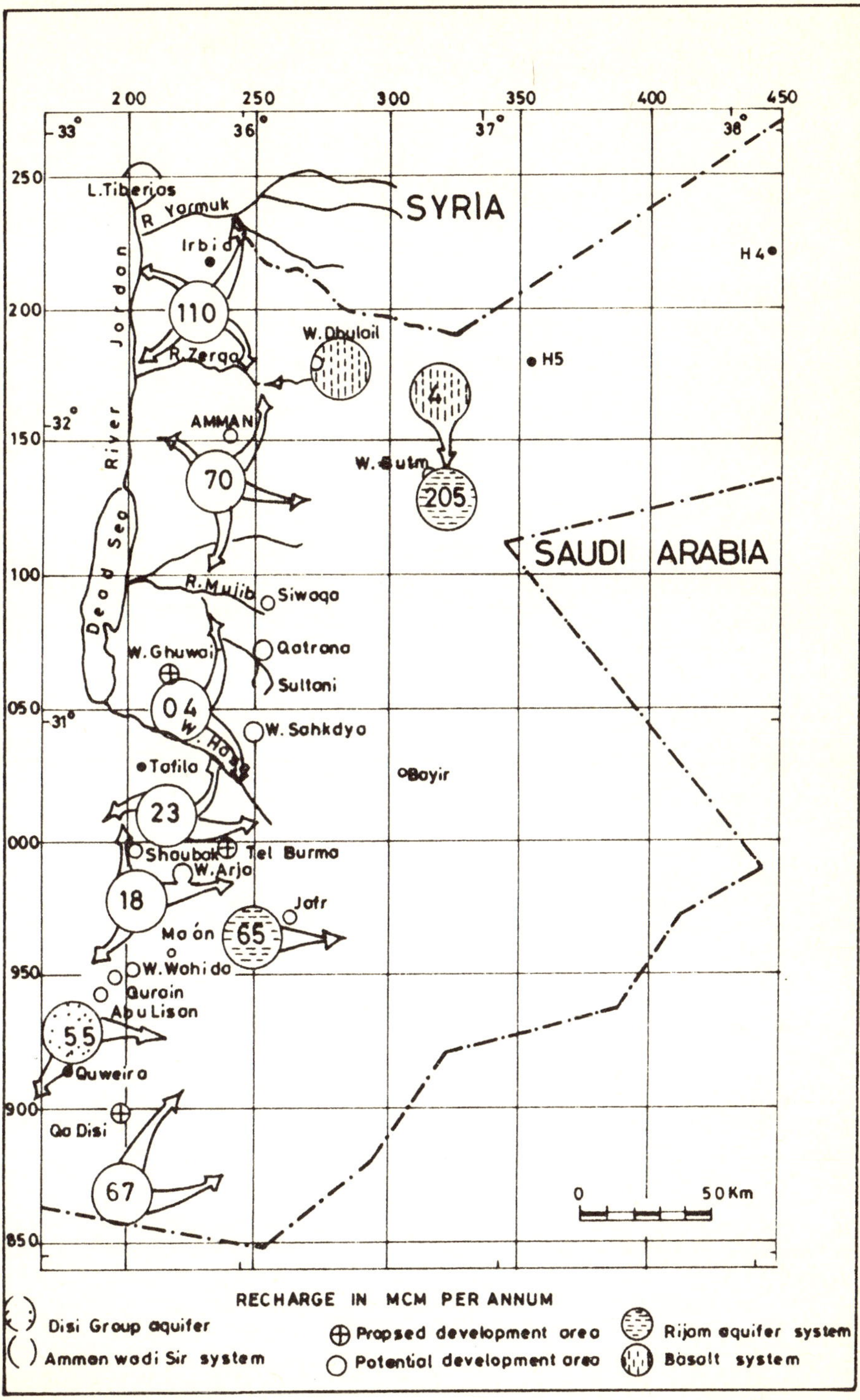

Source: Rashid 1980.

Humphreys (1978) gave the potential recycling in the northern districts of Jordan as in Table 7.12.

Table 7.12: Projection of Recycling of Water in North Jordan, 1977—2002 (million cubic metres per year)

Year	Water re-use
1977	5
1987	35
2002	75

Source: Humphreys 1978.

The NWMP (1977) gives the projections for M&I consumption and waste water effluent as in Table 7.13.

Table 7.13: Projection of M&I Consumption and Waste Water, 1975—2000 (million cubic metres per year)

Year	M&I consumption	Waste water
1975	45.68	21.72
1985	189,83	109.28
2000	299.36	189.24

Source: NWMP 1977.

In Table 7.13, the percentages of return flow from M&I are 48, 58 and 63 per cent for 1975, 1985 and 2000 respectively. Eighty per cent return flow is possible in the future and should be aimed at (Marelius 1977).

Return flow from irrigation, due to low efficiencies of water conveyance, distribution and application, either goes to drains or percolates to the ground water. No data is available on the subject to allow quantitative evaluation.

Water Quality

With the exception of very few springs, the quality of surface water resources is acceptable for agricultural purposes. This water is usually acceptable for domestic purposes after treatment for bacterial and biological contamination. Salinity of the ground water ranges from 300 ppm to over 2,500 ppm (FAO, 1974). In areas west of a line running from Mafraq to Ma'an the ground water salinity is less than 1,000 ppm (FAO 1974). The sodium hazard in both surface and ground water is low.

Overpumping and percolation of return flow to the aquifer have resulted in increases of salinity of ground water in some areas, such as around Jafr.

Percolation from cesspits reaches and pollutes ground water in some areas (NWMP 1977). Return flow from M&I uses in the Amman—Zarqa area increases salinity and causes deterioration in quality of water. Because of this situation, management of water quality becomes as important as that of quantity.

Table 7.14 gives the water quality of the Yarmouk and Zarqa Rivers. Table 7.15 gives the water analysis of the Yarmouk River and the east side wadis. Table 7.16 gives the water analysis and classification of the water resources of the Southern Ghors. Figure 7.3 gives the classification (for irrigation) of ground water in the several aquifers in Jordan.

Regional Divisions

Surface water resources in Jordan are divided into the following 10 drainage basins:

1. Jordan River basin
2. Mediterranean Sea basin (West Bank)
3. Dead Sea basin
4. Wadi Araba North basin
5. Wadi Araba South—Red Sea basin
6. Azraq basin
7. Qaa' al-Jafr basin (Central Desert)
8. Wadi Ibn Hammad basin (North-Eastern Desert)
9. Wadi Sirhan basin (South-Eastern Desert)
10. Southern Desert basin

Ground water resources are divided into 12 ground water basins, as listed previously in Table 7.10. The boundaries of ground water basins and surface water basins do not necessarily coincide.

For the present study, four main geographical divisions are suggested because of interconnections of surface water resources, ground water resources, water exportation-importation and return flows.

1. The Jordan River—North Jordan Region. This region includes the areas from the Yarmouk River and the northern border to Madaba—Zarqa Ma'in and the Dead Sea in the south. Water resources in this area are interconnected naturally through the side wadis of the Jordan Valley, and artificially through the East Ghor Main Canal, the Deir Alla—Amman domestic water supply pipeline, and the planned Yarmouk River—Irbid—Zarqa—Amman domestic water supply line. Return flow from uses in the uplands in this region goes to irrigation in the Jordan Valley.
2. The Wadi Mujib and Southern Ghors region. Water resources in this area will be connected by a planned Mujib carrier from Wadi Mujib to Wadi Fifa.
3. The upland area, which includes several irrigated areas i the plateau and the desert, using springs or wells.
4. The Wadi Araba area, which is characterised by having very limited water resources.

Demand for Water

The main uses for water in Jordan are M&I and agricultural (irrigation). Other uses such as power generation, fish production and recreation are insignificant and will continue to be of minor importance in the future.

Demand for M&I Water: The NWMP (1977) estimated water production for M&I at 45.7 mcm per year, as detailed in Table 7.17.

Table 7.14: Surface Water Quality[1]

Location	Date	Mean daily discharge (m^3/sec)	Specific conductance (mhos/cm)	Total dissolved solids (mg/l)	Chloride (mg/l)	Sulphate (mg/l)	Sodium absorption ratio	Boron (ppm)	pH	Temperature (°C)
Yarmouk River at Adasiya	21 Apr. 71	—	—	320	44.3	101	1.60	—	7.4	—
	27 Dec. 71	12.20	—	349	61.0	65.0	1.50	—	8.2	15
	14 Jan. 72	3.82	675	432	75.2	105	2.01	—	8.4	17
	31 Jan. 72	6.20	—	390	65.9	92.2	1.64	—	8.3	13
	1 Mar. 73	4.35	675	432	85.4	77.7	1.80	—	8.0	—
Yarmouk River at Maqaren	14 Mar. 71	5.60	630	402	60.3	86.4	1.51	—	7.6	—
	8 June 71	4.80	740	474	78.3	141	2.41	—	7.0	—
	9 June 71	4.80	460	294	31.6	29.1	0.42	—	8.0	—
Zarqa River at Deir Alla	27 Dec. 71	1.26	—	730	160	250	1.95	—	7.6	—
Zarqa River at Jerash	12 Mar. 73	1.12	1,080	691	145	223	1.87	—	7.8	—
	Nov. 76	—	—	922	—	—	—	0.38	—	—
Zarqa River at New Jerash Bridge	2 Oct. 70.	0.92	1,300	832	205	223	2.53	—	—	—
	2 Dec. 70	1.52	1,150	736	157	194	1.75	—	—	—
	13 May 71	0.44	1,020	653	123	187	2.50	—	—	—
	9 June 71	0.32	1,120	717	148	204	1.88	—	—	—

Note: [1] Unpublished records from NRA.
Source: Harza and Dar al-Handasah 1978.

Table 7.15: Results of Water Examination, Yarmouk River and East Side Wadis

		Wadi al-Arab		Wadi Ziglab		Wadi Jurm		Wadi Kufranja		Wadi Shueib		Wadi Hisban	
Constituent	Unit	1.64	3.61	11.63	7.63	9.56	10.60	3.69	7.60	3.64	12.73	9.56	8.60
Electric conductivity	(mmhos/cm)	0.350	0.910	0.380	0.760	0.720	0.880	0.470	0.720	0.390	0.680	0.800	1.390
Total dissolved solids	(mg/l)	224	582	243	486	446	363	301	461	250	435	508	890
Calcium (asCa)	(mg/l)	22.0	80.0	46.0	62.0	40.0	98.0	54.0	46.0	44.0	64.0	62.0	80.0
Magnesium (asMg)	(mg/l)	10.9	37.7	12.2	37.7	60.8	60.8	17.0	42.6	12.4	26.8	34.0	48.6
Sodium (asNa)	(mg/l)	34.5	48.1	16.5	32.9	39.1	39.1	14.7	44.9	13.3	28.8	70.6	143.2
Potassium (asK)	(mg/l)	—	—	—	—	—	—	—	—	—	4.7	—	—
Chloride (asCl)	(mg/l)	56.8	102.9	16.0	85.2	71.0	74.6	39.1	81.7	32.0	65.6	127.8	234
Sulphate (asSO_4)	(mg/l)	57.5	47.5	7.2	42.2	92.2	57.6	23.5	88.8	—	38.4	92.1	170.4
Carbonate (asCO_3)	(mg/l)	12.0	—	4.5	6.0	9.0	3.0	0	3.0	3.0	3.0	6.0	1.5
Bicarbonate (asHCO_3)	(mg/l)	18.3	323.3	158.6	259.3	244.0	378.2	228.1	213.5	29.3	192.0	201.3	247.0
Total cations		3.50	9.19	3.80	7.96	8.70	9.60	4.74	7.75	3.68	6.77	8.97	14.25
Sodium cations	(%)	42.8	22.7	13.16	19.60	20.00	17.71	13.50	25.20	15.76	18.46	34.00	43.90
Sodium absorption ratio		1.30	1.20	0.50	0.70	0.80	0.70	0.40	1.10	0.40	0.70	1.80	3.10
pH		8.8	7.9	8.3	8.2	7.9	8.2	7.9	8.1	8.0	8.1	8.0	9.0

Table 7.15 continued

Constituent	Unit	River Yarmouk at Maqaren		River Yarmouk at Adasiya		Wadi Wala		Wadi Mujib		Wadi Hassa Tannour		Wadi Hassa Ghor Safi	
		11.68	11.69	3.67	11.64	3.64	6.64	4.65	6.65	12.73	4.68	10.67	3.76
Electric conductivity	(mmhos/cm)	0.270	0.840	0.235	1.670	0.310	1.180	0.930	1.480	1.500	1.800	0.660	0.920
Total dissolved solids	(mg/l)	173	537	150	1,096	198	755	595	947	960	1,152	422	588
Calcium (asCa)	(mg/l)	18.8	70.4	33.8	54.0	44.0	90.0	83.2	112.0	8.40	95.0	48.6	79.2
Magnesium (asMg)	(mg/l)	10.9	20.4	2.4	53.5	7.3	47.4	36.1	58.4	73.0	86.2	25.9	28.9
Sodium (asNa)	(mg/l)	13.0	71.3	9.2	225.4	7.8	58.0	80.5	115.0	128.8	139.2	46.0	62.1
Potassium (asK)	(mg/l)	5.5	9.0	5.5	—	—	—	—	—	4.7	6.2	7.4	7.0
Chloride (asCl)	(mg/l)	16.7	77.4	14.9	298.2	10.7	142.0	113.6	213.0	203.4	265.5	88.4	101.9
Sulphate (asSO_2)	(mg/l)	17.3	85.4	21.6	189.6	15.8	110.4	—	—	187.2	235.2	64.8	160.3
Carbonate (asHCO_3)	(mg/l)	0	0	0	9.0	3.0	0	0	0	14.4	0	0	0
Bicarbonate (asHCO_3)	(mg/l)	109.2	272.7	101.9	259.3	152.5	341.6	274.5	289.8	363.6	344.0	163.5	184.8
Total cations		2.58	8.53	2.43	16.90	3.14	11.90	10.63	15.40	15.92	18.05	6.75	9.12
Sodium cations	(%)	23.26	36.34	16.46	57.99	13.31	29.41	32.90	32.40	36.00	33.52	29.63	29.60
Sodium absorption ratio		0.66	1.92	0.40	5.10	0.35	1.70	1.85	2.20	2.50	2.48	1.32	1.50
pH		7.75	7.20	8.05	8.20	8.1	7.8	8.21	7.73	8.30	7.40	7.90	7.66

Source: NWMP 1977.

Table 7.16 Water Analysis of Southern Ghors Waters

	Unit	Wadi Mujib	Wadi Hidan	Ain Maghara	Wadi Ibn Hammad	Ain Sikkin	Wadi Karak	Wadi Dharaa'	Wadi 'Isal	Wadi Hassa	Wadi Fifa	Wadi Khanzira
Electric conductivity	(mmhos/cm)	1,700.00	1,480.00	1,390.00	865.00	1,080.00	940.00	800.00	1,010.00	650.00	720.00	1,000.00
Total dissolved solids	(ppm)	1,088.00	947.00	890.00	554.00	691.00	602.00	512.00	646.00	416.00	460.00	640.00
pH		7.8	8.0	7.3	8.55	7.5	7.93	8.2	8.25	8.4	8.1	8.2
Ca	(me/l)	4.3	3.7	3.35	1.52	3.55	2.8	2.8	2.76	2.04	2.3	3.36
Mg	(me/l)	2.99	2.46	3.89	1.82	2.78	1.56	3.09	3.09	1.61	2.44	2.94
Na	(me/l)	8.5	7.3	6.10	4.65	4.15	4.0	3.40	4.15	2.5	2.3	3.40
K	(me/l)	0.39	0.39	0.25	0.24	0.22	0.28	0.15	0.12	0.12	0.12	0.14
SO_4	(me/l)	9.74	8.52	7.4	3.85	4.28	3.75	3.10	4.04	2.65	2.15	3.92
CO_3	(me/l)	3.27	2.54	2.65	2.0	2.6	2.31	2.10	3.15	1.63	2.06	2.45
HCO_3	(me/l)	0.0	0.17	0.0	0.35	0.0	0.0	0.30	0.34	0.22	0.12	0.21
Na%	(me/l)	3.05	2.61	3.55	1.97	3.84	3.19	2.32	2.59	1.82	2.7	3.22
Sodium absorption ratio	(%)	52.5	52.7	44.9	56.5	38.8	43.5	43.0	40.5	47.4	32.1	34.5
Irrigation class		4.45	4.15	3.2	3.6	2.34	2.55	2.30	2.42	1.85	1.5	1.78
		C_3S_1	C_3S_1	C_3S_1	C_3S_1	C_3S_1	C_3S_1	C_3S_1	C_3S_1	C_2S_1	C_3S_1	C_3S_1

Source: Binnie and Jouzy 1979.

Figure 7.3: Water Quality of the Ground Water

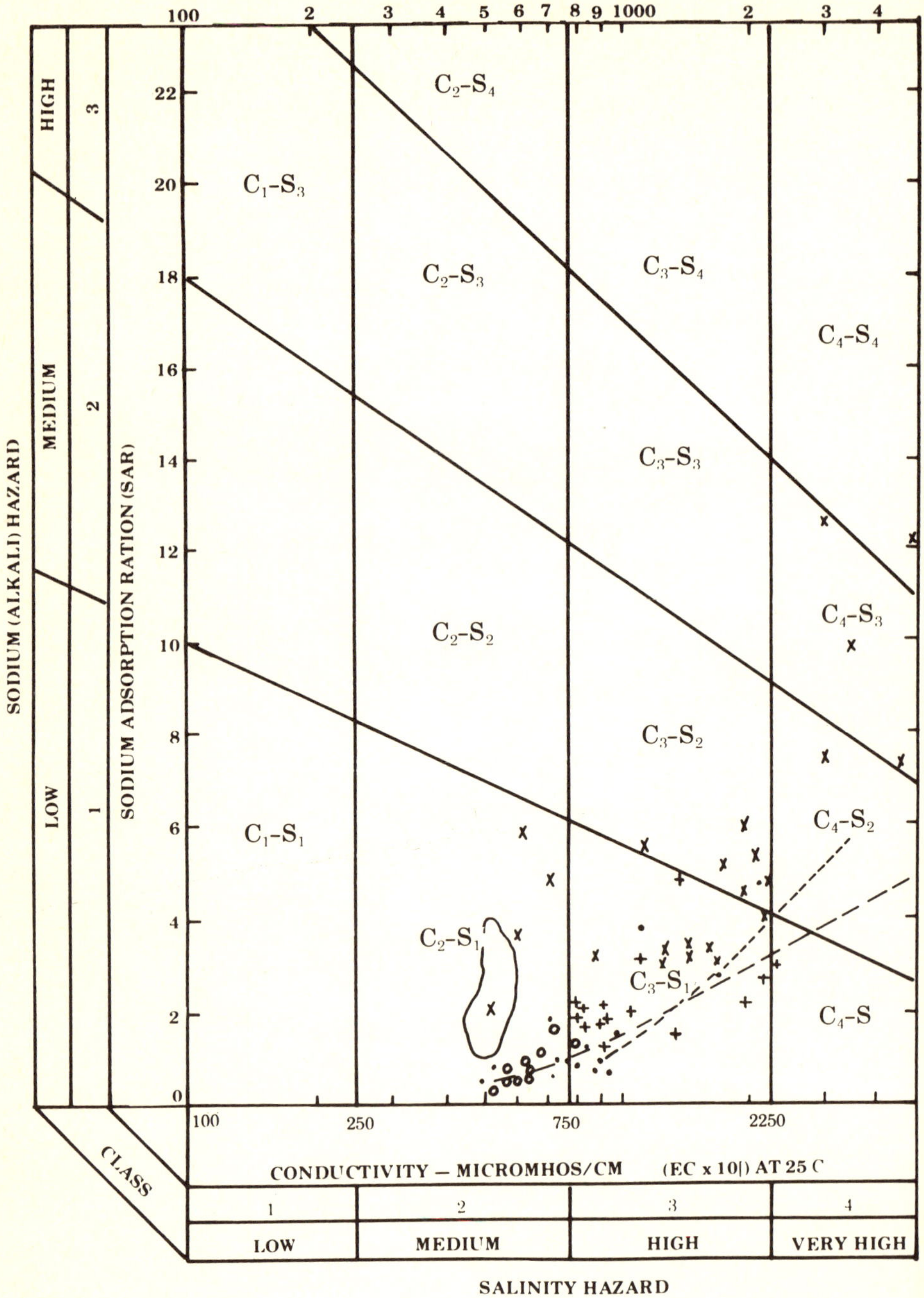

- • Kurnut — Zarqa groups
- o Hummar formation
- \+ Jafr region. Rijam formation
- x Azraq region, shallow aquifers
- ◡ Wadi Dhuleil
- — — Trend line of Amman — Wadi Sir aquifer
- - - - Trend line of Jordan Valley. Rama — Kufrein region

Table 7.17: Water Production for M&I, 1975 (million cubic metres per year)

User	
Cities & towns	22.9
Small communities	5.8
Various other consumers	1.2
Industry	10.0
For population not connected to pipe systems	5.3
TOTAL	45.7

Source: NWMP 1977.

This represents only 55 per cent of demand if enough water had been available. Average *per capita* consumption, according to these figures, was 51 l *per capita* per day l/c/d gross and 40 l *per capita* per day net. The 1975 demand would be around 82 mcm per year (NWMP 1977).

The water supply to the Amman—Zarqa area alone in 1978 was estimated at 30 mcm per year or 50 per cent of the actual demand (Humphreys 1978), for the North Jordan area (north of Madaba), in which about 80 per cent of the population of Jordan reside, as shown in Table 7.18.

Table 7.18: Projections of M&I Water Demand, 1977-2002 (million cubic metres per year)

	1977	1987	2002
Demand for M&I for North Jordan	80	146	289

Source: Humphreys 1978.

Water demand projections for Amman alone were given by the VBB (1977) and are shown in Table 7.19. A more recent estimate of future demands for Irbid, Zarqa and Amman was given by Humphreys in 1983 in their ongoing study of a national M&I carrier pipeline. These figures are shown in Table 7.20. M&I projections are given in Tables 7.21-7.23 (taken from NWMP 1977).

Table 7.19: Municipal Water Demand for Amman, 1980-2005

Year	Population (000)	Average water demand (1/c/day)	Total (mcm/year)
1980	780	50	25
1985	950	60	33
1990	1,129	70	44
1995	1,320	80	59
2000	1,550	90	75
2005	1,800	100	94

Source: VBB 1977.

Table 7.20: M&I Demand for Irbid, Zarqa and Amman, 1982-2005 (million cubic metres per year)

		M&I demand		
		1982	1990	2005
Amman:	Municipal	39.52	68.28	133.01
	Industrial	1.3	7.31	17.06
	TOTAL	40.65	75.59	153.07
Zarqa:	Municipal	11.14	17.54	34.17
	Industrial	7.30	10.08	11.77
	TOTAL	18.44	27.62	45.94
Irbid:	Municipal	14.36	22.98	42.71
	Industrial	0.0	0.50	0.50
	TOTAL	14.36	23.48	43.21
GRAND TOTAL		73.45	126.69	242.22

Source: Ongoing study by Humphreys 1983.

Table 7.22: Daily Municipal Unit Water Demand, 1975-2000 (litres *per capita* per day)

		Years		
Consumer groups	Demand	1975	1985	2000
Cities of Amman Zarqa Irbid	Domestic use	85.0	110.0	120.0
	Commerce & small industries	11.9	15.4	16.8
	Sub-total	96.9	125.4	136.8
	Water losses	33.9	31.4	20.5
	TOTAL	131.0	157.0	158.0
Town of Aqaba	Domestic use	120.0	150.0	170.0
	Commerce & small industries	16.8	21.0	23.8
	Sub-total	136.8	171.0	193.8
	Water losses	41.0	34.2	29.1
	TOTAL	178.0	205.0	223.0
Towns and commu-nitites with a population of 3,000–50,000 inhabitants	Domestic use	65.0	90.0	110.0
	Commerce & small industries	6.5	9.0	11.0
	Sub-total	71.5	99.0	121.0
	Water losses	21.5	19.8	18.1
	TOTAL	93.0	119.0	139.0
Rural communi-ties with less than 3,000 inhabitants	Domestic use	65.0	85.0	100.0
	Commerce & small industries	3.3	4.3	5.0
	Sub-total	68.3	89.3	105.0
	Water losses	20.5	17.8	15.8
	TOTAL	89.0	107.0	121.0

Source: NWMP 1977.

Table 7.21: Annual Municipal Unit Water Demands, 1975-2000

Name/Size of communities	Domestic unit water demand (1/c/day)			Demand for commerce & small industries (%)			Water losses (%)			Total annual municipal water production (m^3/c/year)		
	1975	1985	2000	1975	1985	2000	1975	1985	2000	1975	1985	2000
Cities of Amman Zarqa Irbid	85	110	120	14	14	14	35	25	15	47.8	57.3	57.7
Town of Aqaba	120	150	170	14	14	14	30	20	15	64.9	74.8	81.4
Communities with a population of 3,000-50,000 inhabitants	65	90	110	10	10	10	30	20	15	34.0	43.5	50.7
Communities with a population of less than 3,000 inhabitants	65	85	100	5	5	5	30	20	15	32.5	39.9	44.2
Consumers of unpiped water supply systems	35	35	35	0	0	0	0	0	0	13.0	13.0	13.0

Source: NWMP 1977.

Table 7.23: Municipal and Industrial Supply, Consumption and Demand, 1975-2000 (thousand cm)

			Population No x 10³	Consumption/Demand per capita and year cm	Consumption/Demand total per year	Consumption/Demand factor	Consumption/Demand max per month	Origin of water ground	Origin of water surface	Waste Water factor	Waste Water total per year	Waste Water average per month
Consumption 1975								8 + = 5				
	City/Town of	WSC supply, piped	—	—	22,920	1.3	2,480	22,920	—	0.6	13,750	1,150
	communities	WSC supply, piped	—	—	5,830	1.3	630	5,830	—	0.4	2,330	195
	other consumers	piped	—	—	1,640	1.3	180	1,640	—	0.4	660	55
	industry	piped	—	—	9,960	1.0	830	9,960	—	0.5	4,980	415
	unpiped supply		410.3	13.0	5,330	1.0	440	5,330	—	0.0	—	—
	TOTAL				45,680	1.20	4,560	45,680	—	0.47	21,720	1,815
Demand 1975	City/Town of	piped	925.8		44,530	1.4	5,195			0.6	26,720	2,230
	communities 3,000-50,000	piped	373.7	34.0	12,710	1.5	1,590			0.4	5,080	425
	communities <3,000	piped	242.2	32.5	7,870	1.6	1,050			0.4	3,150	260
	other consumers	piped	—	—	1,680	1.5	210			0.4	670	55
	industry	piped	—	—	9,960	1.0	830			0.5	4,980	415
	unpiped supply		410.3	13.0	5,330	1.0	440			0.0	—	—
	TOTAL		1,952.0		82,080		9,320			0.49	40,600	3,380

Demand 1985	City/Town of (see below)			1,588.5		91,380	1.4	10,660	0.65	59,400	4,950
	comm-	3,000–50,000	piped	561.1	43.5	24,410	1.5	3,050	0.55	13,420	1,120
	unities	< 3,000		371.1	39.0	14,470	1.6	1,930	0.55	7,960	660
	other consumers			—	—	1,680	1.5	210	0.6	1,000	80
	industry			—	—	55,000	1.0	4,580	0.5	27,500	2,300
	unpiped supply			222.3	13.0	2,890	1.0	240	0.0	—	—
	TOTAL			2,743		189,830	1.30	20,690	0.57	109,280	9,110
Demand 2000	City/Town of (see below)			2,948.4		170,740	1.4	19,920	0.7	119,520	9,960
	comm-	3,000– 50,000	piped	830.8	50.7	42,120	1.5	5,260	0.6	25,270	2,110
	unities	3,000		475.5	44.2	21,020	1.6	2,800	0.6	12,610	1,050
	other consumers			—	—	1,680	1.5	210	0.7	1,180	100
	industry			—	—	61,320	1.0	5,110	0.5	30,660	2,560
	unpiped supply			190.6	13.0	2,480	1.0	210	0.0	—	—
				4,445.3		299,360		33,510		189,240	15,780

Demand City/Town	1975 m³/c per year	1975 inhabitant (x 10³)	1975 total (m³x10³)	1985 m³/c per year	1985 inhabitant (x 10³)	1985 total (m³x10³)	2000 m³/c per year	2000 inhabitant (x 10³)	2000 total (m³x10³)
Amman	47.8	606.6	28,995.5	57.3	1,040.8	59,637.8	57.7	2,014.2	116,219.3
Zarqa	47.8	184.0	8,795.2	57.3	333.2	19,092.4	57.7	558.3	32,213.9
Irbid	47.8	119.1	5,693.0	57.3	194.3	11,133.4	57.7	349.9	20,189.2
Zarqa	64.9	16.1	1,045.0	74.8	20.2	1,511.0	81.4	26.0	2,116.4
TOTAL		925.8	44,528.7		1,588.5	91,374.6		2,948.4	170,738.8

Source: NWMP, 1977.

Demand for Irrigation Water: Water is the limiting factor in crop production in Jordan. There is more land that can be beneficially irrigated than there is water to irrigate it. Table 7.24 shows water demands for irrigation of crops until the year 2000.

Table 7.24: Water Demands of Jordan, 1975-200 (million cubic metres per year)

Year	Demand for M&I	Demand for irrigation	Total demand
1975	46	405	451
1985	190	730	920
2000	300	800	1,100

Source: NWMP 1977.

Demands for irrigation represent water requirements of anticipated irrigation projects under assumed cropping patterns. These demands are, of course, constrained by a shortage of water in most areas of Jordan. Water requirements for individual crops and individual projects will be discussed later under the irrigation sub-system.

Water Loss and Efficiency of Water Use

Efficiency of water use in general is low in both the M&I sub-system and the irrigation sub-system. About 40 per cent of M&I water in Amman is lost in the distribution network. Table 7.21 above gave estimates of the percentage water loss in M&I water for the years 1975, 1985 and 2000 as 30-35 per cent, 20-25 per cent and 15 per cent, respectively. The Amman Water and Sewage Authority (AWSA) (1978) estimated losses in the city of Amman for 1975-76 as shown in Table 7.25

Irrigation efficiency is not routinely measured or evaluated in irrigation projects and there is very little data published on this subject. Dar Al-Handasah and NEDECO (1969) reported the results of a single field monitoring of three crops during May 1967 as shown in Table 7.26.

Water Allocation: The allocation of water to different sectors is made by government decree or results as a by-product of development activities. Allocation within the M&I sector is made by the government agencies responsible for M&I water. Allocation within the irrigation sector is made through water rights systems in the area not included in public irrigation projects, and by ordinances in public irrigation projects.

Table 7.25: Water Losses in the City of Amman, 1975-76

	% of total
Leakages, bursts, etc.	17
Over estimation	5
Overflowing reservoirs	1
Illegal connections	1
Inaccurate meters	24
Water consumed	76
Water charged	52

Source: AWSA 1978.

Table 7.26: Irrigation Efficiency in the Jordan Valley, 1967

Crop	Range of observations %	Average of observations (%)
Sweet peppers	45-69	59
Bananas	30-52	44
Citrus	1 observation	52

Source: Dar al-Handasah and NEDECO 1969.

A higher priority is given to M&I use by planners in Jordan. A well-studied water allocation policy does not exist (NWMP 1977). The system at present is to meet the requirements of the M&I users first, and then to allocate the remaining water to irrigation.

Recycled water (return flow from M&I use or from irrigation) is used mostly for irrigation. Some industries use recycled water but domestic users do not (although some return of unintentional recycling may occur).

Development of the Water Resources System

The M&I Sector

The M&I water for North Jordan (Madaba and north) represents about 80 per cent of the M&I water sub-system. The supplies for the Amman—Zarqa area in 1975 were as given in Table 7.27. Humphreys (1978) gave approximately similar figures for the situation in Amman—Zarqa in 1975. The northern district (Irbid and the northern villages) in 1975 was fed from sources and in quantities as shown in Table 7.28.

Table 7.27: Municipal and Industrial Water Usage in Amman-Zarqa, 1975 (million cubic metres per year)

Amman basin User	(mcm/year)	Zarqa basin User	(mcm/year)
AWSA	15.7	Zarqa municipality	3.9
Water Supply Corporation	1.7	Factories	4.0
Factories	0.3	Private domestic water	2.4
Private domestic water	0.3	Irrigation	10.0
Irrigation	2.0		
TOTAL	19.7	TOTAL	20.3

Source: VBB 1977.

Table 7.28: Municipal and Industrial Water Usage in Irbid and North District, 1975 (million cubic metres per year)

Water source	(mcm/year)
Sama Sdoud Wells	2
Dhuleil Wells	4
Minor local wells	1
Irbid—Ramtha	
Azraq Springs	2
TOTAL	9

Source: Humphreys 1978.

The plans for Amman, as given in the NWMP (1977) included the development of local ground water up to a maximum of 25 mcm per year and the importation of water from King Talal Dam (existing) and upper Wadi Wala Dam (planned). This plan has changed since then. Importation from King Talal Reservoir has been abandoned. A pipeline has been constructed to transport 10 mcm per year from Azraq to Amman, the Wadi Wala Dam has not been started, and 4 mcm per year are transported from Qastal and Swaqa to the Madaba area. Plans to transport Yarmouk River water to Amman have been completed, and construction has begun.

Major On-Going M&I Water Projects

Deir Alla—Amman Domestic Water Supply Systems: A project is under construction at present, which, when completed early in 1985, will supply Amman with 45 mcm per year from the East Ghor Main Canal which draws its water from the Yarmouk River.

Wadi al-Arab-Irbid Domestic Water Supply System: Construction was started in 1983 on a project which will transport 20 mcm per year of water from the Wadi al-Arab well field to Irbid. Completion is scheduled for 1985.

Qaa' al-Disi—Aqaba Domestic Water Supply Project: A project is under construction at present to pump 18 mcm per year of water from Qaa' al-Disi Wells to Aqaba.

Future Plans for M&I Water

Despite the new Deir Alla—Amman project, the Amman—Zarqa area will have a shortage of M&I water starting in 1987. The future plan calls for the construction of another pipeline from Deir Alla to Amman similar to the one being built at present. A further 40 mcm per year is planned from the Mukheiba well field to Irbid and Amman. Two studies are in progress now to formulate new projects to meet the increasing demands of Amman, Zarqa and Irbid. One study will evaluate the feasibility of transporting 80 mcm per year of water from the Euphrates River as a first stage. The second study is for a national M&I carrier pipeline between Irbid—Zarqa—Amman to complete the loop with the East Ghor Canal—Deir Alla—Amman branch.

Ground Water Development

Development of ground water for domestic use and irrigation by the public and private sectors followed the public exploration programme so rapidly that many ground water basins were overpumped and then restricted in a short time.

In 1973 there were 737 production wells in East Jordan, 31 of them being part of public irrigation projects (National Planning Council (NPC) 1973). These projects and the number of wells in each are given in Table 7.29.

In 1975 there were 859 wells in East Jordan producing 65 mcm per year. Only 48 of these were in public irrigation projects. Table 7.30 gives the area and number of wells, as reported by the NPC in the 1975-80 Development Plan. Between 1976 and 1980, 170 more wells were added to the total, bringing an additional 800 du under well irrigation in public projects. Major areas irrigated by wells in the private sector are given in Table 7.31.

Table 7.29: Wells in Public Irrigation Projects, 1975

Project	Number of wells	Production (m^3/hour)	Irrigated area (du)
Jafr	14	2,222	3,000
Dhuleil	7	2,142	6,000
North Qatranah	3	360	1,000
South Qatranah	3	360	700
Tell Burma	2	360	1,000
Wadi Abyad	1	170	500
Ghuweir	1	170	500
TOTAL	31	5,784	12,700

Source: NPC 1975.

Table 7.30: Production Wells on the East Bank, 1975

Area	Wells drilled up to 1975	Total production capacity (m^3/hour)
Jordan Valley	282	22,200
Amman—Zarqa	166	11,350
Ma'an	84	6,100
Wadi Dhuleil	78	10,300
Karak	76	5,600
Ajloun	66	2,950
Azraq	42	2,600
Jafr	22	2,650
Salt	21	1,100
H-4	14	200
Wadi Araba	8	150
TOTAL	859	65,200

Source: NPC 1975.

Table 7.31: Major Private Areas Irrigated by Wells (dunums)

Region	Irrigated area
Sama Sdoud	2,800
Qasr al-Hallabat/Dhuleil	22,500
Baqa'a	770
Mafraq	590
Qastal	600
Azraq	3,575

Source: NWMP 1977.

Development of Dams and Reservoirs
A number of small to medium dams have been constructed for water conservation and irrigation purposes. Table 7.32 gives a brief listing of them.

Table 7.32: Completed Dams in Jordan

Dam	Location	Height (m)	Type	Capacity (mcm)	Date completed
King Talal	R. Zarqa	92	Earth	56.0	1978
Shurahbeel	R. Ziglab	30	Earth	4.2	1967
Shueib	R. Shueib	?	Earth	2.5	1969
Kafrein	R. Kafrein	30	Earth	4.5	1969
Sama Sdoud	Desert	8	Rock	1.5	1965
Al-Ghadir al-Abyad	Desert	?	Earth	0.85	1967
Buweida	Desert	9	Concrete	0.25	?
Qatranah	Desert	?	Earth	4.5	1963
Sultaneh	Desert	?	Earth	1.25	1962
Um al-Jimal	Desert	?	Earth	1.8	?

Source: NPC 1975.

Two dams are under construction at present. These are:

1. Wadi al-Arab Dam, which is an earth-rock fill dam 80.5 m high and which will store 20 mcm.
2. King Talal Dam, rising by an additional 15 m. The new capacity of the reservoir will be 86 mcm.

The following dams are planned:

1. Maqaren Dam on the Yarmouk River. This dam will be earth-rock fill, 148 m high and with a capacity of 470 mcm.
2. Tannour Dam on Wadi Hassa, which will be 58 m high with a gross capacity of 12 mcm, an active capacity of 10.95 mcm and an annual release of 3.0 mcm per year.
3. Rumeil Dam on Wadi Wala, with a gross capacity of 50 mcm and an annual release of 7.3 mcm per year.
4. Nukheila Dam on Wadi Mujib, with a gross capacity of 8.5 mcm per year and an annual release of 6.3 mcm per year.
5. Wadi Lihfi Dam, in Dhuleil area.
6. Wadi Mallaha Dam, in the Jordan Valley, to store uncontrolled winter flows from the Yarmouk and Zarqa Rivers.

The Irrigation Sub-System

The irrigation sub-system is the major water user in Jordan. In most of Jordan, rainfall is inadequate for crop production, and irrigation is needed to make it possible.

Brief History

Irrigation on a limited scale around springs was probably practised in Jordan before recorded history. Archaeological findings in the Jordan Valley and in other parts of the country confirm the existence of irrigated farming communities as old as this.

Public irrigation projects in modern times started in the Jordan Valley in the late 1940s and early 1950s when the Government built small irrigations schemes on some of the side wadis. These schemes consisted of concrete diversion weirs on the wadis with gates, spillways, short lengths of concrete-lined canals, division boxes, and earth canals to divert and distribute the base flow of the wadis to their command areas. These command areas were always larger than what could be irrigated by the base flows, so partial irrigation was the rule. These schemes provided water shares for 226,000 du and were built on the following side wadis on the East Bank:

1. Wadi al-Arab
2. Wadi Ziglab
3. Wadi Jurm
4. Wadi Yabis
5. Wadi Kufranja
6. Zarqa River
7. Wadi Shueib
8. Wadi Kafrein
9. Wadi Hisban

In 1952 Jordan employed Michael Baker and Harza Engineering Companies to study the feasibility of irrigating the Jordan Valley using the Jordan River, the

Yarmouk River and the side wadis. The Baker-Harza plan called for the following:

1. A storage dam and reservoir on the Yarmouk River at Maqaren and/or Wadi Khalid
2. A concrete diversion dam on the Yarmouk River at Adasiya
3. An East Ghor Main Canal 113 km long from Adasiya to the Dead Sea, with a capacity of 20 m^3 per second at the intake
4. A West Ghor Canal 47 km long for the West Bank
5. A syphon across the Jordan Valley from the East Ghor Main Canal to the West Ghor Canal near Deir Alla
6. Four large hydroelectric power plants on the Yarmouk River
7. A distribution system of open canals to serve 514,000 du of irrigable land, 104,000 du of which were to receive water by pumping

In 1957 Jordan decided to implement the first stage of this development plan without storage on the Yarmouk River and using only the base flows of the Yarmouk and the side wadis. The East Ghor Canal Project was constructed between 1958 and 1963. The Project brought under full irrigation 120,000 du distributed into farm units of about 30 du each. Each farm unit was served by a concrete-lined lateral canal and had a measuring farm turnout. The Main Canal was extended by an additional 8 km, and 14,000 du were added to the irrigated area between 1966 and 1968. During the same period the Main Canal capacity was increased to its full design capacity of 20 m^3 per second by raising its lined side slopes.

Public development of ground water for irrigation in the uplands and the desert started in 1963/64 with development of the Jafr Irrigation Pilot Project, followed by the Wadi Dhuleil Pilot Irrigation Project in 1966-70.

Private irrigation with spring water has continued throughout Jordan's history. Private ground water development gained momentum in the 1950s and 1960s.

Irrigated Areas

Table 7.33 gives the extent of irrigated areas grouped according to water source, crops grown, surface drainage basins and private or public sector. Table 7.34 gives public irrigation projects in 1975 and the planned areas for 1985.

Irrigation Development in the Jordan Valley

Completed Projects: Development in the Jordan Valley was interrupted by the 1967 War and it was not started again until 1972 when the Jordan Valley Commission (JVC) started an integrated socio-economic development plan in the Valley. Between 1974 and 1978 four irrigation projects were completed, adding 84,000 du of irrigated land and converting 10,000 du from irrigation by open canals to irrigation by a pressurised pipe system. The new projects utilised this pipe system to enable farmers to use more efficient sprinkler and drip irrigation methods. The four irrigation projects are listed in Table 7.35.

On-Going Projects: Construction of the Wadi al-Arab irregation scheme was started in October 1983. This project will add 8,300 du of new land and will convert 4,200 du from open canals to pressurised pipes.

Future Projects: In the 1981-1985 Development Plan, the following projects are scheduled:

1. Extension of the East Ghor Main Canal by 14 km to irrigate 60,000 du by

pressurised pipe networks. The area will be divided into farm units of about 40 du each.

2. Conversion of about 43,000 du from open distributary canals to pressurised pipe networks. These are located between Wadi Rajid and the end of the 8 km extension.

Development of the remaining irrigable areas in the Jordan Valley will depend on water storage on the Yarmouk and the Wadi Mallaha Dam. If water is made available, an additional 50,000 du can be irrigated. Conversion of the remaining 77,000 du of land served by open distributary canals will also be implemented in the future.

When the development of the East Jordan Valley is completed, the system will be a complex one comprising the following:

1.	360,000 du of irrigated land	
2.	Reservoirs and reservoir discharge controls	
	Maqaren on the Yarmouk River	1
	King Talal on the Zarqa River	1
	Balancing Reservoir on the Zarqa River	1
	Wadi al-Arab Reservoir	1
	Wadi Ziglab Reservoir	1
	Wadi Shueib Reservoir	1
	Wadi Kafrein Reservoir	1
	Wadi Mallaha Reservoir	1
3.	Diversion dam intakes	5
4.	East Ghor Main Canal intake	1
5.	Main Canal	
	Length	*110 km*
	Check gates	*34*
	Sluice gates	*2*
	Spillways	*3*
	Wasteways	*11*
	Lateral turnouts	*22*
6.	Pipe distribution network	*1,857 km*
7.	Main isolation valves	*420*
8.	Gravity carrier pipelines	*17*
9.	Low-head pumping stations	*16 with 48 pumps*
10.	Medium-head pumping stations	*9 with 27 pumps*
11.	High-head pumping stations	*5 with 15 pumps*
	Total irrigation pumping stations	*30 with 90 pumps*
12.	Wadi al-Arab M&I pumping station	*1*
13.	Amman M&I pumping station at Deir Alla	*1*
14.	Farm units with controls and water meters	*9,277*

Figures 7.4 and 7.5 show schematically the configuration of the Jordan Valley—North Jordan water resources system when the present plans are implemented.

Irrigation Development in the Southern Ghors

Along the eastern and southern coasts of the Dead Sea, old irrigation diversion schemes irrigated about 25,000 du (by partial irrigation). Modern irrigation by drip method is presently being introduced.

On-Going Projects: The first stage of irrigation development in the Southern Ghors was started in October 1982 and will finish in 1984. This project will utilise the following water resources:

Table 7.33: Irrigated Areas, 1975

	(du)	(%)
According to water sources:		
Rivers	214,000	64
Wells	72,000	21
Springs	50,000	15
	336,000	
According to crops grown:		
Vegetables	185,600	55
Trees	86,400	26
Field crops	64,000	19
	336,000	
According to surface drainage basins:		
Jordan River basin	253,700	75.5
Dead Sea basin	43,800	13.0
Wadi Araba basin North	23,400	7.0
Wadi Araba basin South & Red Sea	4,900	1.5
Azraq basin	4,400	1.3
Jafr basin	5,800	1.7
	336,000	
According to private or public schemes:		
Private schemes	190,155	56.6
Public project areas	145,845	43.4
	336,000	
Existing public project areas:		
Jordan Valley Project areas	130,445	
Wadi Dhuleil	6,300	
Qatranah (North & South)	1,440	
Wadi Abyad	560	
Tell Burma	500	
Wadi Arja	1,125	
Jafr	1,050	
Ghuweir	200	
Qaa'al-Disi	4,225	
	145,845	

Source: NWMP 1977.

1. Wadi Ibn Hammad
2. Ain Maghara
3. Ain Sikkin
4. Wadi Karak
5. Wadi Dhraa'
6. Wadi Numeira
7. Wadi Hassa
8. Wadi Fifa
9. Wadi Khanzira

An area of 46,000 du will be irrigated by pressurised pipe networks and drip irrigation. The land will be divided into farm units of around 30 du each. Each

...ual Situation and Projected Situation for 1985) (dunums)

	Public projects (1975)	Private projects to be transferred to public projects (1975)	New physical developments (1975-85)	Expected total of public projects (1985)
Existing East Ghor Canal	117,000	—	—	117,000
North-East Ghor Irrigation & Rural Development Project	10,000	8,880	8,720	27,600
Zarqa Triangle Irrigation Project	—	11,525	3,675	15,200
East Ghor Canal Extension Project	3,445	3,150	28,405	35,000
Kafrein—Hisban Project	—	9,340	6,160	15,500
Stage II of Jordan Valley Irrigation Project	—	12,830	136,870	149,700
Mujib and Southern Ghors Irrigation Project	—	31,745	17,825	49,570
Siwaqa	—	—	—	—
Qatranah	1,440	—	6,500	7,940
Sultaneh	—	—	7,000	7,000
Wadi Abyad	560	—	4,440	5,000
Hassa	—	—	—	—
Tell Burma	500	—	3,500	4,000
Wadi Arja	1,125	—	—	1,125
Jafr	1,050	—	1,400	2,450
Qaa' al-Disi	4,225	—	2,775	7,000
Wadi Dhuleil	6,300	—	—	6,300
Shaubak	—	—	—	—
Abu Makhjtoub	—	—	4,000	4,000
Udruh	—	—	2,100	2,100
Wuheida	—	—	1,710	1,710
Qurein	—	—	4,300	4,300
Abu Lisan	—	—	1,170	1,170
Ghuweir-Adir	200	—	6,800	6,800
TOTAL	145,845	77,470	247,350	470,465

Source: NWMP 1977.

Figure 7.4: Plan Showing Trunk Conveyance System for Alternative Schemes

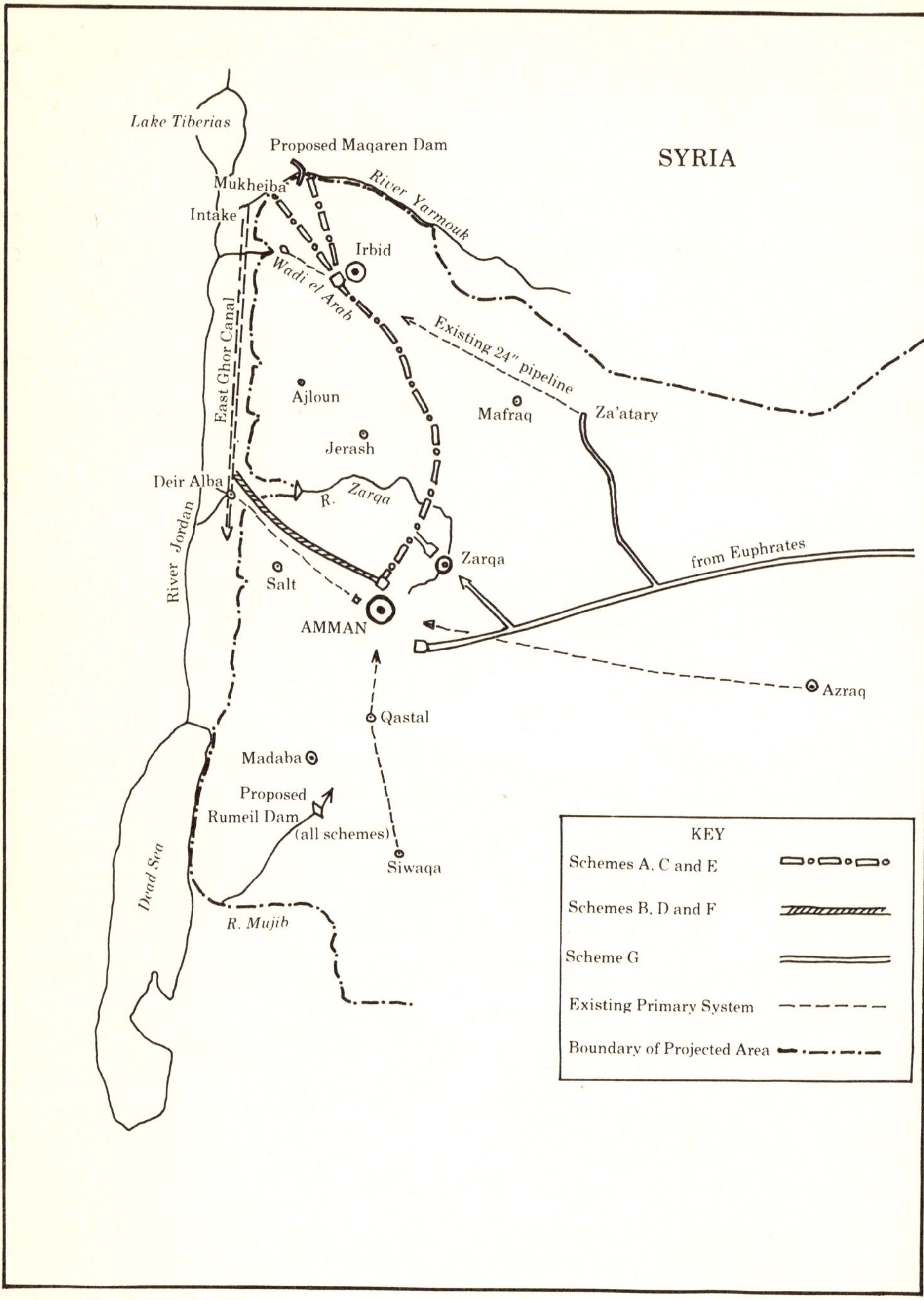

Source: Humphreys 1983

Figure 7.5: Schematic of Irrigation Systems in the Jordan Valley

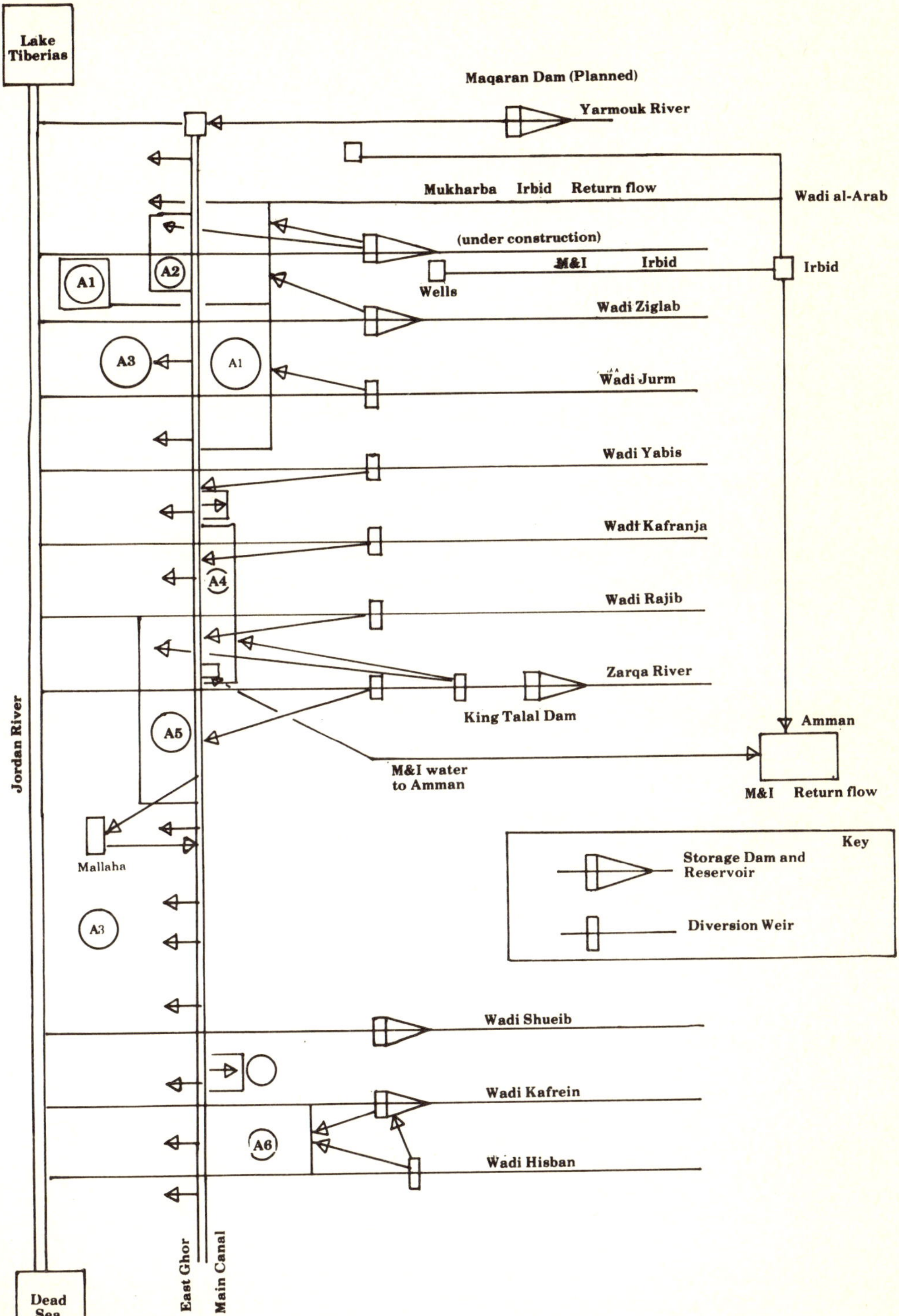

Table 7.35: Pressurised Pipe Irrigation Projects in the Jordan Valley, 1978 (dunums)

Project	New land	Converted land
East Ghor Canal Extension 18 km	36,000	
Zarqa Triangle Irrigation Project	15,500	
North-East Ghor Irrigation Project	17,500	10,000
Hisban-Kufrein Irrigation Project	15,000	
TOTAL	84,000	10,000

Source: JVA files.

unit will have its own turnout with controls and water meter.

Future Projects: The second stage of the irrigation of the Southern Ghors is being studied at present. This stage will include diversion of the Mujib water through a main pipe carrier 72 km long, building Tannur Dam on Wadi Hassa (12 mcm capacity), and a pipe network to irrigate an additional area of 46,000 du divided into 30-du farm units as in the first stage.

The third stage of the irrigation development will include building Rumeil Dam on Wadi Wala and Nukheila Dam on Wadi Mujib and pipe networks to irrigate an additional area of 17,600 du.

Irrigation Projects in the Plateau: Pilot irrigation projects were established in connection with the programme of ground water investigation to test the aquifers under long-term pumping, to investigate potential irrigation, crops and technologies, and to plan for future expansion.

Table 7.36 shows the completed public projects in the plateau and the desert area. Table 7.37 gives the planned projects.

Table 7.36: Public Irrigation Projects Completed in the Jordan Plateau and Desert

Project	Date completed	Area (du)	No. of wells	Capacity of wells (m^3/hour)	Depth to water meter
Jafr	1964	3,000	14	2,222	20-25
Wadi Dhuleil	1968	6,250	7	2,142	20.106
Qatranah	1973	1,700	6	600	97
Wadi Abyad	1972	500	1	170	
Tell Burma	1973	1,000	2	350	57
Arja		1,200	3	400	15
Ghuweir	1973	500	1	170	95
Qaa' al-Disi		7,000	14	2,500	
Abu al-Lasan		1,200			
Sultaneh		2,700			
Wuheida		3,000			

Source: NPC 1975.

Table 7.37: Planned Irrigation Projects in the Plateau

Project	Area (du)
Shaubak	4,500
Siwaqa	2,700
Abu Makhtoub	2,100
Udruh	1,500
Qurein	2,700
Qatranah Extension	6,500
Wadi Abyad Extension	4,500
Tell Burma Extension	3,000
Ghuweir-Adir Extension	2,000

Source: NPC 1975.

Irrigation in Wadi Araba: The private sector developed ground water for irrigation in Wadi Araba some time before the public sector. In 1980/81 the Jordan Valley Authority (JCA) established four pilot irrigation projects in Wadi Araba. Each pilot project consisted of 500 du divided into 15-du farm units irrigated by drip systems. Thirteen wells, each providing 40-75 m^3 per hour were used. The projects were at:

1. Um Metla
2. Wadi Musa
3. Qaa' Saa'diyyine
4. Rahma

Future irrigation developments in Wadi Araba will depend on the availability of ground water in the area. A programme of exploration is going on at present.

Irrigation Methods

Surface on-farm irrigation methods predominate in Jordan. The main surface methods used are the zig-zag furrow method for vegetables, the basin irrigation method for citrus, the big border method for bananas and alfalfa, and the wild flooding method for cereals. Some of the pilot irrigation projects of the plateau used surface methods (borders and furrows), as in Qatranah and Jafr, while others used sprinkler methods, as in Tell Burma, Wadi Abyad and Ghuweir projects.

The more recent irrigation projects in the Jordan Valley were designed to use sprinkler irrigation to attain better on-farm irrigation efficiency. Adoption of this method by farmers has been much less than expected. Drip irrigation is being used more, both in public projects and by private farmers. The public irrigation projects in the Southern Ghors are designed to use drip irrigation. There are no data available on areas under each method of irrigation.

Irrigation Efficiency

Irrigation efficiency is generally low. Very little data on this subject are available such as that reported in Table 7.26 above. Seventy per cent efficiency is attainable with a well-designed surface method, 80 per cent with a well-designed sprinkler system and 90 per cent with a good drip system. However, efficiency attained by farmers depends on how they operate the systems. The writer estimates that the average efficiency in the Jordan Valley is 40-45 per cent.

Table 7.38: Actual Cropping Patterns in the Jordan Valley, 1953-80

Crop	1953 Pre-project (%)	1966 (%)	1980 (%)	Crop	1953 (%)	1966 (%)	1980 (%)
Field crops	75.0	42.8	17.5	Vegetables	19.0	51.0	71.26
Wheat	48.0		12.7	Tomatoes	6.0		22.9
Barley	13.0		4.4	Eggplants	3.0		9.5
Berseem			0.13	Peppers			2.3
Sorghum	8.0			Potatoes	0.5		2.5
Corn	1.0		0.22	Squash	1.0		10.4
Sesame	5.0			Cucumbers	1.0		6.7
Other	—		0.04	Green beans	—		3.1
				Broad beans	1.0		3.8
				Cabbages	0.5		1.1
Fruit	6.0	14.5	16.3	Cauliflowers	0.5		0.56
Citrus	1.0			Onions/Garlic	0.5		1.8
Bananas	2.5			Watermelons	5.0		1.1
Grapes				Melons	—		2.1
Other	2-5			Other	—		3.5
		TOTAL (%)			100.0	108.3	105.15
		AREA (du)				120,000	238,000

Source: JVA files.

Conveyance efficiency ranges from 80 to 92 per cent in the Jordan Valley, depending on the time of the year and the demand for water.

Cropping Patterns

Cropping patterns in irrigation projects are very important at all stages, from planning to design to operation. During the planning stage, cropping patterns are basic for defining irrigated areas and for feasibility studies. At the design stage they are important for assessing the project components. At the operational stage they are important in water allocation, although they have not been used for that purpose in Jordan.

There are numerous suggested cropping patterns for irrigation projects. Between 1953 and 1982 more than 22 different cropping patterns were recommended for the Jordan Valley. Table 7.38 shows the actual cropping pattern in the Valley in 1953, 1966 and 1980. Five suggested cropping patterns for the Southern Ghors are given in Figure 7.6.

Crop Water Requirements

Evaluation of crop water requirements is very important in the planning, design and operation of irrigation projects. It is also important in irrigation scheduling. There are many empirical formulæ to estimate crop water requirements from climatic data such as temperature, humidity and length of daylight hours. There are two formulæ commonly used in Jordan. The Blaney-Criddle formula has been used in the Jordan Valley and the Pennman equation for the Southern Ghors. A few experiments were undertaken by the Natural Resources Authority (NRA) (Hanbali 1977) to check the actual evapo-transpiration measured by drainage-type lysimeters. The data collected are not sufficient yet to evaluate which of the two empirical equations is more applicable to Jordan. Tables 7.39, 7.40 and 7.41 give the net crop irrigation requirements in the North Jordan Valley, the South Jordan Valley and the Southern Ghors, respectively. To calculate the gross irrigation requirement, these figures are divided by the irrigation efficiency. Table 7.42 shows a comparison of the annual crop water requirements of the Jordan Valley, the desert areas and the highlands.

Irrigation Water Management

On-farm irrigation management is left to the farmer within general rules of beneficiary use and water availability limitations. This also applies to privately owned irrigation projects. Off-farm irrigation management of public projects is carried out by government organisations. The JVA operates and maintains irrigation projects in the Jordan Valley, the Southern Ghors and Wadi Araba. The Ministry of Agriculture (MOA) and the NRA operate and maintain irrigation projects in the highlands and the desert area.

The day-to-day management will be described in the section entitled 'The System Analysis' below.

Drainage of Irrigated Lands

Planning for agricultural land drainage should be made at the time of planning for irrigation. In newly irrigated areas, a drainage problem may or may not occur, depending on the sufficiency of the natural drainage of the soil. If natural drainage is insufficient, artificial drainage must be installed. If it is not drained, crop production will be impaired and salinisation of the soil will build up.

At present, there are 30,345 du of land in the Jordan Valley and over 38,000 du in the Southern Ghors, all requiring drainage. The JVA, through its Land Drainage Division, monitors the water table in the irrigation projects. If the need for drainage becomes apparent through this monitoring programme, then

Figure 7.6: Examples of Crop Rotations, Southern Ghor

Source: Binnie and Jouzy 1979.

Table 7.39: Net Crop Irrigation Requirements, North Zone, Jordan Valley (millimetres)

Crop	Month												
	Jan.	Feb.	Mar.	Apr.	May	June	July	Aug.	Sept.	Oct.	Nov.	Dec.	Total
Wheat/Barley	33	62	75	69	17	—	—	—	—	—	15	52	323
Berseem	41	46	72	58	—	—	—	—	15	84	53	35	404
Maize	—	—	—	15	85	124	182	181	142	58	—	—	787
Sorghum	—	—	—	15	82	108	156	168	149	64	—	—	742
Tomatoes I	13	—	—	—	—	—	15	95	108	128	81	28	468
II	48	38	31	—	—	—	—	—	15	81	51	48	312
III	—	—	—	15	89	123	182	181	131	51	—	—	772
Eggplants I	50	53	71	95	64	—	—	15	88	80	81	57	664
II	—	15	65	84	156	189	182	166	134	70	28	—	1,089
Peppers I	45	43	61	48	—	—	15	95	108	128	91	48	682
II	15	56	54	121	167	173	173	156	124	54	—	—	1,093
Squashes/Cucumbers I	13	—	—	—	—	—	—	15	88	90	74	23	303
II	15	56	54	99	128	72	—	—	—	—	—	—	424
Cabbages/Cauliflowers I	—	—	—	—	—	—	15	95	108	118	71	17	424
II	15	56	54	121	151	81	—	—	—	—	—	—	478
Beans I	13	—	—	—	—	—	—	15	88	96	71	38	321
II	15	56	59	111	136	72	—	—	—	—	—	—	449
Melons	—	15	65	90	108	153	77	—	—	—	—	—	508
Potatoes I	22	—	—	—	—	—	—	15	88	103	81	41	350
II	15	56	65	124	159	79	—	—	—	—	—	—	498
Onions/Garlic	3	—	—	—	—	—	15	99	115	122	66	10	430
Other vegetables	13	7	—	—	—	—	—	15	92	96	76	33	332
Bananas	35	53	80	130	186	226	248	238	206	153	88	43	1,686
Citrus	11	29	37	77	102	123	130	124	107	83	45	8	876
Other fruit	—	—	17	63	111	144	152	133	98	62	29	—	809

Source: Harza and Dar al-Handasah 1978.

Table 7.40: Net Crop Irrigation Requirements, South Zone, Jordan Valley (millimetres)

Crop	Month												
	Jan.	Feb.	Mar.	Apr.	May	June	July	Aug.	Sept.	Oct.	Nov.	Dec.	Total
Wheat/Barley	53	81	96	75	18	—	—	—	—	—	15	55	393
Berseem	62	64	93	62	—	—	—	—	15	87	69	61	513
Maize	—	—	—	15	89	126	185	185	143	61	—	—	803
Sorghum	—	—	—	15	85	110	159	171	151	68	—	—	759
Tomatoes I	23	—	—	—	—	—	15	96	108	135	98	55	530
II	68	55	41	—	—	—	—	—	15	84	67	75	405
III	—	—	—	15	91	126	184	184	132	55	—	—	787
Eggplants I	69	71	91	104	67	—	—	15	89	96	98	84	784
II	—	15	74	92	163	193	184	168	134	109	36	—	1,168
Peppers I	65	61	81	52	—	—	15	96	108	135	108	75	796
II	15	64	73	130	179	176	176	158	124	55	—	—	1,150
Squashes/Cucumbers I	23	—	—	—	—	—	—	15	89	96	77	50	350
II	15	64	73	107	135	74	—	—	—	—	—	—	468
Cabbages/Cauliflowers I	—	—	—	—	—	—	15	96	144	125	88	29	497
II	15	64	73	120	159	82	—	—	—	—	—	—	513
Beans I	23	—	—	—	—	—	—	15	89	103	88	55	373
II	15	64	79	120	143	74	—	—	—	—	—	—	495
Melons	—	15	74	98	153	166	78	—	—	—	—	—	584
Potatoes I	31	—	—	—	—	—	—	15	89	109	98	68	410
II	15	64	84	133	167	91	—	—	—	—	—	—	554
Onions/Garlic	12	—	—	—	—	—	15	100	116	129	83	37	492
Other vegetables	32	15	—	—	—	—	—	15	92	103	93	60	410
Bananas	54	71	100	139	194	230	252	242	207	161	105	70	1,825
Citrus	31	37	56	85	108	125	131	126	108	89	61	34	991
Other fruit	—	—	25	68	118	145	153	161	99	67	9	—	845

Source: Harza and Dar al-Handasah 1978.

Table 7.41: Net Crop Water Requirements for Crops Planted on Typical Date (average)

Crop	Planting date	Aug.	Sept.	Oct.	Nov.	Dec.	Jan.	Feb.	Mar.	Apr.	May	June	July	Total
Vegetables														
Tomatoes I	1/10			44.6	47.6	54.9	57.6	72.4	85.0					362.1
Tomatoes II	1/11				29.3	32.2	54.9	72.5	121.2	120.4				430.5
Tomatoes III	1/12					19.2	32.2	69.1	121.2	170.9	154.5			567.7
Eggplants/Peppers I	1/9		60.0	81.3	79.5	52.2	52.2	58.2						383.4
Eggplants/Peppers II	1/10			44.6	52.9	52.2	52.2	65.5	90.6					358.0
Eggplants/Peppers III	1/11				29.3	35.7	52.2	65.5	109.6	137.8				430.1
Potatoes I	1/12					21.2	49.0	72.5	92.7					239.9
Potatoes II	1/1						21.2	63.3	121.2	131.1				336.8
Green Beans/Broad Beans/Peas I	1/9		67.8	100.5	77.2									245.5
Green Beans/Broad Beans/Peas II	1/10			50.2	65.5	50.3								166.0
Green Beans/Broad Beans/Peas III	1/11				32.8	43.3	50.3							126.4
Green Cowpeas	1/6	200.7										87.6	175.2	463.5
Cucumbers/Squashes I	1/11				33.0	43.0	47.6	12.8						136.4
Cucumbers/Squashes II	10/2							17.7	76.6	146.4	100.8			341.5
Melons/Okra	1/3								46.9	128.8	179.6	51.1		406.4
Cabbages/Cauliflowers	20/10			13.2	40.4	49.4	49.4							152.4
Onions/Garlic I	1/10			52.1	73.0	52.2	51.3	54.7						283.3
Onions/Garlic II	10/3	54.8							28.8	124.1	197.3	208.0	186.2	799.2
Lettuces	10/11				17.8	32.4	50.0	33.3						133.6
Mallow	20/3								15.3	77.7	173.4	189.8		456.2
Beetroot/Turnips	1/11				32.7	43.9	52.2	14.6						143.5
Sweet Corn	20/9		18.3	72.0	87.9	38.4								216.6
Radishes - 3 successive crops	1/10			72.1	49.8	35.7	40.3							197.9
Fodder crops														
Berseem Clover	1/10			54.0	71.1	46.7	46.7	58.7	98.1					375.3
Sudan Grass	1/6	186.1	145.6	108.3	71.2							94.9	186.1	792.2
Alfalfa/Rhodes Grass	Perennial	186.1	145.6	108.3	71.2	46.7	46.7	58.7	98.1	138.3	176.6	186.1	186.1	1,448.4
Fruit Crops														
Citrus — Young	Perennial	98.6	77.1	63.7	41.9	27.5	30.2	38.0	57.7	81.4	103.9	98.6	98.6	817.2
— Intermediate	Perennial	120.5	94.2	70.1	50.2	32.9	35.7	44.9	69.2	97.6	124.6	120.5	120.5	980.9
— Mature	Perennial	142.4	111.4	89.2	58.6	38.4	41.2	51.8	80.8	113.9	145.4	142.4	142.4	1,157.9
Dates — Young	Perennial	120.5	94.2	70.1	50.2	32.9	35.7	44.9	69.2	97.6	124.6	120.5	120.5	980.9
— Intermediate	Perennial	153.3	119.9	95.6	62.8	43.9	46.7	58.7	92.3	130.2	166.2	153.3	153.3	1,276.2
— Mature	Perennial	175.2	137.0	114.7	75.3	52.2	54.9	69.0	109.6	146.4	186.9	175.2	175.2	1,471.6
Bananas — First year	Perennial	186.2	162.7	127.4	83.7	54.9	54.9	69.0	75.0	97.6	114.2	131.4	153.3	1,310.3
— Mature	Perennial	218.0	179.9	133.8	83.7	54.9	54.9	69.0	115.4	162.7	166.2	186.2	197.1	1,622.8
Grapes	Perennial	153.3	94.2	70.1	29.3				28.9	73.2	135.0	164.3	164.3	912.6

Source: Binnie and Jouzy 1979.

Table 7.42: Crop Water Requirement. Situation 1977

Crop	Potential evapo-transpiration (mm)			Net irrigation requirement (mm or m³/du)			Gross irrigation requirement (mm or m³/du)		
	Desert	Jordan Valley	High-lands	Desert	Jordan Valley	High-lands	Desert	Jordan Valley	High-lands
Tomatoes	1,335	912	553	1,265	712	273	2,420	1,362	520
Eggplants	1,335	987	553	1,265	787	273	2,420	1,362	520
Peppers	1,335	912	553	1,265	712	273	2,420	1,362	520
Cucumbers	1,252	828	1,100	1,182	628	820	2,257	1,202	1,570
Squashes	1,200	913	1,140	1,130	713	860	2,150	1,361	1,640
Melons	1,312	900	1,140	1,242	700	860	2,370	1,337	1,640
Okra	1,100	822	—	1,030	622	—	1,970	1,190	—
Potatoes	521	616	640	451	416	360	860	796	680
Onions	751	724	513	681	524	223	1,300	1,003	425
Cabbages/ Cauliflowers	674	596	270	604	396	200	1,150	785	380
Beans/Peas	770	577	324	700	377	200	1,340	722	380
Broad beans	950	489	—	880	289	—	1,680	553	—
Wheat/Barley	726	654	520	656	454	260	1,250	869	500
Maize	—	730	—	—	530	—	—	1,014	—
Alfalfa	1,935	2,050	1,707	1,865	1,850	1,427	3,560	3,540	2,725
Annual fodder	—	1,412	—	—	2,212	—	—	2,315	—
Citrus	—	1,305	—	—	1,105	—	—	2,115	—
Bananas	—	2,181	—	—	1,981	—	—	3,791	—
Other fruit	2,066	—	1,606	1,996	—	1,326	3,810	—	3,530

Source: Humphreys 1978.

drainage investigation and studies are made, and a drainage system is designed and constructed under the powers of the JVA.

The Institutional Environment

The development and management of water resources involves collective action by the beneficiaries or by society. Water resources management is usually a public responsibility. The objectives of water utilisation have increased over the past years. More users and more uses are involved, especially as big water-transfer projects over long distances are realised. Decision-makers must integrate and reconcile conflicting and competing demands for water by different water users. Institutional aspects are thus very important in water management.

The institutional machinery for water resources are:

1. Legal:
   a. system of water rights
   b. water laws and regulations
2. Organisational:
   a. institutions in charge of water resources, studies, development or use

The Legal Aspects

The System of Water Rights: Water rights in Jordan started as 'appropriative rights'. Users first started by diverting water from streams for their uses (mainly irrigation) and thus establishing a water right. This continued until these rights were fixed and registered in modern times through land and water ownership registration based on Land and Water Settlement Laws. In modern public projects, water rights were expropriated and the water rights rest with the Government. Ground water is considered the property of the nation and its use is regulated by laws and regulations. At no time have prior irrigation rights excluded drinking water taken by a pail or a pot.

Laws and Regulations

1. *Laws Establishing Water Rights*

Land Settlement Law No. 9 of 1937, Water Settlement Law No. 38 of 1946, Water Settlement Ordinance No. 1 of 1949, and Land and Water Settlement Law No. 40 of 1952 settled matter and conflicts related to rights of use or rights of ownership of land and water. The value of these Settlements was to give security of water right tenure. New appropriations were stopped; the water rights were attached to the land and not to the owner; and if water was found to be in excess of the user's requirements, the excess was reassigned to other users.

2. *Water Control Laws*

Water Control Law No. 31 of 1953, Water Law No. 2 of 1954, and Water Law No. 22 of 1957 established water right registers, water control procedures, rules of operation and maintenance of irrigation projects by the beneficiaries, and rules for protection of irrigation projects against unauthorised use and pollution.

3. *Laws for Government Agencies Dealing With Water*

a. Municipalities Law No. 29 of 1955 enables municipality boards to own and control water resources and to set up water projects.
b. Organisation of the Natural Resources Law No. 12 of 1968 and its amendments—Law No. 15 of 1969 and Law No. 25 of 1976—superceded Law No. 37 of 1966 which established the NRA. This Law entrusted the NRA (among other non-water related activities) with responsibilities for:
   i) Planning, designing, constructing, operating and maintenancing irrigation projects, land reclamation and applying government water policies in irrigation projects.

ii) Recommending a water policy for the Kingdom and submitting this policy for approval by the Council of Ministers
iii) Controlling the development of ground water

Until 1984 the NRA was responsible for the following activities:

i) All hydrological and hydrogeological studies and data collection (except in the area under the JVA)
ii) Drilling of water wells for exploration and for public use
iii) Design, construction, operation and maintenance of irrigation projects and dams in the Plateau
iv) Development of springs
v) Control and licensing of private wells, drilling rigs and rig operators

c. The Water Supply Corporation (WSC) Law No. 56 of 1973. The WSC is entrusted with planning, design, construction, operation and maintenance of water supply projects for domestic water, except in the Jordan Valley, where the JVA is responsible for these activities. According to its Law, the WSC does the following:
   i) Draws up the policy for water supply for domestic purposes
   ii) Provides drinking water to cities, villages, tribes and groups of users
   iii) Cooperates with the Ministry of Health (MOH) in matters related to water quality and public health

d. The Jordan Valley Development Law No. 18 of 1977. This Law established the JVA and entrusted it with the overall integrated development of the Jordan Valley Region. This region was later extended to include the Southern Ghors and Wadi Araba. Among its other functions, the JVA took over the water functions of the NRA in this region. This puts the JVA in control of most of the water resources for irrigation at present.

e. The Amman Water and Sewage Law No. 18 of 1973, and the Amman Water and Sewage Authority Law No. 48 of 1977 established the AWSA and gave it the following responsibilities:
   i) Supply and distribution of domestic water for the population of the AWSA's region.
   ii) Protection of water supplies against pollution
   iii) Prevention of water wastage
   iv) Planning and execution of schemes for water and sewage
   v) Collection and drainage of sewage water in the Amman region
   vi) Location of the water basin boundaries and location of restricted use areas for water protection

4. *Ordinances for Irrigation Water and Ground Water*

Ordinances are regulations based on laws, but they are more detailed and specific. The following relevant Ordinances are in operation at present:

a. *Water Ordinances for the Jordan Valley*: Water Ordinance No. 89 of 1966 and its amendment Water Ordinance No. 35 of 1973 are still in use for the Jordan Valley area. These are the culmination of a long series of improvements in the Water Ordinance for the East Ghor Canal Project No. 48 of 1961, the East Ghor Canal Project Water Ordinance No. 93 of 1965, Ordinance No. 131 of 1965 and Ordinance No. 13 of 1966. The main concepts of these Ordinances are as follows:
   i) The JVA controls the allocation and distribution of irrigation water.
   ii) The farmer applies for water weekly, and is then assigned his turn to receive it.
   iii) Water is given on a semi-demand basis to the farmer and the water delivered is measured. The farmer pays water charges on the basis of the volume consumed.
   iv) The JVA decides on the cropping pattern for the irrigation project area. (This has not yet been enforced.)

v) Perennial crops and high water-consuming crops like rice and sugar cane are not grown in the Jordan Valley. Licensing of citrus and bananas was practised for a number of years but this was discontinued in 1978; since then, irrigable unit areas of land have been given equal amounts of water available, thus leaving the farmer to decide (at his own risk) whether or not to plant trees, and on what percentage of his land. Cropping patterns are surveyed monthly by the Operation and Maintenance (O&M) Department, but this information is not utilised for water management.

vi) Licensed trees are ensured of their water requirements at times of water shortage.

b. *Water Ordinance for Wadi Dhuleil Project No. 38 of 1969*: This Ordinance is very similar to the Water Ordinance for the Jordan Valley except that the NRA controls the project. Similar ordinances are used for the Plateau and desert irrigation projects.

c. *Ground Water Control Ordinance No. 88 of 1966*: Authority is given to the Vice-President of the NRA in all matters of control of ground water. The area within the jurisdiction of the JVA was later excluded. The main concepts of this Ordinance are:

i) All drilling and development of ground water are subject to licensing by the Vice-President of the NRA.

ii) All drillers have to be licensed by the NRA.

iii) Ownership of drilling rigs has to be licensed by the NRA.

iv) The procedure of licensing is given. The applicant applies to the Vice-President of the NRA. If the area in which the well is intended is not restricted and drilling is not prohibited, the Vice-President of the NRA announces his intention of licensing the well in the newspapers for two weeks. If there are no objections the licence is given.

v) A water extraction licence is also needed. It gives peak allowable extraction, purpose or use of the water extracted and the measuring device that has to be installed to measure water extraction.

vi) The Vice-President of the NRA can cancel licences and can alter the quantities of water licensed for extraction.

vii) All information required from the well owner should be provided to the NRA.

The Organisational Aspects

The following organisations have major dealings with the water resources system:

1. Municipalities, as users of water and sources of effluent.
2. The NRA, which deals with studies, planning, design and construction for water resources development.
3. The WSC, as a user of M&I water, and supplier of the same to municipalities and communities.
4. The JVA, which is entrusted with water control, planning, design, construction, operation and maintenance of irrigation and M&I development projects in the Jordan Valley, the Southern Ghors and Wadi Araba.
5. The AWSA, which is in charge of water supply and sewage disposal in the Amman area.
6. The MOA, which is taking care of operations and maintenance of some of the irrigation projects in the Plateau and the desert. Research and extension in irrigation for the whole Kingdom are undertaken by the MOA.
7. The Agricultural Credit Corporation (ACC), which has been highly instrumental in on-farm irrigation development through providing credit for

the introduction of modern irrigation methods. It has also been most active in providing credit for the private sector development of ground water.

8. The MOH, which controls drinking water quality.
9. The Ministry of Municipal, Rural Affairs and Environment, which controls pollution and shares in monitoring water quality.
10. The University of Jordan and Yarmouk University, through their research and extension in irrigation and water quality.
11. The Jordan Cooperative Organisation (JCO), which also extends credit for on-farm irrigation development.
12. The Jordan Valley Farmers' Association (JVFA).
13. The Royal Scientific Society (RSS), which is undertaking research on water quality.

Comments on the Institutional Environment

The main institutional deficiency is the lack of communication and coordination between agencies in charge of water development and supply to the different users. Each organisation is authorised according to its laws to plan, design, construct and operate water projects for its own purposes. This has resulted in conflicts, the absence of a national water policy and a less than optimum control over water resources. In addition to the above, Nimry (1978) reported the following deficiencies:

1. Centralised management of the organisations
2. Conflicts between the diversified functions of some of the organisations, such as the NRA
3. Lack of manpower due to the brain drain and loss of skilled manpower to neighbouring oil-rich countries

The Future Outlook

One of the recommendations of the National Water Symposium of 1978 was to create a 'National Water Board' to be entrusted with policy-making for national water management. It was also recommended that a comprehensive water act be enacted to organise the conflicts between the various organisations. Law No. 34 of 1983 established the National Water Authority (NWA) to replace the AWSA, the WSC and the water functions of the NRA. This new Authority is to be responsible for water policy and will gradually take over the water functions of the other departments.

The Economic Environment

The economics of water supply development, transport, distribution and use vary according to space, project and time. Money value changes rapidly and it is difficult to give quantitative values which will hold for a long time. The Tables and Figures given are meant only as illustrative examples tied to their times. Table Table 7.43 gives examples of the cost of production of M&I water. Table 7.44 gives the unit cost of ground water for different well spacings and depths to water. Table 7.45 gives the cost of irrigation development in the Jordan Valley and the Southern Ghors. Table 7.46 gives a comparison of the cost of water storage by dams constructed in Jordan. Table 7.47 gives the 1979 estimated cost of water for the Maqaren Dam and the Jordan Valley Irrigation Project, Second Stage. These costs depend on the size and configurations of the particular project.

Very little data in the published literature can be found on costs of on-farm irrigation systems. Binnie and Jouzy (1979) gave 1979 price estimates of the cost of three different irrigation methods as illustrated in Table 7.48.

Table 7.44: Unit Cost of Water for Different Well Spacings at Different Depths to Water Table, Varying the Discount Rate

Depth to water table (m)			15.0	20.0	25.0	30.0	35.0	40.0	45.0
Discount rate (%)	Number of wells	Well spacing (m)	Unit cost of water (JD/1000 m^3)						
	8	250	8.16	9.02	9.89	10.76	11.62	12.49	13.36
	9	185	8.15	9.06	9.98	10.90	11.819	12.73	13.64
	10	145	8.28	9.24	10.21	11.18	12.14	13.11	14.08
	11	117	8.45	9.47	10.48	11.50	12.52	13.53	14.55
	12	98	8.75	9.81	10.88	11.95	13.01	14.08	15.14
	13	84	8.92	10.04	11.16	12.27	13.39	14.50	15.62
10	14	73	9.22	10.39	11.56	12.72	13.89	15.05	16.22
	15	65	9.45	10.67	11.89	13.10	14.32	15.53	16.75
	16	58	9.76	11.03	12.29	13.56	14.82	16.09	17.35
	17	52	10.04	11.35	12.67	13.99	15.30	16.62	17.93
	18	48	10.37	11.73	13.10	14.47	15.83	17.20	18.56
	19	44	10.64	12.06	13.47	14.89	16.30	17.72	19.14
	20	41	10.95	12.42	13.89	15.35	16.82	18.28	19.75
	21	38	11.25	12.77	14.28	15.80	17.31	18.83	20.34

Source: Rashid 1980.

Table 7.43: Cost of Production of M&I Water, 1978 (thousand JD per million cubic metres)

Project	Cost (present worth of capital & running costs)
Supply to Bani Hasan area	2,451.0
Water from Wadi Sir to Amman	1,493.1
Water to Hashimiya village	809.0
Qastal and Siwaqa Wells	4,161.23
Water to Wadi Sir villages	1,924.0
Yarmouk—Irbid water	696.78
Qaa' al-Disi—Aqaba water	848.45

Source: NWMP 1977.

Table 7.45: Cost of Irrigation Development (JD per dunum)

Project	Construction date	Cost
East Ghor Canal Project	1958-66	58.3
8-km Extension Project	1966-68	71.4
18-km Extension Project	1974-78	152.8
Zarqa Triangle Project	1975-78	206.0
Hisban-Kafrein Irrigation Project	1976-79	186.0
North-East-Ghor Project	1976-79	163.0
Wadi al-Arab Scheme	1983-85	360.0
Southern Ghors, Stage I	1982-84	397.0

Source: JVA files.

Table 7.46: Cost of Water Storage by Dams

Dam	Dates	Live capacity (mcm)	Cost (JD mn)	Cost (JD/m^3)
King Talal Dam	1972-78	48	16.25	0.34
Wadi al-Arab Dam	1982-86	17	17.0	1.0
King Talal Dam raising	1983-87	30	20.0	0.67
Maqaren Dam (estimated)	1980-87	430	200.0	0.46
Tannour Dam (estimated)	1983-87	12	12.0	1.0

Source: JVA files.

Table 7.47: Estimated Cost of M&I Water, 1979 Prices, Maqaren Dam (fils per cubic metre)

Cost component	Cost of water	
	to Irbid	to Amman—Zarqa
Cost at dam site	82.8	82.8
Cost of conveyance	272.3	229.5
Cost of distribution	83.9	102.2
Unaccounted for water	86.1	81.3
TOTAL	525.1	495.8

Source: JVA files.

Table 7.48: Estimated cost of On-Farm Irrigation Methods, 1979 Prices

	Furrow	Sprinkler	Drip (biwall)
Capital cost (JD/du)	42.25	108.43	56.5
Total variable cost (JD/hour)		0.461	0.938

Source: Binnie and Jouzy 1979.

A cost comparison was made by the Food and Agriculture Organisation of the United Nations (FAO) (1974) and is given here for comparison purposes only, as the prices are rather out-of-date.

Table 7.49: Comparison of Costs of On-Farm Irrigation Methods, 1974 Prices

	Desert schemes		Highland schemes	
	Basin	Sprinkler	Basin	Sprinkler
Initial capital (JD/du)	42.0	42.0	36.0	45.0
Direct annual (JD/du/year)	5.7	7.6	4.7	7.3
Indirect annual (JD/du/year)	15.1	11.0	4.2	3.3
Labour (man-hour/du/year)	18.0	10.0	52.0	7.0

Source: FAO 1974.

The writer's experience suggests that 1983 capital costs in the Jordan Valley are around JD 100 per du for drip systems and about JD 130 per du for a solid set sprinkler system.

The cost of drainage is about JD 75-85 for a field drain spacing of 160 m as practised in the Jordan Valley. This cost can change from one locality to another as it depends on the spacing of the drains, which in turn depends on soil characteristics that vary widely from one place to the next.

Value added by irrigation water was estimated by Humphreys (1978), as illustrated in Table 7.50. The same reference also gave the net output per unit of water for the different regions and before deduction of water charges, as shown in Table 7.51. The Economic Commission for Western Asia (ECWA) (1976) gave the gross value of water for the Jordan Valley as illustrated in Table 7.52.

Table 7.50: Value Added by Irrigation Water, 1978 (fils per cubic metre)

Region	1977	1987	2002
Desert irrigation schemes	25	52	93
Jordan Valley irrigation	25	107	222
Highland irrigation schemes	62	107	180

Source: Humphreys 1978.

Table 7.51: Net Output per Unit of Water, 1978 (JD per thousand cubic metres)

		Net output	
	1977	1987	2002
Desert irrigation schemes	41.0	71.5	113.8
Jordan Valley irrigation	49.8	129.9	249.6
Highlands irrigation schemes	93.0	145.3	224.7

Source: Humphreys 1978.

The Financial Environment

The development of water resources has enjoyed an increasingly high priority in financing by the public sector in Jordan. Table 7.53 shows a comparison of budgets planned for water, irrigation and agricultural development in the three Development Plans since 1973. Table 7.54 provides details of water and irrigation projects in the current 1981-85 Development Plan.

The Tables show the public expenditure. Although private-sector expenditure in water development (especially in ground water development and on-farm irrigation facilities) is substantial, there are very few published data to show this. The NPC (1980) reported that the actual investment in agriculture and irrigation during the 1976-80 Development Plan was JD 47.0 mn by the private sector and JD 31.4 mn by the public sector.

The Informational Environment

The Data Base

Apart from certain specific project areas where fairly detailed studies have been carried out, the data base of water supply and water use is either lacking or inadequate. The programme of water data collection, storage and updating is weak and a large portion of collected data is not published. Data collection is not evenly distributed and the accuracy of the data is questionable. Inconsistencies in data collected by different working groups and over different time periods occur because of changes in organisations responsible for data collection and because of the shortage of skilled personnel due to the brain drain which took place in the 1970s. The analysis and interpretation of data are also weak. Flood flows in many areas of the country are not measured. Hydrological data on intermittent streams are lacking. Data on ground water abstraction are scarce. The ground water observation network needs to be expanded and the aquifer characteristics need to be re-evaluated. Calculations of the ground water recharge coefficients may be in error from ± 20 per cent to ± 50 per cent (Joudeh and Abu Taha 1978).

Table 7.52: Average Value of Water (Gross), Jordan Valley, 1976 (fils per cubic metre)

		Unit value	
Crop	Season	at producer price	at social cost
Tomatoes	AW	40	40
Tomatoes	ESS	16	18
Eggplants	AW	24	33
Eggplants	ESS	10	21
Hot peppers	AW	29	13
Hot peppers	ESS	16	14
Sweet peppers	AW	18	14
Sweet peppers	ESS	10	11
Broad beans	AW	54	28
String beans	AW	89	36
String beans	ESS	25	13
Cabbages	AW	24	68
Cabbages	ESS	5	13
Squashes	AW	31	23
Squashes	ESS	16	24
Cucumbers	AW	84	21
Cucumbers	ESS	17	14
Watermelons	ESS	4	14
Sweet melons	ESS	13	14
Jew's mallow	ESS	12	15
Potatoes	AW	21	26
Potatoes	ESS	20	25
Green onions	AW	32	53
Green onions	ESS	9	21
Dry onions	AW	18	42
Spinach	AW	21	29
Lettuces	AW	25	34
Citrus	P	14	14
Bananas	P	17	8
Deciduous fruit	P	23	23
Wheat	AW	7	12
Barley	AW	5	8
Maize - grain	ESS	7	9
Winter pulses	AW	9	14

Notes: AW = Autumn/Winter.
ESS = Early Spring and Summer.
Source: ECWA 1976.

Past Studies

Joudeh and Abu Taha (1978) gave a brief listing of some of the major studies in hydrology and geohydrology which have contributed to our present data base in the hydrology and geohydrology of Jordan. These studies are briefly listed below:

Table 7.53: Comparison of Planned Investments for Water in Jordan since 1973

Sector	1973-75 Plan Planned (mn JD)	1973-75 Plan Planned (% of plan)	1973-75 Plan Implemented (mn JD)	1973-75 Plan Implemented (% of plan)	1976-80 Plan Planned (mn JD)	1976-80 Plan Planned (% of plan)	1976-80 Plan Implemented (mn JD)	1976-80 Plan Implemented (% of plan)	1981-85 Plan (mn JD)	1981-85 Plan (% of plan)
Agriculture & cooperatives	13.02	7.3			40.1	5.24	51.7		234.5	7.11
			23	10						
Water & irrigation	14.636	8.2			97.4	12.73	73.8		521.7	15.81

Source: NPC 1972, 1975 and 1980.

1. A hydrological survey of Transjordan was conducted by M. G. Ionidis between 1937 and 1939.
2. The Baker and Harza study of the feasibility of irrigation in the Jordan was conducted between 1952 and 1955.
3. Hydrological data collection was greatly improved in 1961 after the establishment of the Hydrology Division in the Central Water Authority (CWA): in 1964 the CWA published the 30-year rainfall isohyetal map of Jordan.
4. The water resources of East Jordan were studied during the period 1965-69 by Sir M. Macdonald and Partners.
5. M. E. Ibbitt and the NRA studied rainfall intensities in Jordan in 1969.
6. In 1977 a German team from Agrar and Hydrotechnik made a comprehensive review of available data and studies on water and prepared their NWMP for Jordan.
7. Between 1965 and 1970 the NRA and FAO carried out a hydrogeological study of the deep sandstone aquifer in Jordan.
8. A number of computer models were prepared and used for some aquifer systems such as the Amman—Zarqa System, Wadi Dhuleil, Azraq, Qatranah, Qaa' al-Disi and the Shaubak area.

Numerous other studies have dealt with the development of specific projects in irrigation or M&I water. Most of these studies were undertaken by expatriate consultants.

Organisations Dealing with Basic Data

More than one public organisation deals with data collection, processing and interpretation in the water resources system. Coordination and cooperation between these organisations are not always present. Climatic and rainfall data are collected by the Meteorology Department and the NRA. Evaporation and evapo-transpiration studies are performed by these two Departments, but both programmes are much smaller than required. Stream flow measurements are taken by the NRA and the JVA. Spring measurements and ground water monitoring are done by the NRA. Water quality monitoring is undertaken by the NRA, the JVA, the MOH, the RSS, the University of Jordan, the MOA, the Ministry of Municipal, Rural Affairs and Environment, the WSC and the AWSA.

Information to Water Users

In the M&I sub-system, the transmission of information to users is almost non-existent. The transmission of information to farmers in the irrigation sub-system is done by the JVA, the MOA, the NRA and private companies. Training of farmers in water management is done by the MOA and the JVA. New technologies and applications in irrigation are transmitted through this training or extension programme.

The Future Outlook and Research Requirements

Efforts to improve the data base, in the water resources system are continuous. The NRA is at present establishing a computer-based data bank that will greatly help to improve the quality of data processing, storage and retrieval. Hydrological studies are planned, and a Water Atlas is to be compiled on their completion.

Research in the field of water has been scanty and there are many areas that

Table 7.54: Budget for Water and Irrigation Projects (thousand JD)

Summary of water and irrigation projects

	1981	1982	1983	1984	1985	Total
A. Irrigation						
Wadi al-Arab Dam	1,600	2,400	2,500	4,000	5,500	16,000
Raising the King Talal Dam	100	1,000	3,500	3,500	3,900	12,000
Raising the Kafrein Dam	—	—	—	1,500	2,000	3,500
Desert dams	100	150	150	150	50	600
Jordan Valley Irrigation Project phase II	2,000	2,000	4,000	55,000	83,000	200,000
Irrigation of Southern Ghors	1,000	2,500	3,500	12,000	15,000	34,000
Irrigation of Wadi Araba	300	1,000	1,000	1,550	2,150	6,000
Subterranean drainage	500	500	500	1,000	1,000	3,500
TOTAL (A)	5,600	27,550	51,150	78,700	112,600	275,600
B. Development of water resources & water delivery						
Groundwater in the highlands and modernisation of drilling equipment	838	1,000	1,000	1,000	1,279	5,117
Groundwater in the Jordan Valley	500	750	1,000	1,500	2,250	6,000
Training of hydrological technicians	28	70	45	45	12	200
Pump water from Azraq wells	1,000	1,150	—	—	—	2,150
North Jordan domestic water supply:						
a) Deir Alla-Amman pipeline	500	12,000	17,000	14,000	9,000	52,500
b) Irbid Governorate domestic water supply	200	3,000	7,000	7,800	10,000	28,000
Euphrates/Jordan water supply	100	500	1,000	2,000	11,400	15,000
TOTAL (B)	3,166	18,470	27,045	26,345	33,941	108,967
C. Water & waste-water projects in the governorates and districts						
1) Irbid Governorate:						
Water & waste-water in Irbid	—	3,400	4,000	4,000	1,000	12,400
Development of Za'tari/Irbid waters	500	400	545	445	—	1,890
Water supply to the Jerash area	190	150	150	—	—	490
Water supply to the Ajloun area	50	150	150	150	170	670
Water supply to the villages of the northern Badia area and H4	390	110	—	—	—	500
Improvement of water network in the Irbid Governorate	210	600	550	540	600	2,500
Water and waste-water for Ajloun, Anjara, Ain, Janneh, Ramtha and Mafraq	—	—	2,000	5,500	7,500	15,000
Sumayya water to Irbid, Ramtha and Yarmóuk University	—	1,800	1,800	—	—	3,600

	1981	*1982*	*1983*	*1984*	*1985*	*Total*
2) Amman and Balqa Governorates:						
Water and waste-water at Zarqa and Ruseifa	—	4,000	5,000	7,500	8,500	25,000
Improvement of water distribution network & drilling of wells in the Amman and Balqa Governorates	240	400	610	800	1,000	3,050
Water & water distribution networks in the Amman Municipality	1,000	2,000	3,000	3,000	3,500	12,500
Greater Amman waste-water						
a) Amman city waste-water	—	3,000	6,000	6,000	3,000	18,000
b) Waste-water areas around Amman	—	1,000	2,000	2,000	2,000	7,000
Storm-water drainage in Amman	—	500	750	1,000	1,500	3,750
Water supply to southern Amman and the Madaba district	1,500	1,000	800	800	—	4,100
Water supply to areas west and north of Amman, Balqa and Beirein	435	500	600	735	—	2,270
Rumeil Dam	—	—	—	—	1,000	1,000
3) Karak and Ma'an Governorates:						
Sultaneh to Karak water supply	—	1,300	—	—	—	1,300
Water supply to remote villages	115	350	350	450	335	1,600
Improvement of water distribution networks in the Karak Governorate	30	150	250	215	200	845
Qaa'al-Disi/Aqaba water supply	2,260	—	—	—	—	2,260
Aqaba waste-water	—	2,500	1,000	625	—	4,125
Water supply to villages in the Ma'an Governorate	110	80	150	200	200	740
Water network improvement in the Ma'an Governorate	50	100	150	125	125	550
Water and waste-water in Madaba, Karak, Tafila and Ma'an	—	—	3,000	4,000	3,000	10,000
4) Jordan Valley:	250	250	500	500	500	2,000
TOTAL (C)	7,330	23,740	33,355	38,585	34,130	137,140
GRAND TOTAL	16,096	69,760	111,550	143,630	180,671	521,707

Sources of financing for water and irrigation projects

	1981	1982	1983	1984	1985	Total
General budget	8,874	26,360	37,075	48,465	56,297	173,071
Self-financing	500	2,500	4,750	5,000	4,000	16,750
Assistance & foreign loans	1,016	2,650	6,325	8,190	13,924	32,105
Loans	5,706	38,250	63,400	81,975	106,450	295,781
TOTAL	16,096	69,760	111,550	143,630	180,671	521,707

need future exploration. Some of these fields for study, with their research requirements, are listed below:

1. Improvement of methods of hydrological studies
2. Water budgeting for Jordan and for the hydrological basins.
3. Improvement of methods of exploration for ground water
4. Research on using solar energy for desalinisation of sea water and for pumping
5. Research on artificial rain
6. Research on methods of irrigation
7. Research on crop water requirements and climate/soil/crop/water relations
8. Research on efficient water conveyance
9. Research on additional plant nutrients under irrigation
10. Research on methods of decreasing labour requirements of irrigation
11. Research on drainage and re-use of drainage return flow in irrigation, with or without blending with other water
12. Research on methods of use of treated sewage water for irrigation
13. Research on water quality-crop response
14. Research on modern methods for projection of M&I water requirements
15. Research on reducing non-beneficial uses of M&I water
16. Research on water quality of drinking water and its relation to public health
17. Research on efficient methods of treatment and use of sewage effluent
18. Research on water requirements for industrial development
19. Research on water pollution caused by industry and in methods of treatment of this pollution
20. Research on water quality standards for the different industries and possibilities of recycling within one industry and between industries

The System Analysis

Introduction

The water resources system may mean different things to different people. To the hydrologist it is a system of natural processes which produce water. To the engineer it means the physical system of rivers, dams, barrages, aqueducts, canals and hydraulic structures. To the economist it is a system of economic activities or human behaviour related to water. To the lawyer it is a system of laws, regulations and water rights. For our purposes in the present chapter the water resources system is considered as an integrated engineering/economic activity system.

Every activity that is performed to modify the natural hydrological system to serve a human objective is part of our system, as is every activity performed to bring water to the root zone of the crops. Activities not directly related to and far removed from the crop production system will eventually have their effect on it. An activity like the rehabilitation of an old municipal water supply network, although not directly related to the crop production sub-system, will have an effect on it by making more water available to crops through increased efficiency in the M&I sub-system. A decision to provide sewage systems to all M&I users will affect the quantity and quality of water available for crops. Decisions in watershed management and in industrial water recycling, although far removed from the crop production system, will affect it. If all activities are taken up in detail, the presentation of the water resources system will become too long, greatly complicated and very difficult to follow and comprehend. Because of this it has been decided to aggregate activities into major sub-systems and to go into detail only where activities become more directly related to the farm system.

The system model will be a verbal one rather than a 'mathematical model', and a performance model rather than an optimisation one.

Characteristics

In brief, the water resources system in Jordan is characterised by the following features:

1. A limited data base. As mentioned earlier, 60 per cent of the catchment areas are ungauged. The density of meteorological stations is very low.
2. Limited water resources
3. Inadequate water storage facilities
4. The absence of a well-defined water policy
5. A low efficiency of water use
6. A high degree of water loss in old water networks
7. Insufficient enforcement of regulations governing water utilisation
8. Duplication and lack of coordination and cooperation between organisations dealing with water
9. A shortage of trained and skilled technical personnel to meet the requirements of the efficient management of the system

The Sub-Systems

The water resources system can be divided into sub-systems in many different ways: into physical sub-systems or components such as river basins or projects; into sub-systems of use, such as M&I and agriculture; into regional sub-systems and into surface and ground water sub-systems; and also into sub-systems of activity, such as planning, allocation, design, construction, operation and maintenance. The following list gives only examples of sub-systems:

1. Hydrological sub-system
2. Surface water sub-system
3. Ground water sub-system
4. Irrigation sub-system
5. M&I sub-system
6. Project level sub-system (off-farm sub-system)
7. Farm-level sub-system
8. Sub-system of conjunctive use of ground water and surface water
9. Soil and land classification sub-system
10. Crop rotation patterns as sub-systems
11. Soil-plant-water sub-system
12. Crop production sub-system
13. Economic and financial sub-systems
14. Conveyance sub-system
15. Organisational sub-system
16. Management sub-system/water planning sub-system
17. Water quality sub-system
18. Land drainage sub-system
19. Technology sub-system
20. Institutional sub-system

Activity Analysis

A very long list of activities can be identified in the water resources system. The major activities will be listed below, divided into activity sub-systems. Each of the activities listed is actually a sub-sub-system of activities in itself consisting of many individual activities and decisions.

1. Identifying the hydrological system
   a. Data collection and interpretation
   b. Investigations and studies
   c. Analysis of rainfall-run-off relationships
   d. Ground water exploration and analysis of ground water yields
2. Water resources policy formulation
   a. Obtaining data: hydrological, geological, engineering, agricultural, demographic and socio-economic
   b. Defining the objectives of planning
   c. Defining the scope of planning
   d. Defining all uses and users
   e. Identifying all sources of supply
   f. Defining priorities
   g. Defining water allocations policy between users and between regions. Identifying complementary and competing uses and regions
   h. Deciding the overall water policy and the development strategy
3. Definite planning in water resources
   a. Identifying long-range plans
   b. Identifying particular resource projects, and developed projects
   c. Performing feasibility studies of projects
   d. Selecting priority development projects
   e. Deciding on timing and sequencing of development projects
   f. Drafting of action programmes
4. Design of projects
   a. Analysis of design alternatives
   b. Analysis of joint use of ground water and surface water
   c. Selection of reservoir sizes and locations
   d. Sizing, sequencing and timing of the project elements
   e. Aqueduct route optimisation
   f. Sizing of irrigation command area
   g. Design optimization
5. Programming implementation
   a. Deciding on optimum timing and sequence of construction
   b. Financing and budgeting
   c. Institutional arrangements. Institution building, staffing, administration and materials procurement
   d. Construction of the project
6. Operation of the project
   a. Forecasting of water supplies
   b. Assessment of actual water supplies
   c. Allocation of water supplies
   d. Investigation of reliability of water supplies at the farm level
   e. Investigation of water management problems at the farm level
   f. Water release and diversion
   g. Water distribution to regions, laterals and farms
   h. Direct control of gates
   i. Recording and monitoring of water supplies and their use
   j. Estimation of water losses
   k. Estimation of future requirements and supplies
   l. Investigation of new sources of water supply
7. Fiscal management
   a. Setting water charges
   b. Deciding subsidies in water charges
   c. Collection of water charges
   d. Deciding differential water charge incentives

8. Maintenance and rehabilitation
   a. Restoration of irrigation facilities: distribution systems, surface drainage systems, sub-surface drainage systems, project roads, equipment, buildings, telecommunication and power transmission lines
   b. Facility improvement: design and construction of new canals, drains, roads, etc. in the project
9. Drainage programme implementation: investigation of drainage problems, design and construction of drainage systems
10. River bank protection programmes: investigation, design and construction
11. Land distribution programme and land tenure reform: studies, plans and administration
12. Development of water users' organisations
13. Research
14. Training and extension
   a. Staff training
   b. Users' training and extension work

Decisions in the Water Resources System

Almost every activity undertaken in the water resources system is usually preceded by a process of evaluating alternative courses of action available and decision-making. Decisions in the water resources system are therefore numerous. Decision-making is widely distributed between the private and the public sectors, and between many levels and many agencies in the public sector. The following list of decisions is meant only to give examples and not to be complete or exhaustive. Example decisions are to answer the following questions:

1. Is water available?
2. Is the source of water satisfactory? Criteria are the availability of water, water quality, water rights, distance of source, pumping head, etc.
3. What is the optimum use of water?
4. What is the optimum allocation of water for M&I and for agriculture?
5. What is the optimum allocation of water within the agricultural sub-system, timewise, regionwise, and cropwise?
6. Is irrigation needed?
7. Is irrigation profitable?
8. Is the land fit for irrigation?
9. What is the optimum area to be irrigated? With or without water deficit?
10. What is the optimum mix of projects?
11. What is the optimum design of the projects?
12. What is the optimum construction programme in scale, size, sequence and methods?
13. What is the best layout of farm units?
14. What is the best land tenure system?
15. Which irrigation method should be used?
    Project's decision
    Farmer's decision
16. Which crops to irrigate?
    Project's decision
    Farmer's decision
17. When to irrigate?
    Project's decision
    Farmer's decision
18. How much water to apply to each crop?
19. Is drainage required?

20. What is the best drainage system to install?
21. Decision to apply for water from the project: farmer's decision.
22. Decision to give water to the farmer: project's decision.
23. Which source of water to utilise? project's decision.
24. Which reservoir to release from? project's decision.
25. Choice of energy source for pumping.
26. Decision to automate irrigation control facilities.
27. Decision to reclaim saline lands.
28. Decision to re-use return flow for irrigation.
29. Choice of plant for deep well pumping.
30. Decision to leach and decision on leaching requirements.

Policy Instruments in the Water Resources System

Policy instruments are variables in the input to, or the environment of, the water resources system that can be modified or controlled by the policy-maker to reach an objective. Most activities present an amount of choice for the policy-maker, and the policy instruments are therefore numerous. Some examples are the following:

1. Development decisions. Most development decisions are policy instruments such as whether or not a certain water resource is developed, for which uses, when and how. This includes irrigation projects, as well as water supply projects for M&I and sewage projects.
2. Water allocation. Most decisions regarding water allocation are policy instruments. Allocation between regions, uses and sectors, within sectors, and for each type of crop are included.
3. Water distribution activities. Time and rate of water distribution can act as policy instruments.
4. Land tenure system for irrigated lands.
5. Choice and enforcement of selected cropping patterns.
6. Institutional instruments:
   a. Laws and regulations related to water
   b. Institution building
   c. Training of personnel and users
7. Economic incentives or controls:
   a. Water pricing and water charges
   b. Differential taxation
   c. Decisions to recover what percentage of the capital costs from the beneficiaries

Activity Flow Chart

Figure 7.7 shows the hydrological physical system which serves as the basis of the activity system in water resources. Any activity made to modify the hydrological system is part of the water resources system. Figure 7.8 shows the schematic of the water use system which includes engineering works to facilitate the use of water. Figure 7.9 gives a flow chart of major aggregate activities—or sub-systems of activities—in the water resources sub-system. The activities in Figure 7.9 can be very briefly described as follows:

1. Data collection is done by the NRA, the JVA and the Meteorology Department.
2. Ground water exploration is done by the NRA and the JVA.
3. Interpretation of data and assessment of water supply is carried out by the NRA and the JVA.

Figure 7.7: Hydrological System Schematic for a Typical Watershed

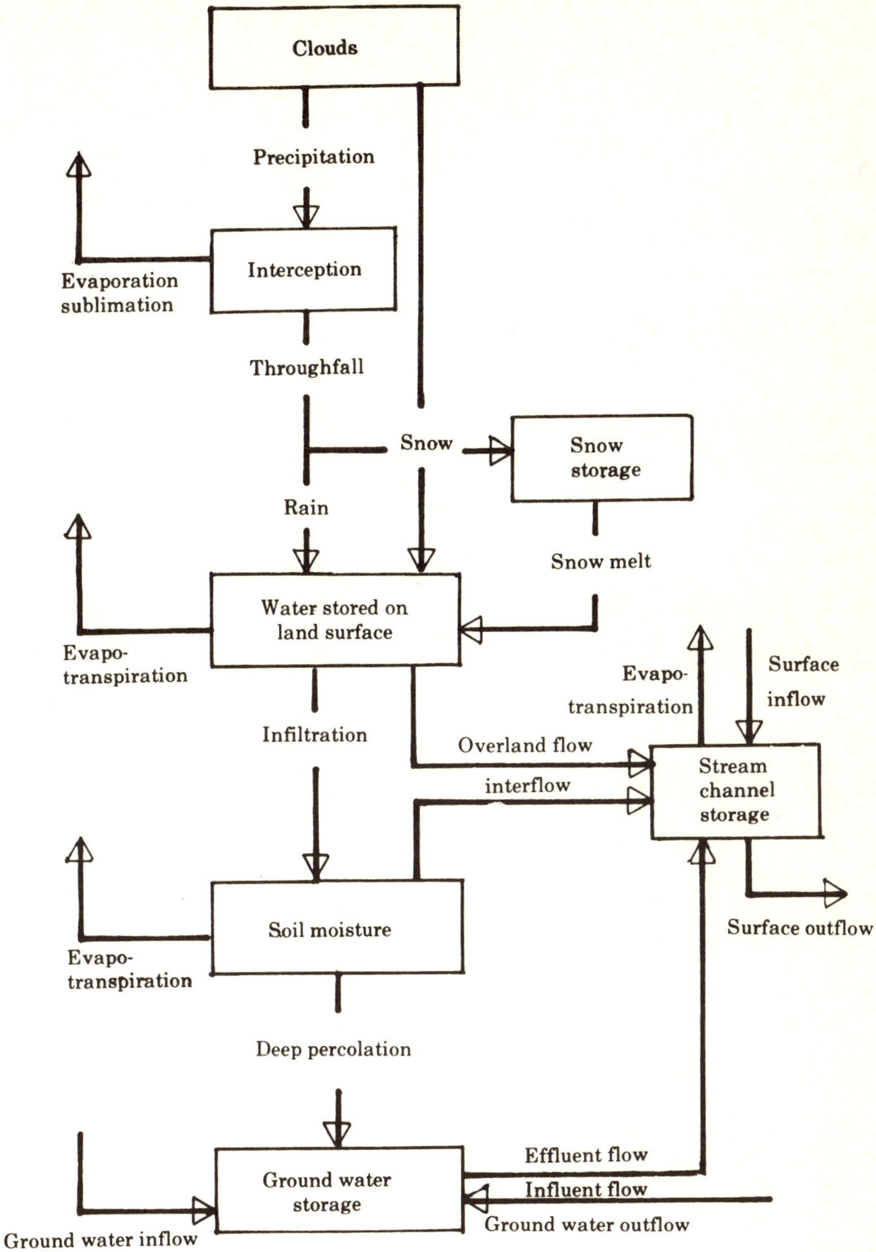

Figure 7.8: Water Use Sub-System

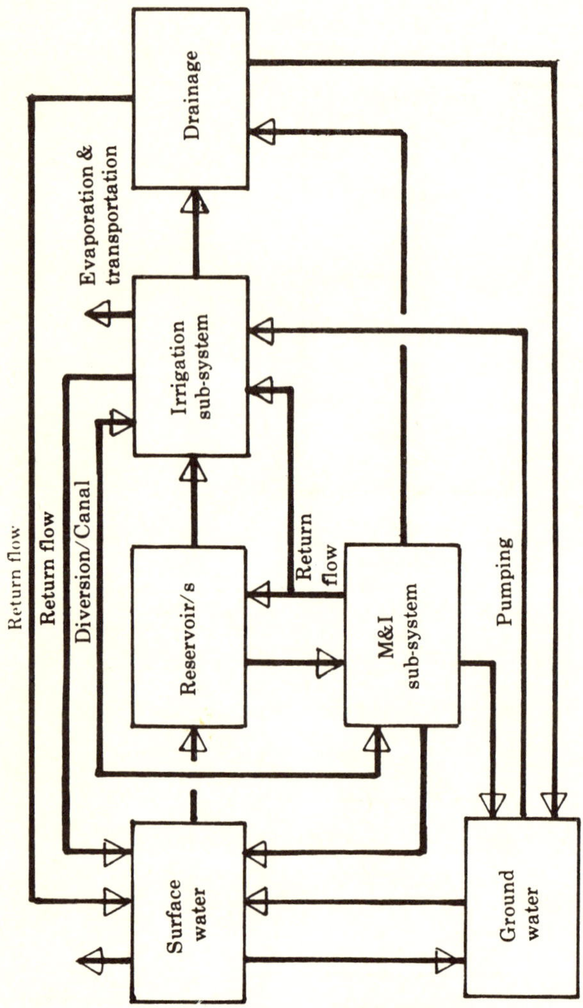

Figure 7.9: Flow Chart for Major Aggregated Activities in the Water Resources System

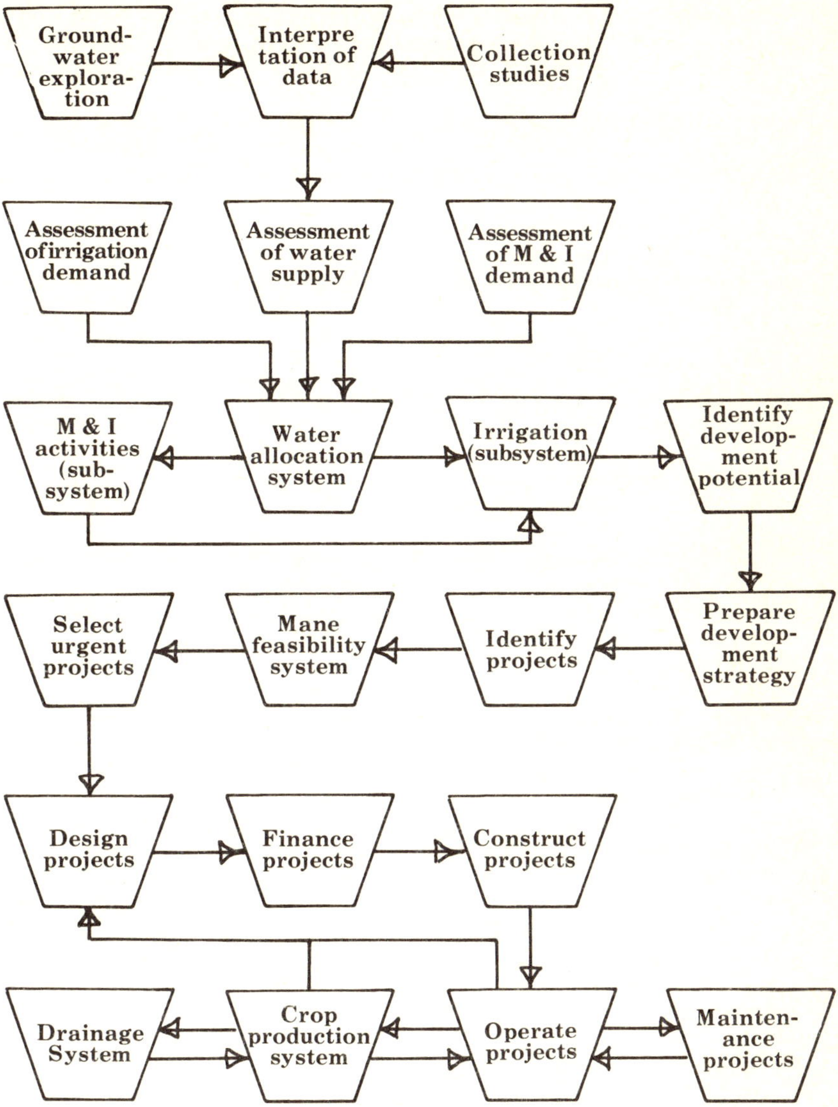

4. Assessment of M&I demand is made by organisations serving M&I water. Assessment of irrigation water demand is also made by organisations dealing with irrigation, such as the JVA, the NRA and the MOA. Consultants are often employed in both sub-sectors.
5. With the assessment of water supply and water demands, allocation of water to different uses and regions is made. This activity is not refined and it could benefit from optimisation. The rule is that M&I requirements are given priority.
6. In the irrigation sub-system, a long list of activities in project development planning follows. There is no central unit that carries out this work. Each department is left to do its own project planning. This does not lead to optimum results.
7. Design, financing, construction, operation and maintenance of the projects are carried out by the department dealing with that project. The Government exercises some control over planning and implementation through project budgeting.

Figure 7.10 shows details of the operation sub-system, whose activities are briefly described below:

1. The farmer decides whether irrigation is required, depending on the rainfall in the area, the soil moisture storage and the evapo-transpiration rate of the crops.
2. The farmer decides whether irrigation is economic or not, depending on costs and benefits of irrigation.
3. If surface water is not available and the farm lies in a region where ground water is available at economic depths, in acceptable quantity and quality, and if the drilling of wells is permitted, the farmer may choose to apply for a licence to drill a well. An application is submitted to the NRA, and the Vice-President of the NRA (if he and the farmer's neighbours have no objections) may issue a drilling licence. The farmer does not pay for the licence, but he does pay a deposit of JD 30 to ensure that he will provide the NRA with information about the drilling.
4. The farmer contracts with a private drilling rig for the well to be dug.
5. If the well is successful, the farmer applies to the Vice-President of the NRA for an extraction licence. He does not pay for extracting ground water.
6. The Vice-President of the NRA grants the extraction licence, indicating the total volume per year and the peak rate of extraction allowed. The licence also gives the type of meter and control that the NRA specifies for monitoring, measurement and control. This, however, is not strictly enforced.
7. The farmer then makes his choice of pump and motor installation for the well. Most private farmers choose deep turbine pumps and diesel motors. In public projects, most wells are equipped with submersible pumps and diesel-motor generator sets. Many factors enter into the farmer's choice, including economy, dependability, distance to repair shops, and so on.
8. The farmer then prepares his land, chooses the farm irrigation system, designs and installs it. If he has chosen the sprinkler or drip systems, he seeks the technical assistance of the company supplying it. The farmer's choice depends on many factors, such as availability, service, soil and field conditions, energy cost and availability, water supply, experience of others with whom he is in contact, his own personal experience, cost of the system,

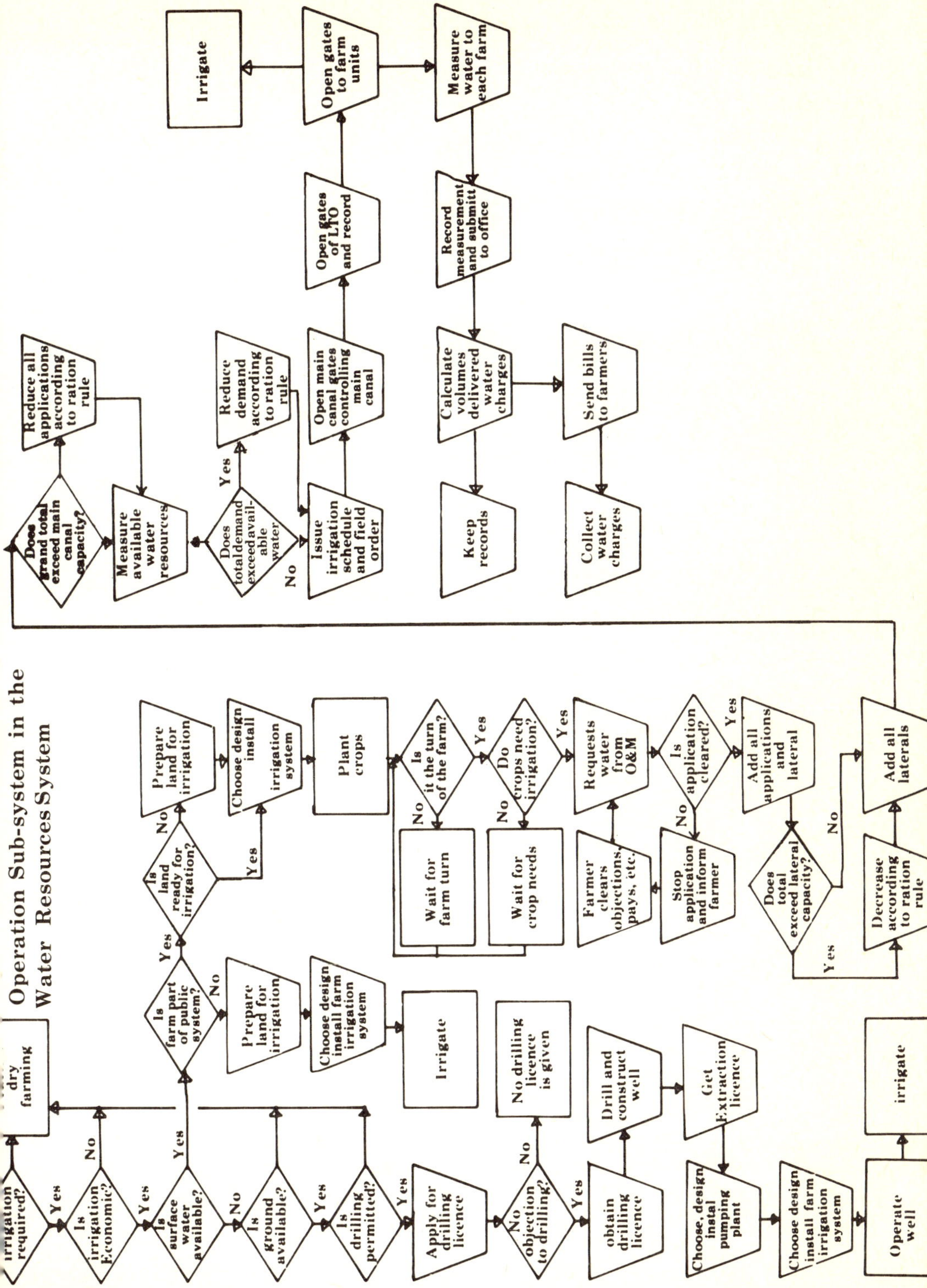

Operation Sub-system in the Water Resources System

labour requirements of the system, crops grown, operating cost of the system, convincing salesmanship and the farmer's ability to manage the system. Choices available to him are: drip irrigation, all types of sprinkler systems, straight furrows, borders and zig-zag furrows.

9. After planting, the farmer can operate the well and irrigate as his crops require.

If the farmer has a private surface water source, then he only has to do activity 8 and irrigate as required by his crops.

If the farmer owns a farm unit in one of the public projects, then he performs activity 8, plants jhis crops and requests water when needed from the project management. The majority of irrigated areas fall into this category, as will most future irrigation development.

10. Forty-eight hours before it is the turn of his farm unit, and if his crops need irrigation, the farmer submits an application to the O&M Department and gives the following required information: name of owner, the farm unit number, the discharge requested in l per second, and the length of time the water is required. This application is governed by the following:
    a. Water will not be given unless requested.
    b. Every farm unit has one day per week when its irrigation can start. The farmer has to apply for water two days ahead of this. Small seedlings may have two turns.
    c. The farmer can ask for any discharge he wants, within reasonable limits and within the physical limits of the irrigation system. If the project is using open distributaries and laterals, the farmer can get flows of between 5 and 30 l per second. In projects where water is distributed by pressurised pipes he can get only one rate of discharge, which is either 4, or 8 l per second.
    d. In the open canal systems the farmer can request water delivery for6, 12, 18 or 24 hours for the farm unit. This is modified in two cases:
        i) in drip installations where 5 l per cent second are given for longer times
        ii) in area not accessible at night-time

       In the piped irrigation system the farmer is given the set discharge for any continuous length of time up to six days per week.
    e. The farmer can cancel or alter his application until 11.00 am the day before his turn if he does so at the Water Master's Office.
11. All irrigation applications are collected by the O&M staff and checked for the following:
    a. Is the date requested for irrigation the turn of the farm unit?
    b. Has the owner paid his water charges?
    c. Are there any notes on the file of the farm unit preventing delivery?
12. If not cleared, the application is stopped, the farmer is informed and no water is given.
13. If the application is cleared, the quantities of water requested by all farms on each lateral are added up.
14. If total water requested on the lateral exceeds its capacity, then all the water requests are decreased. Licensed permanent crops on the lateral are ensured of their full requirements and the remaining capacity is divided between requests for annual crops, according to the cropped areas.
15. All requested water on the main canal is added up.
16. If the total for the main canal exceeds its capacity then all requests are reduced according to activity 14 above.

17. All available water flows to the project are measured by the O&M Department.
18. If total demand on the main canal exceeds the available water, then the water requests are cleared according to activity 14 above.
19. A written irrigation schedule or order is issued to all ditch riders and main canal inspectors, giving times, discharges and durations of flow for each farm unit, each control gate, lateral turn outs and the main canal intake gates.
20. Field staff of the O&M Department implement this irrigation schedule in the field:
    a. The main canal inspector opens the main canal intake gates, the control gates of incoming water from storage dams and the diversion weirs to the main canal. He also inspects flow in the main canal during the day.
    b. The ditch rider's adviser opens the lateral turnouts, giving water to the laterals according to the irrigation schedule. He also inspects the flow of water in the laterals and checks on the measurements of the ditch riders.
    c. The ditch rider opens the gates of the farm unit, giving water to the farmer on time and according to the irrigation schedule issued to him. The ditch rider measures the water delivered to the farmer at least twice a day and these measurements should promptly be entered in ink on a form carried by the ditch rider called the field report form. This record should be presented for checking by the ditch rider adviser or for any surprise check by the Water Master.
21. The field report is sent to the office, where volume delivered to each farm unit is calculated. The farmer is charged for water delivered to him on the basis of volume delivered. The water charge in the Jordan Valley is 3 fils per m^3. This is highly subsidised, as the cost of O&M alone, without capital cost recovery, was around 24 fils per m^3 in 1982.
22. Bills are issued to farmers at the end of the month and they are required to pay the water charges due within one month.
23. Water will not be given to a farmer who has not paid after that month.
24. Records of all flows and charges are kept by the O&M Department.

Planned Optimisation

Although this chapter is not intended for optimisation purposes, as this function will come at a later stage in the study, a very brief mention will be made of the plans to improve the operation sub-system by the JVA which controls most of the irrigated areas. A computer-based water management system is planned. This system will be coupled with remote control, automation and an on-line computer control system for the irrigation facilities. The system is described below:

Seasonal Planning

1. Two to three months before the start of the cropping season in the Jordan Valley (preferably in June) the O&M Department in the JVA will make an assessment of the availability of water at any time in the coming season (monthly). Data on stored water and inflows in all streams at the beginning of the season are the starting-point. Forecasts of the stream flows (monthly) are made, using forecasts of data such as precipitation (using stochastic methods of hydrology).
2. Data needed for the optimisation of a cropping pattern are collected. This includes, *inter alia*, prospects for marketing, forecasts of prices of produce and prices of inputs, and availability of inputs, especially labour and energy. With this data and water availability data in hand, an optimum cropping pattern is developed. Linear programming is used to produce optimisation.

3. The optimum cropping pattern is publicised to farmers, together with water availability and market data forecasts for the following season. Farmers are invited to submit their cropping plans for the season on special forms giving crops, areas and dates of planting. Farmers are encouraged to consider the information made available to them and to stay within the limits of the optimum cropping pattern.
4. The requested cropping pattern is compiled from the requests of farmers. The water requirements of this pattern are calculated and the pattern is compared with the optimum one. The optimum cropping pattern should not be regarded as sacred and inviolable, but only as a guide. This is because the optimisation process depends on uncertain stochastic or forecasted data. If the cropping plan requested by the farmer is acceptable, he is informed of the JVA's approval of his request. If his plan is not acceptable, he is told to make changes and in what manner, until an acceptable planned cropping pattern is evolved.
5. Seasonal allocations of water are made according to the cropping plan, and the plan and allocation are publicised to farmers.

Monthly Check on Plan

1. The farmer is required to get the approval of the operating agency (JVA) for any change in his approved cropping plan.
2. A survey is conducted every month during the season to determine the actual cropping in the project area.
3. The actual cropping pattern is compared to the optimum cropping pattern to note deviations and find remedies.
4. The forecast of water availability for the remainder of the season is updated.
5. The water requirements of the actual cropping pattern are calculated and compared to the updated water availability forecast. If no shortage is forecast, monthly water allocations in the seasonal plan are updated and farmers informed. If a shortage is anticipated, an operational rule to minimise the adverse effects of this shortage is made. This is done through dynamic programming, taking into consideration the relative economic importance of the different crops, their stage of growth and the previous history of water deficits experienced by the crop. The water allocation for the rest of the season is made and farmers are informed of the expected shortages and future operational rules to meet them.
6. From rain fall forecasts, stream flows and stages in the different reservoirs, the storage and release schedule is worked out for the coming month.

Weekly Planning

1. From weekly stream flow forecasts and weekly forecasts of water demands, a comparison is made to indicate whether there will be a water shortage or not. Demands for the coming week are calculated from forecasts based on climatological data of the previous week. If there is a shortage, the shortage optimisation rule will be followed. If no shortage is expected, then the weekly allocation is made.
2. The following schedules are drawn up:
   a. weekly reservoir releases schedule
   b. schedule of operation of main canal control gates
   c. operation schedule of lateral turnouts, pumping stations, sub-lateral gates and headgate unit mains
   d. operation of gates and valves of farm units (farm turnout assemblies)
3. Farmers are informed of the weekly allocation and schedule.

4. The status of the irrigation system is read, monitored and recorded. Inflows, reservoir levels and releases are recorded.

Recommendations, Measures and Actions Required

Informational

Data collection and interpretation, studies and research in the water resources system should be strengthened. Information should be systematically updated. Coordination and cooperation between departments engaged in the collection of data should be strengthened. Information to water users should be made available.

Water Resources Development

1. Jordan should give high priority to the maximum development of its water resources.
2. Storage of flood flows should be given priority. Every drop of water counts in Jordan.
3. Sewage collection, treatment and re-use for irrigation should be stressed, as it is the only way to make ends meet for M&I and irrigation. Water in Jordan should be used at least twice.

Water Management

1. Water conservation and the efficient use of water should be given a very high priority in any actions to be taken. Water distribution losses should be minimised.
2. Water quality conservation and management should be given as much importance as management of water quantity.
3. Water policy and planning should be undertaken in a central organisation that can integrate all water supplies and uses.
4. The enforcement of existing controls on water extraction and use should be strengthened. Adequate legal measures to enforce optimum ground water use should be enacted.

References and Other Works Consulted

Alexandratos, N., A. Condos and P. A. Wardle, 1982, 'A model for the evaluation of forest sector development strategies in peninsular Malaysia', in E. Thorbecke and L. Hall (eds), *Agricultural Sector Analysis and Models in Developing Countries*, Food and Agriculture Organisation, Rome, pp. 47-83.

AWSA (Amman Water and Sewage Authority and Water Supply Authority), 1978, 'Problems of potable water supply in Jordan', in *Proceedings of the National Water Symposium*, pp. 156-71. Natural Resources Authority, Amman, 19-22 March 1978.

Arar, Abdullah A., 1978, 'Some considerations for increasing the supply of and reducing the demand for usable water in Jordan', in *Proceedings of the National Water Symposium,* Natural Resources Authority, Amman, 19-22 March 1978.

Bhattacharjee, Jyoti P., Apostolos Condos and Carlo Cappi, 1982, 'FAO perspective studies for agricultural development', in E. Thorbecke and L. Hall (eds), *Agricultural Sector Analysis and Models in Developing Countries*, Food and

Agriculture Organisation, Rome, pp. 17-24.

Binnie and Jouzy, 1979, 'Mujib and Southern Ghors Irrigation Project', feasibility report submitted to Jordan Valley Authority, Amman.

Burra, Elwood S., 1968, *Operations Management, Problems and Models* (2nd edn), John Wiley & Sons, New York, p. 745.

Cappi, Carlo, Lehman Fletcher, Roger Norton, Carlos Pomareda and Molly Wainer, 1982, 'A model of agricultural production and trade in Central America', in E. Thorbecke and L. Hall (eds), *Agricultural Sector Analysis and Models in Developing Countries*, Food and Agriculture Organisation, Rome, pp. 307-45.

Cirialy-Wantrup, S. V., 1964, 'Water policy', in Ven Te Chow (ed). *Handbook of Applied Hydrology*, McGraw-Hill, New York, pp. 28-1 to 28-25.

Clyde, Calvin G., Alton B. King and Jay C. Anderson, 1971, *Application of Operation Research Techniques for Allocation of Water Resources in Utah*, Utah State University, Logan, Utah, PRWG-73-2, p. 132.

Condos, Apostolos and Carlo Cappi, 1982, 'Agricultural sector analysis: A linear programming model for Tunisia', in E. Thorbecke and L. Hall (eds), *Agricultural Sector Analysis and Models in Developing Countries*, Food and Agriculture Organisation, Rome, pp. 119-48.

Danes, Samuel R. (ed), 1982, 'Colombia agriculture sector analysis I: Methodological summary and sample results', in E. Thorbecke and L. Hall (eds), *Agricultural Sector Analysis and Models in Developing Countries*, Food and Agriculture Organisation, Rome, pp. 189-224.

Dar al-Handasah and NEDECO, 1969, *Jordan Valley Project: Agro- and Socio-Economic Study*, Jordan River and Tributaries Regional Corporation, Amman.

Darley, P. R., J. R. Rydzewski and W. M. Clark, 1972, 'A simulation model for the optimal design and operation of irrigation systems based on yield-water deficit relationships', in Proceedings of Symposium on Water Resources System Planning with Special Reference to Irrigation, Drainage, New Delhi, pp. S 135-S 167.

Department of Statistics, 1972, *Survey of Agricultural Land Irrigated by Artesian Wells, 1971*, Amman.

— 1973, *Social and Economic Survey of the East Jordan Valley*, Amman, p. 157.

Duloy, John H. and Roger D. Norton, 1982, 'CHAC, A programming model of Mexican agriculture', in E. Thorbecke and L. Hall (eds), *Agricultural Sector Analysis and Models in Developing Countries*, Food and Agriculture Organisation, Rome, pp. 227-61.

ECWA (Economic Commission for Western Asia), 1976, 'Water Regimes for Jordan Valley', a study report submitted to Jordan Valley Authority, Amman.

Food and Agriculture Organisation (FAO), 1966, Mediterranean project, country report, Jordan, FAO, Rome.

—— 1974, *Development and Use of Groundwater Resources of East Jordan*, FAO, Rome, p. 114.

Fletcher, Lehman B., 1982. 'Agricultural sector analysis and sector lending by the U S Agency for International Development (USAID)', in E. Thorbecke and L. Hall (eds), *Agricultural Sector Analysis and Models in Developing Countries*, Food and Agriculture Organisation, Rome, pp. 25–45.

Goreux, Louis, 1982, 'Multi-level planning and decentralized decision making: On the problem of linkages between projects, sector and economy-wide analysis', in E. Thorbecke and L. Hall (eds), *Agricultural Sector Analysis and Models in Developing Countries*, Food and Agriculture Organisation, Rome, pp. 287-305.

Hall, Warren A. and John A. Dracup, 1970, *Water Resources Systems Engineering*, McGraw-Hill, New York, p. 372.

Hanbali, Mohammad, Mohammad Abu-Hyaht, Hani Rashid and Munther Kharraz, 1977, *Consumptive Use of Water Requirements for Jordan—Interim Report*, Natural Resources Authority, Amman, p. 25.

Hargreaves, George H., 1977, *World Water for Agriculture (Climate, Precipitation, Probabilities and Adequacies for Rainfed Agriculture)*, Utah State University, Logan, Utah.

Harza and Dar al-Handasa, 1978, 'Jordan Valley Irrigation Project, Stage II', feasibility study, Vol. 1, main report submitted to Jordan Valley Authority, Amman.

Hirshleifer, Jack, James C. DeHaven and Jerome W. Milliman, 1963, *Water Supply Economics, Technology and Policy*, University of Chicago Press, Chicago, Ill., p. 378.

Hotes, Frederick L. and Richard Phillips, 1972, 'System management and planning for irrigation, drainage and flood control', in Proceedings of Symposium on Water Resources System Planning with Special Reference to Irrigation, Drainage and Flood Control, International Commission on Irrigation and Drainage, Varna, 1972, pp. 51-548.

Humphreys, Howard, 1978, 'Water use strategy, North Jordan', summary report submitted to Jordan Valley Authority, Amman.

—— 1983, 'Water supply project from the Euphrates River, water demand projections', discussion draft submitted to Jordan Valley Authority, Amman.

International Commission on Irrigation and Drainage (ICID), 1980, *The Application of System Analysis to Problems of Irrigation, Drainage and Flood Control*, International Commission on Irrigation and Drainage, New Delhi, p. 203.

Johnson, Glen L., 1982, 'The general systems — science simulation approach to agricultural sector analysis', in E. Thorbecke and L. Hall (eds), *Agricultural Sector Analysis and Models in Developing Countries*, Food and Agriculture Organisation, Rome, pp. 87-95.

Joudeh, Omar M., and Mohammad Abu Taha, 1978, 'Present and needed information on water resources in Jordan', in Proceedings of the National Water Symposium, Natural Resources Authority, Amman, 19-22 March 1978, pp. 17-50.

King, Alton B., 1972, 'Development of regional supply factors and a least cost model for allocating water resources in Utah: A parametric linear programming approach', unpublished PhD dissertation, Utah State University, Logan, Utah, p. 231.

Marelius, Kenneth, 1977, 'Amman water resources masterplan', in Proceedings of the National Water Symposium, Natural Resources Authority, Amman, 19-22 March 1978, pp. 140-55.

Marsh, Albert W., 1967, *Applied Irrigation Research*, Food and Agriculture Organisation, Rome, p. 54.

McGaugheyl, Stephen E., 1982, 'Usefulness of sector analysis to a large lending organization', in E. Thorbecke and L. Hall (eds), *Agricultural Sector Analysis and Models in Developing Countries*, Food and Agriculture Organisation, Rome, pp. 349-63.

NPC (National Planning Council), 1972, *Three-Year Development Plan, 1973-75*, Amman, p. 327.

—— 1975, *Five-Year Development Plan, 1976-80*, Amman, p. 325.

—— 1980, *Socio-Economic Development Plan for 1981-85*, Amman, p. 360.

Natur, Fahd S., 1969, *Land and Water Use in the Hashemite Kingdom of Jordan — Part IV: Irrigation in Jordan*, Natural Resources Authority, Amman, p. 22.

NRA (Natural Resources Authority), a. Irrigation Department, 1974, *Irrigation Development in East Jordan*, NRA, Amman, p. 62.

—— b. 1974, 'Water resources for irrigation in Jordan', NRA, Amman (mimeo; in Arabic), p. 60.
—— 1977, *National Water Master Plan of Jordan* (7 vols.), Agrar-Und Hydrotechnik, NRA, Amman.
—— 1978, Proceedings of the National Water Symposium, Amman, 19-22 March 1978.
NWMP (National Water Master Plan of Jordan), 1977, Vols. I to VI, Natural Resources Authority, Amman, GTZ, Hanover.
Nimry, Y. E., 1978, 'Present water legislation and institutional framework in Jordan', in Proceedings of the National Water Symposium, Natural Resources Authority, Amman, 19-22 March 1978, pp. 231-49.
Onigkeit, Dietmar, Christian Kim and Bent Schmid, 1969, 'Optimal design of an irrigation system', in Proceedings of Symposium on Application of Computers in the Analysis of Various Problems Relating to Irrigation and Drainage Systems, ICID, New Delhi, pp. S 95-S 110.
Pricop, G. V. Ionesen, U. Bevar and I. Ghizelen, 1978, 'Water management in an irrigation system', in Proceedings of the Tenth Congress on Irrigation and Drainage, International Commission on Irrigation and Drainage, New Delhi, pp. 35.181-35.200.
Rangeley, W. R. and D. G. Knott, 1969, 'Application of computers to compare developments in water demand and supply in complex irrigation systems', in Proceedings of Symposium on Application of Computers in the Analysis of Various Problems Relating to Irrigation and Drainage Systems, International Commission on Irrigation and Drainage, New Delhi, pp. S.1—S.24.
Rashid, Hani Abdul-Hamid, 1980, *Optimization of Conjunctive Use of Surface and Ground Water for Irrigation in Jordan*, PhD dissertation, University of Southampton (UK), p. 316.
Rossmiller, George E., 1982, 'A capacity for agricultural analysis for planning and policy: The Republic of Korea', in E. Thorbecke and L. Hall (eds), *Agricultural Sector Analysis and Models in Developing Countries*, Food and Agriculture Organisation, Rome, pp. 97-116.
Soltesz, J. *et al.*, 1975, 'Application of computer technology and automatic control in planning, designing and operation of irrigation systems in Czechoslovak Socialist Republic', in Proceedings of Symposium on Introduction of Automatic Facilities for the Needs of Collecting and Processing of Information Planning and Managing in the Fields of Irrigation and Drainage, International Commission on Irrigation and Drainage, New Delhi, pp. S.113—S.142.
Stoecker, Arthur, Kanok Khaliharn, Earl O. Heady and Somnuk Sriplung, 1982, 'Development and application of a rational linear programming model for development planning of Thailand agriculture', in E. Thorbecke and L. Hall (eds), *Agricultural Sector Analysis and Models in Developing Countries*, Food and Agriculture Organisation, Rome, pp. 149-89.
Thomas, Jimmie L., J. Paul Riley and Eugene K. Israelsen, 1974, *A Computer Model of the Quantity and Chemical Quality of Return Flow*, Utah State University, Logan, Utah, p. 94.
Thorbecke, Erick and Lana Hall, 1982, 'Nature and scope of agricultural sector analysis: An overview', in E. Thorbecke and L. Hall (eds), *Agricultural Sector Analysis and Models in Developing Countries*, Food and Agriculture Organisation, Rome, pp. 3-13.
USAID (US Agency for International Development), 1979, *Environmental and Natural Resources Management in Developing Countries*, Vol I, USAID, Washington, DC, p. 184.
—— 1983, 'Water systems and services management', project identification documents (mimeo), p. 25.

VBB (Vappen Byggnas Byran), 1977, *Amman Water and Sewage Works. Water Resources, Investigations and Feasibility Studies for Amman Water and Sewerage Facilities*, Vols I & II, Amman Water and Sewage Authority, Amman.

World Bank Staff, 1978, 'Agricultural sector planning models: A selected summary and critique', *World Bank Working Paper No. 297*, World Bank, Washington, DC, p. 61.

8. THE SUPPLY OF PURCHASED AGRICULTURAL INPUTS

M. Haitham el-Hurani

Introduction*

This chapter aims to analyse the operations and activities of the Jordanian agribusiness input market, and the supply of purchased inputs. It will analyse the domestic production of agricultural inputs, describe the role of the Jordanian Government in the supply of such inputs and discuss the overall operations of the private agribusiness input market. The research was based on personal interviews and meetings with managers in both the private and public sectors.

General Description of Jordanian Supply of Purchased Agricultural Units

The availability of improved agricultural inputs on the market in Jordan is a vital requirement to speed up the adoption rate by farmers and thereby increase the growth of agricultural production. There are two sources of purchased inputs: domestically produced, and imported. Certain inputs, such as fertilisers, plastic materials and pipes for agriculture, concentrated feeds and veterinary medicaments, are produced at home, but Jordan depends on imports for the majority of its agricultural inputs. (See Tables 8.1 to 8.8.) The institutional set-up which deals with the supply of purchased agricultural inputs in Jordan can be divided into two principal groups: first, the private firms which constitute the agribusiness input market; and second, the semi-public organisations which serve the agricultural sector and deal with the supply of purchased agricultural materials among their other activities. This group includes two major organisations, the Jordan Cooperative Organisation (JCO) and the Jordan Valley Farmers' Union (JVFU).

* I wish to extend my gratitude and sincere appreciation to all officials at the Ministry of Agriculture, the Department of Agricultural Research and Extension, and the Department of Animal Health and Production, and to all managers of agribusiness companies interviewed and to commissioners, who kindly provided the required information. I owe a special debt of gratitude to the General Manager of the Jordan Fertiliser Industry Company, the General Manager of the Jordan Valley Farmers' Union, and the Deputy General Manager of the Jordan Cooperative Organisation for their warm reception and cooperation. I also grateful to all farmers and managers of local stores in the Jordan Valley who made the tedious task of travelling and interviewing a delightful and rewarding experience.

Table 8.1: Quantities of Pesticides Imported 1974-81 (tonnes)

Year	Pesticides
1974	738
1975	621
1976	664
1977	1,451
1978	1,350
1979	1,534
1980	1,850
1981	1,705

Sources: DOS 1979: 136.
DOS 1981: 76.

Table 8.2: Value of Pesticides Imported, 1970-80 (thousand $US)

Year	Pesticides
1970	481
1971	424
1972	710
1973	925
1974	1,180
1975	1,861
1976	1,982
1977	3,092
1978	3,575
1979	3,578
1980	8,457

Sources: FAO a. 1976: 292.
FAO a. 1980: 296.
FAO a. 1981: 302.

Table 8.3: Quantities of Chemical Fertilisers Imported, 1974-81 (tonnes)

Year	Nitrogenous	Phosphatic	Potassic	Other fertilisers
1974	8,185	2,712	310	2,743
1975	1,107	4,384	961	5,855
1976	15,168	3,375	4,425	987
1977	13,205	2,723	7	773
1978	18,437	6,490	1,850	670
1979	15,180	19,784	994	998
1980	17,841	10,502	143	2,863
1981	9,425	11,545	1,774	1,691

Sources: DOS 1979: 136.
DOS 1981: 76.

Table 8.4: Value of Manufactured Fertilisers Imported, 1970-80 (thousand $US)

Year	Manufactured fertilisers
1970	667
1971	847
1972	890
1973	440
1974	1,358
1975	757
1976	3,779
1977	2,528
1978	4,586
1979	5,847
1980	7,577

Sources: FAO b. 1976: 289.
FAO b. 1980: 293.
FAO b. 1981: 299.

Table 8.5: Value of Crude Fertilisers Exported, 1970-80 (thousand $US)

Year	Crude fertilisers
1970	6,263
1971	6,272
1972	9,792
1973	12,182
1974	60,742
1975	61,319
1976	57,968
1977	52,554
1978	63,478
1979	87,441
1980	158,027

Sources: FAO b. 1976: 287.
FAO b. 1980: 291.
FAO b. 1981: 297.

Table 8.6: Quantities of Veterinary Medicaments Imported, 1974-81 (tonnes)

Year	Veterinary medicaments
1974	58
1975	108
1976	107
1977	140
1978	235
1979	159
1980	96
1981	194

Sources: DOS 1979: 136.
DOS 1981: 76.

Table 8.7: Quantities of Principal Agricultural Machinery Imported, 1976-81

Year	Tractors	Soil preparation machinery	Harvesting machinery
1976	437	460	156
1977	278	656	217
1978	344	189	397
1979	152	832	54
1980	218	107	948
1981	238	129	61

Sources: DOS 1981: 76.

Table 8.8: Quantities of Tractors and Harvesters in Service, 1969-80

Year	Total tractors	Total harvester-threshers
Average 1969-1971	2,759	139
1975	3,748	n.a.
1976	3,914	n.a.
1977	4,074	210
1978	4,223	220
1979	4,343	230
1980	4,561	240

Source: FAO a. 1980: 262.

The Government of Jordan carries out several functions in organising the supply of agricultural inputs. The first group of measures aims to reduce market prices in granting full exemption of import tax on all agricultural materials. The second group aims to regulate and control the importing of certain inputs to protect farmers from poor-quality inputs and potential health risks. Analysis of the Jordanian agribusiness input market shows a steady development during the last 10 years as new agribusinesses have entered the market alongside the JCO and the JVFU. A comprehensive analysis of the size and location of agribusiness input firms, with their importing activities, distribution and transport services, provides a useful insight into the operation of this sub-sector. Finally, a study of the price mechanisms, terms of payment, and role of central market commissioners in the provision of agricultural inputs, sheds light on farmers' decisions with regard to purchasing inputs and the current problems which face agribusinesses and the central market middlemen in their dealings with Jordanian farmers. Nevertheless, this chapter concludes that agribusiness firms are functioning efficiently and are responsive to the demands of Jordan's agricultural sector.

Domestic Production of Agricultural Inputs

Jordan depends largely on imports to satisfy the demands of the agricultural sector. However, domestic production of agricultural inputs is developing at a satisfactory rate. The most important newly emerging industry in this respect is the production of chemical fertilisers. Two large companies were established in the mid-1970s and started production in 1982.

The first of these, the Jordan Fertiliser Industry Company (JFI), was established in 1975 to produce chemical fertilisers, and specially phosphatic fertilisers. JFI's Complex, which started in September 1983, will be a major contribution to Jordan's industrial development and technological base. Its stated goals (JFI 1982) are to help improve the country's balance of trade, to create employment opportunities and to improve regional and international economic cooperation. The Complex utilises 1.3 mn t per year of rock phosphate from Jordan Mines in the al-Hassa area, situated about 200 km north of Aqaba. The other two major raw materials used, sulphur and ammonia, are imported from Arab and international sources. According to the JFI's Bulletin, this complex is considered one of the largest single-stream plants in the world for the production of phosphoric acid and DAP. It is estimated that the marketable final products will total 740,000 t of DAP and 105,000 t of phosphoric acid per year. The same plant can also produce two grades of MAP. To guarantee distribution of this large output of fertilisers, the JFI has concluded marketing agreements for a period of three years with two international organisations. One of these is the Mitsubishi Corporation of Japan, which will market 40 per cent of JFI production of DAP in areas east of Suez; the second is Woodward and Dickerson of the US, which will market 35 per cent in areas west of Suez; and the remaining 25 per cent of DAP will be marketed direct by the JFI. It is expected that this Fertiliser Complex will have a substantial impact on the national economy, and particularly the southern region of Jordan. The JFI will become one of the major foreign-exchange earners in the country, and its fertiliser production will help to develop and improve domestic agriculture and industry. The company plans to undertake special extension programme in cooperation with the Ministry of Agriculture (MOA) to encourage Jordanian farmers to use this new kind of fertiliser in place of the traditional types such as MAP, which farmers have been purchasing from the local market. The General Manager of the company has indicated that the

JFI sells DAP to Jordanian agricultural organisations at very reasonable prices and payment conditions. The first sale to the local market, for 1,700 t, was bought by the JVFU in 1983.

The second company is the Arab Potash Company (APC), established in 1956 as a result of the recommendation taken by the Arab Economic Council. The establishment of this company arose from Jordan's strong interest in benefiting from the magnificent resources of the Dead Sea. But political circumstances in the region over the last two decades, together with administrative difficulties, made it impossible for the company to achieve its objectives, and its activities have ceased. However, the Jordanian Government has given priority to the exploitation of potash in setting up the Three-Year (1973-75) and first Five-Year (1976-80) Development Plans. In 1982 commercial production of potassium chloride started, and a new APC plans to utilise all the by-products from the potash operation in order to expand Jordan's chemical industry. The main projects for the future are the production of potassium sulphate, dicalcium phosphate and magnesium oxide (APC 1982: 20). In 1983 the company produced 160,000 t of potassium chloride. The Marketing Manager reports that it exported 150,000 t to Iraq, France and other countries. As to the domestic use of potassium chloride, it is generally believed that Jordanian soils are rich in potassium. Nevertheless, the company has been requested by the MOA to conduct soil tests in order to investigate the possible promotion of this kind of fertiliser in Jordan.

In addition to this fertiliser industry, Jordan produces plastic materials and pipes for plastic houses and irrigation systems. A study of these plastic factories shows that there are six companies which produce several kinds of plastic materials for agricultural use. An interview with the manager of one factory revealed that the total production of these factories satisfies the market demand for all plastic materials and pipes for drip irrigation systems. Jordanian private companies also produce poultry feed. There are five major factories which produce different types of feed, in addition to many small feed mills which operate on several poultry farms. The manager of one feed factory interviewed reported that all Jordanian factories depend mainly on imports for the necessary raw materials for processing and producing feed. With respect to the production of veterinary medicaments and vaccines, there are two factories operating which produce several kinds of veterinary medicaments and depend completely on importing the required raw materials. Finally, the Jordanian Government takes responsibility for providing locally treated and cleaned wheat seeds and, through the JCO, distributes them to all wheat farmers in Jordan.

The Role of the Jordanian Government in the Supply of Agricultural Inputs

The Jordanian Government has a multi-purpose role in the supply of purchased agricultural inputs. The various aspects of government intervention can be classified into two categories: government measures taken to support the availability of agricultural inputs at reasonable prices; and government regulations and laws which aim to regulate and control the importing and selling activities for certain inputs. The following is a brief discussion of these two categories.

Government Measures to Support the Availability of Inputs

The Government places a heavy emphasis on the development of the agricultural sector. On the production side, it believes that the adoption of innovations and improved agricultural inputs is sound practice by farmers and must be supported. Studies in the diffusion and adoption of innovations have concluded that the variables affecting the farmer's decision to adopt a new technology can be divided into two types: first, the situational and structural aspects of the environment; and second, individual knowledge and perception (Barghouti 1980: 115). Among the elements of the first type are the high cost of necessary inputs such as improved seeds, fertilisers and pesticides, and the scarcity and inefficiency of public services and necessary infrastructure such as markets, roads and transportation. The Government has taken several steps to tackle these problems to increase the adoption rate of improved agricultural inputs. Among these measures are the following:

Government Exemption from Import Tax: The Government grants full exemption from import tax for all imported agricultural inputs. This policy should assist the provision of inputs to farmers at reasonable prices. However, problems sometimes arise between the Customs Department and agribusiness firms with respect to imported materials. Company managers interviewed indicated that the problem arises because the Customs Department lacks technical experts capable of determining whether or not certain imported materials are used in agriculture and deserve full exemption from import tax.

Direct Government Intervention: In 1975 the Government directly intervened in the agribusiness input market by importing a large quantity of fertiliser and selling it at a reduced price. The purpose was to encourage farmers to use fertiliser more systematically (El-Hurani 1975: 340). Since that time the Government has not repeated such direct action, but it tries to help farmers indirectly through the financial support and encouragement it extends to specialized organisations such as the JCO and the JVFU.

The Role of the JCO in the Supply of Inputs: The JCO, which is government-controlled and -funded, plays an important role with respect to the provision of inputs to Jordanian farmers in all areas except the Jordan Valley. Its activities here can be classified into two groups:

i) The JCO takes on the responsibility on behalf of the Government to provide certain inputs in sufficient quantities to satisfy all Jordanian farmers' needs. These inputs are clean and treated seeds of wheat, barley, chick-peas and feeds.

ii) The JCO provides farmer members with requirements such as fertilisers, pesticides and concentrated feeds, in addition to certain services such as chemical spraying, tillage and seeding. The JCO purchases inputs either by direct importing or by soliciting tenders if the quantity required is large; it may buy from the Amman agribusiness input market if the amount is small. After purchasing inputs, the JCO distributes them to farmer members on a loan basis. The selling prices are lower than the retail market price level by a margin of 5-10 per cent because the JCO makes bulk purchases (JCO 1981: 37).

The JCO serves a total farmer membership of about 42,000 distributed across 167 agricultural associations. Table 8.9 shows the size of total purchases and sales of agricultural inputs in 1981.

The Role of the JVFU in the Supply of Inputs: The JVFU's main objective is to carry out the task of developing agricultural production in the Jordan Valley. One of the major departments in this organisation is responsible for purchasing all kinds of agricultural materials and loaning them to Jordan Valley farmers. The JVFU, which is non-profit-making, actively studies the world agricultural input markets and tries to purchase at the best possible prices in order to provide

Table 8.9: Jordanian Cooperative Organisation, Size of Total Sales and Purchases of Agricultural Inputs, 1981 (JD)

Inputs	Sales	Purchases
Fertilisers	648,144	336,218
Agricultural equipment	24,023	7,025
Feed materials	891,522	963,249
Plastic	454,541	432,106
Seeds & pesticides	302,166	203,578
Poultry concentrated feeds	118,756	126,512
Other materials	672,770	642,669
TOTAL	3,111,924	2,711,365

Source: JCO 1981: 38.

inputs at cheaper prices than the Amman market level. The organisations also purchases its requirements either through direct imports or by soliciting tenders. The JVFU serves a total of 4,500 members by extending agricultural loans of about JD 3,000 for the preparation and planting of farm units of 30-44 du. This loan consists of 25-50 per cent in cash and the rest in goods, which include all the inputs the farmer requires. Table 8.10 shows the size of total purchases and sales of agricultural inputs during 1978-82.

Table 8.10: Jordan Valley Farmers' Union: Size of Total Sales and Purchases of Agricultural Inputs, 1978-82 (JD)

Year	Sales				Purchases			
	Fertilisers	Seeds	Pesticides	Total sales	Fertilisers	Seeds	Pesticides	Total sales
1978	109,783	20,776	17,122	147,681	122,533	226,928	22,963	172,423
1979	124,588	39,640	32,123	196,351	116,576	50,893	57,562	225,031
1980	297,563	104,287	49,653	451,503	428,720	80,268	10,354	519,342
1981	407,124	123,371	86,361	616,856	167,788	107,924	134,116	409,828
1982	711,211	322,484	153,763	1,187,458	377,389	262,867	42,807	783,063

Source: JVFU 1982: 6.

Government Regulations to Organise the Importing of Certain Inputs

The Government has issued certain regulations to watch over and organise importing and selling activities, as well as to protect farmers from potential health hazards and to ensure that the private sector imports a high standard of the three basic agricultural inputs, namely, fertilisers, pesticides and veterinary medicaments. The following are summaries of these regulations.

Regulation of Fertilisers: The Government issued a ministerial decree in 1975 setting out all the regulations and conditions which organise and control the importing and selling of all kinds of fertilisers. This decree consists of 44 articles, which define all the steps that must be followed by agribusiness firms wishing to deal in fertilisers. The main points of the decree are as follows:

i. All kinds of fertilisers must be registered at the MOA prior to import. The purpose of this advance registration is to ensure good-quality imports of all kinds of fertilisers, and thus to protect farmers from purchasing and using poor ones. To register, the applicant agribusiness must submit several documents showing the high quality of the desired brand of fertiliser.

ii. After the application for registration is approved, it must be renewed every five years. Registration may be cancelled if the firm does not import any of the registered fertiliser during the first year and a half from the date of registration.

iii. In addition to this registration requirement, the firm must acquire import permission from the MOA. In order to sell fertilisers, the firm must also obtain a sales licence from the MOA.

Regulation of Pesticides: The Government issued five ministerial decrees in 1983 which strictly regulate the import and sale of all chemicals. The following are the main points of these decrees.

Decree No. 11 consists of 16 articles setting out the conditions and procedures required to obtain an import licence from the MOA. This pesticide import licence is valid for one year and must be renewed at the end of every December. The company selling pesticides must keep a record of the total quantity imported, the date of importation and the daily sales.

Decree No. 12 consists of 21 articles and sets out the conditions and procedures for the registration of pesticides. All kinds of chemicals must be registered at the MOA before they may be imported. To register, the applicant agribusiness must submit several documents showing the effectiveness of the desired brand of pesticide. The registration of a pesticide may be cancelled if the material is not imported during the first eighteen months from the registration date, or if the material is not imported during a period of two years from the date of the last import, or if the materials are proved ineffective, or if they have harmful side-effects, or if it is disclosed that the materials are not allowed for use in the country of origin, or if registration was made with false information. Finally, registration must be renewed every five years.

Decree No. 13 consists of 22 articles and sets out the conditions for importing pesticides. It is not permitted to import a pesticide unless it has been registered at the MOA and the agribusiness has a valid import licence and sales permission. When the chemical materials arrive at the port of Aqaba, the firm must get clearance from the MOA.

Decree No. 14 consists of 17 articles and sets up the conditions for dealing with the sales of pesticides. Any firm which wishes to sell chemical materials must obtain a sales licence in advance from the MOA. The people in charge of sales must be qualified and possess a degree in agriculture, or at least a high school diploma in agriculture, and the store may not sell any kind of food for human consumption or animal feeds.

Finally, Decree No. 15 consists of six articles and sets out the pricing regulations for pesticides. The MOA has the authority to decide the ceiling on profit margin which should not exceed 30 per cent. This 30 per cent is to be shared between the wholesale firm (20 per cent) and the retail firm (10 per cent).

Managers interviewed complained that the two-year period is too short for the company either to import or to be faced with the cancellation of registration. They suggested that this period be extended to four years. They also reported that registration for new brands of fertilisers and pesticides usually takes a long time.

Regulation of Veterinary Medicaments: The Government issued an ordinance in 1967 which consists of 36 articles and sets out the requirements and conditions for dealing with veterinary medicaments. All veterinary medicaments must be registered at the MOA before inport. The firm applying for registration must submit all the necessary documents which certify the effectiveness of these materials and the company must obtain a sales licence for selling them. The MOA determines the wholesale and retail prices of all veterinary medicaments.

The Jordanian Agribusiness Input Market

The Jordanian agribusiness input market has shown healthy signs of development during the last 10 years, as many new firms have entered the market and made a positive contribution to competition. Major input selling activities are centred on Amman, where all large agribusinesses have their headquarters. There are about 30 firms dealing with four major agricultural inputs, namely, seeds, pesticides, fertilisers, and small agricultural tools and equipment. There are about 10 firms dealing with manufacturing and selling plastic materials and drip irrigation equipment, and there are also about 24 firms dealing with importing and distributing concentrated feeds and veterinary medicaments and vaccines.

In addition, there are about 10 big companies which deal with importing and selling large agricultural machinery such as tractors, grain drills, combines and tillage equipment. Some of these firms concentrate their business activities on agricultural equipment, while others deal with importing and selling agricultural equipment alongside their automobile businesses, such as Volvo, Ford and General Motors car sales.

Size and Location of Agribusiness Input Firms

In general, there are three main sizes of agribusinesses which deal with the major inputs such as seeds, fertilisers, pesticides and small agricultural tools. Large companies are the principal operators in the importing activities of the input market. There are eight companies of this size, and it can safely be stated that they import most purchased agricultural inputs. Five of these firms have been in the Jordanian agricultural input market for over 20 years, have a high level of professional expertise and are well known to farmers with respect to certain brands of vegetable seeds. As previously mentioned, all large firms are located in Amman, with headquarters in the city centre. Some of them have branches in the Amman Central Wholesale Market (ACWM) area, called Wahdat and considered a most strategic location for marketing and distributing inputs, as farmers bring their produce there to sell and purchase the inputs they require.

Company managers interviewed said that their firms conduct experiments to test new varieties of seeds and new brands of pesticides. Most of these testing activities are carried out in the Jordan Valley. More elaborate discussion of this activity will be included in a later section on informational environment.

Medium-sized companies are the second type of supplier. There are about 20 agribusiness firms located near the ACWM which sell direct to farmers. Most of them depend on importing companies to supply them with inputs. All these agribusinesses reported that they do not themselves conduct experiments but depend on large firms for testing new varieties. Third, there are a few small firms located in agricultural areas such as the Jordan Valley and cities such as Zarqa, Irbid, Jerash and Mafraq. They depend entirely on the large companies to supply them with all agricultural inputs. These local stores sell direct to farmers, who usually buy seeds and pesticides in small quantities. Table 8.11 summarises the activities of these firms.

Table 8.11: Numbers of Private Agribusiness Firms Dealing with the Supply of Agricultural Inputs

Activities	No.
Large companies dealing with importing & distributing major agricultural inputs (seeds, pesticides, fertilisers, small tools)	8
Medium-sized firms dealing with selling major agricultural inputs (seeds, pesticides, fertilisers, small tools)	20
Companies dealing with importing & selling concentrated feeds, veterinary medicaments, vaccines	24
Companies dealing with manufacturing & distributing plastic materials & pipes & installing drip irrigation systems	10
Large companies dealing with importing, distributing & selling large agricultural machinery	10
Companies deaing with manufacturing plastic packing containers	4
Companies dealing with tree seedlings	8

Source: Information collected from several interviews with agribusinesses.

Imports of Agricultural Inputs
Jordan depends mainly on imports to provide its required agricultural inputs. Importing activities are undertaken largely by private agribusiness companies and, of late, by some semi-government organisations, such as the JCO and the JVFU. Importing activities by private firms take the following pattern:
Input Items which are Familiar and have been Well Known to Farmers for Many Years: These materials are imported by the large companies after they have studied the market and estimated their share in it.
New Brands of Input: These are usually introduced to the importing companies through catalogues and brochures sent by exporting companies. One manager interviewed said that a sales representative of an exporting company sometimes visits agribusinesses to promote new varieties and brands. With respect to new inputs, all managers interviewed gave the following sequence of actions. First, they order a trial sample to test the new materials as well as to introduce them to selected and trusted farmers to use and report results. If the new materials give good results, the companies order a small quantity for the first few cropping seasons and increase the amount gradually as farmers become accustomed to them.

Most agricultural inputs are imported from Western Europe and the US and shipped by sea to the port of Aqaba. However, some shipments have been sent lately by truck from Europe. Some kinds of fertiliser imported from Arab countries are also transported by truck. One manager interviewed reported that a few items might occasionally be shipped by air, for example, some varieties of seeds from Holland. After imported materials are released from the port of

Aqaba, they are transported by truck to the firm's storage place, which is usually located in Wahdat near the ACWM.

Informational Environment

One of the major factors affecting the farmer's decision to adopt a new input technology is his own knowledge and perception of this improved input. Information available to Jordanian farmers on the advantages of different improved inputs is generally meagre and erratic. As for the role of the private sector, large agribusiness companies provide some technical information to farmers on the results of using different brands of seeds, pesticides and fertilisers. Interviews with the managers of these firms showed that their technical departments conduct some experiments and demonstrations on farmers' land, especially with seeds and pesticides. In addition, agricultural engineers and technicians from these companies make scheduled daily visits to farmers' fields to discuss problems encountered and to give technical advice on specific problems, such as plant disease. After the field trips, agricultural engineers submit their daily reports to the head of the technical department, and a weekly meeting is usually held to discuss the results. Managers interviewed reported that their companies concentrate extension activities on Jordan's principal irrigated areas, such as the Jordan Valley, Jerash, Baqa'a and the newly developing irrigated lands in Balqa and Amman districts.

Other agribusinesses which are smaller, like those in Wahdat and the local store in agricultural areas, also provide useful information to farmers visiting these places and requesting specific advice on the best kind of pesticides. It is noted that merchants in these stores provide very careful instructions and advice on how to use chemicals. It is also noted that farmers trust the information and advice directed at them through the daily agricultural radio programme broadcast from Amman Radio Station in the morning. This programme, called 'With the Farmer' has been able to reach farmers for over 15 years and has become a useful reference for farmers making decisions on the quantity of inputs required according to their own experience, as well as in consultation with the local merchants. All farmers interviewed asserted that they buy seeds, fertilisers and pesticides at the time of application, which means immediately they decide to prepare the land and plant the seeds. Because the local stores try to make all necessary inputs available at the right time, farmers feel that there is no need to buy before the season and store the inputs themselves.

Distribution and Transportation of Agricultural Inputs

Large agribusinesses are the principal distributors and transporters of inputs. These companies distribute their materials via three main channels. Local stores in rural areas are considered by some major importing companies to be one of the most important channels for distributions. Companies usually provide a transportation service for agricultural inputs free of charge to local stores in rural areas. This free service is one way of promoting and increasing sales of the large importing firms. However, managers interviewed reported that such companies usually charge a transportation fee for fertilisers, being heavy and bulky, and because their profit margin on them is narrow. Large firms give local stores a discount of 10-15 per cent for their profit margin.

The second channel of distribution for these large firms is provided by the small- and medium-sized agribusiness located in Wahdat near the ACWM. Most of these firms depend on the importing companies to supply them with agricultural inputs. The third channel of distribution is provided by the semi-public agricultural origanisations, namely, the JCO and the JVFU, which buy certain inputs from companies in quantities which do not merit being imported direct or even by tender.

The Pricing Mechanism of Agricultural Inputs

Despite the Government's previous intent to fix the prices of certain agricultural inputs such as pesticides and fertilisers through ministerial decree, it does not enforce these regulations, which set the profit margin as 20 per cent to wholesalers and 10 per cent to retailers. It seems that the Government has realised that it is better to leave the input market free of intervention and to allow increasing competition to determine the equilibrium price to farmers. In this respect, the Governments has adopted a liberal market policy and is trying to increase and encourage more private agribusinesses to strengthen the degree of competition among themselves. The Government has also pulled out of direct intervention, and supports semi-public organisations such as the JCO and the JVFU competing with private firms on the import market.

Although this government policy has benefited farmers generally; a close investigation of the input market indicates that a few large agribusinesses still control, to some extent, the imported input market in Jordan. This is especially the case with respect to certain brands of input which have become very popular with farmers who insist on purchasing the ones which have proved to be most effective. This pattern still exists, but is decreasing as new inputs develop continuously and the JCO and JVFU work extremely hard to convince farmers to use them and thereby diversify the sources of inputs. The results of these efforts have been very successful with certain varieties of seeds and have led to marked price reductions.

The prices of veterinary medicaments and vaccines are still laid down by the MOA. According to the ordinance issued in 1967, the MOA has the authority to determine both the wholesale and retail prices of all veterinary medicaments. Managers of firms which sell these materials complained when interviewed that world prices sometimes rise and they have difficulty in persuading the Government to allow price increases.

Methods of Payment and the Role of Central Market Middlemen in the Provision of Inputs to Farmers

Generally speaking, Jordanian farmers are unable to finance their farming activities from their own resources, and seek payment concessions from agribusinesses until the harvest season. However, all managers interviewed who deal direct with farmers pointed out the acute problems of delay in payment and false cheques issued by farmers which have been increasing over the last few years. This problem may be attributable to some poor seasons and bad markets, both of which depress farmers' revenues.

After close investigation it can be stated categorically that the middlemen who are stationed at the ACWM in Wahdat play a pivotal role in providing necessary inputs to a large number of farmers. It has been customary for many years for vegetable farmers in particular to depend on middlemen to finance their farming operations by the extension of loans in cash and in kind, these commission agents arranging with their company clients to supply farmers with all their necessary inputs. The usual arrangement is for the middleman to write a purchasing order and send it with the farmer. At the end of each month the agribusiness company will collect from the middleman.

Questions were asked to find out whether the middleman receives a discount from the agribusiness companies for dealing with them on behalf of the farmers. Some middlemen interviewed asserted that they did not get any discount, and that they paid the company exactly the amount that farmers promised to pay at harvest; others admitted that they received a 10 per cent discount.

In the harvest season, farmers are expected to market their produce through the middlemen who financed them during the planting season. Middlemen are faced with the risk that the harvest might be bad or prices low, which means that farmers will be unable to pay off all their debts and will have to carry them over to another season. The middlemen interviewed also pointed out another problem. Some farmers do not fulfil their promises at harvest time but market their produce through other middlemen in order to avoid paying their debts. Because this bad behaviour has been on the increase in the last few seasons, there has been a tendency on the part of the middlemen to reduce sharply their lending services to farmers.

References

APC (Arab Potash Company), *Annual Report, 1982*, Amman.

Barghouti, Shawki, 1980, (The role of agricultural cooperatives in introducing new technology in wheat production', in *Improving Dryland Agriculture in the Middle East and North Africa*, Food Research Institute, Stanford University and the Ford Foundation, Middle East Regional Office, Cairo.

DOS (Department of Statistics), 1979, *Statistical Yearbook 1974-75*, Vol. 30, Amman.

—— 1981, *Agricultural Statistical Yearbook and Agricultural Sample Survey, 1976-81*, Amman.

FAO (Food and Agriculture Organisation), a. 1976, *Production Yearbook, 1970-75*, Rome.

—— a. 1980, *Production Yearbook, 1976-79*, Rome.

—— a. 1981, *Production Yearbook, 1980*, Rome.

—— b. 1976, *Trade Yearbook, 1970-75*, Rome.

—— b. 1980, *Trade Yearbook, 1976-79*, Rome.

—— a. 1981, *Trade Yearbook, 1980*, Rome.

El-Hurani, M. Haitham, 1975, 'Economic Analysis of the Development of the Wheat Sub-sector of Jordan', unpublished PhD dissertation, Iowa State University, Ames, Iowa.

JCO (Jordan Cooperative Organisation) *Annual Statistical Report, 1981*, Amman.

JFI (Jordan Fertiliser Industry Co. Ltd), *Company Bulletin, 1982*, Amman.

JVFU (Jordan Valley Farmers' Union), *Annual Report, 1982*, Amman.

9 AGRICULTURAL CREDIT AND FINANCE

Nureddin A. Taqieddin

I. Introduction

Although the economy of Jordan has been transformed in the last three decades from a predominantly agrarian into a more balanced multi-sectoral one, agriculture still occupies an important place. It provides direct income and employment for over 15 per cent of the local labour force and it has to be relied upon to reduce the substantial and chronic deficit in the economy's agricultural trade account.

The consecutive Development Plans of Jordan have, therefore, placed special emphasis on the development of the agricultural sector. While actual farming is done by the private sector, the Government has taken on the responsibility of providing infrastructure as well as the basic services required to facilitate and accelerate agricultural development. The importance of low-cost agricultural credit as a strategic element in the development process has long been recognised by the Government of Jordan. Due to various risk and uncertainty elements involved in the farming business, privately owned capital does not pass smoothly into medium- and long-term agricultural investments. Commercial banks are hesitant to provide finance, if any, for longer than one year, and then only at a relatively excessive cost to borrowers. These factors, together with the generally low savings of the agricultural communities, and the high cost of modern farming technologies, go to explain the strong demand for official agricultural credit.

Considerable efforts have been made since the late 1950s to organise and upgrade the public agricultural credit system in Jordan. Currently, most credit is provided by public sources. Due to a number of constraints and limitations, however, these sources have not been able to satisfy fully the needs of farmers. Private and commercial sources of credit have had to fill the gap. Hence, agricultural credit systems in Jordan can be divided into two general sub-systems: public and private. A number of credit agencies and sources are included under each sub-system. In what follows, each sub-system together with the relevant agencies and sources of credit will be separately presented and discussed.

II. The Public Sub-System of Agricultural Credit

The public sub-system of agricultural credit consists of three specialised agencies: the Agricultural Credit Corporation (ACC), the Jordan Cooperative Organisation (JCO) and the Jordan Valley Farmers' Association (JVFA). The main characteristics of these agencies are that their capital is provided mainly by the Government for the purpose of extending soft development credit, and that generating profit is not their underlying objective.

In 1980 the balances of outstanding agricultural loans of these credit institutions were JD 14.05 mn (59 per cent), JD 9.18 mn (38.5 per cent) and JD 0.62 mn (2.5 per cent) for the ACC, JCO and JVFA respectively. Each of these agencies will be presented separately below, along with a brief introductory account.

The Agricultural Credit Corporation (ACC)

Introduction

This is the largest specialised agricultural credit agency in terms of the size of loans extended to farmers over the years. It was established in 1959 as a semi-autonomous government body to provide supervised credit for productive agricultural projects. Its current overall financial status is reflected by its balance sheet and revenue and expenditure statements as at the end of the years 1980-1982. (See Tables 9.1 and 9.2.) Its outstanding loans have increased from JD 3.75 mn in 1960 to JD 20.0 mn at the end of 1982 (ACC 1960, 1980, 1981 and 1982).

The authorised capital of the ACC is JD 12.0 mn, of which JD 8.36 mn are already paid up. Total assets by the end of 1982 approached JD 21.0 mn, financed from the ACC's own capital (40 per cent), retained earnings (10 per cent), foreign long-term loans (34 per cent), local loans (13 per cent) and the balance (3 per cent) from other sources. In addition to amounts collected from interest and from outstanding loans, the ACC's capital and borrowed funds represent the major sources from which the Corporation has financed its lending operations.

The ACC provides short-term seasonal loans at 8 per cent interest, and medium-term loans extending to 10 years at 7 per cent interest. Loans for more than 10 years, although permissible, are in practice very rare. Interest on seasonal loans has to be collected in advance. If the seasonal loan is repaid on time, interest of 1 per cent is refunded to the borrower; otherwise a penalty charge of 1 per cent is added. So far, no penalty charges have been imposed on delinquent medium-term loans. Only simple (not compound) interest rates are applicable to the ACC.

Objectives

The ACC aims to provide loans to farmers at a reasonable cost to finance productive agricultural projects which contribute to agricultural development in accordance with national development plans. Although profit is not in itself an objective, the ACC needs to generate enough income to cover its costs. Interest on loans is the major source of such income.

Environmental Characteristics

Physical: The ACC operates through its head office in Amman and its 14 branch offices distributed over the East Bank of the Jordan.* The head office, located in a convenient part of the capital, is easily accessible by both public and private transport. The branch office for Amman district shares premises in the head office. Access by farmers to the ACC has been greatly facilitated by the existence of its branch offices, while at the same time they have made it easier for ACC staff to visit farms all over the country to carry out their supervisory duties. The ACC employed about 240 people in 1982, distributed between the head office and branch offices. Each branch office normally has a branch manager, a loan supervisor, an accountant, a clerk, a loan collector, a driver and a messenger. The most intensive agricultural activity is found in the Jordan Valley area, where virtually all cultivable land is under permanent irrigation. Three branch offices operate here, more than in any other area.

* Four branch offices used to operate in the West Bank of Jordan before it was occupied by Israel in 1967.

Table 9.1: Agricultural Credit Corporation: Consolidated Balance Sheet for the Years 1980, 1981 and 1982 (JD)

Assets	1982	1981	1980	Liabilities, capital & retained earnings	1982	1981	1980
				Current liabilities			
Cash at banks	146,571	379,788	739,703	Deposits	231,625	250,786	178,873
Debit advances	18,262	169,273	149,465	Credit advances	2,739,654	2,352,647	1,375,066
Expenses of loan collection	1,032	664	748	Loans	7,072,597	5,942,872	4,683,578
				Interest on loans	123,582	68,221	30,559
Loans	20,216,954	17,279,724	14,048,550	Total current liabilities	10,167,458	8,614,524	6,270,176
Net book value of the fixed assets	329,108	258,192	224,781	Capital & retained earnings			
				Capital	8,368,801	7,365,068	7,108,360
				Retained earnings	1,972,727	1,747,600	1,429,608
				Total capital and retained earnings	10,341,528	9,112,668	8,537,968
				Staff housing funds & retained earnings			
				Capital	300,000	300,000	300,000
				Total staff housing funds & retained earnings	367,017	360,599	355,103
TOTAL ASSETS	20,876,003	18,087,791	15,163,247	TOTAL LIABILITIES, CAPITAL & RETAINED EARNINGS	20,876,003	18,087,791	15,163,247

Source: ACC 1980-82.

Table 9.2: Agricultural Credit Corporation: Statement of Revenues and Expenditures for the years 1980, 1981 and 1982 (JD)

Revenues	1982	1981	1980
Received interest on loans	902,697	843,536	645,101
Bank interest	47,868	46,392	64,171
Other revenues	7,354	9,581	6,871
TOTAL REVENUES	957,919	899,509	716,143
Expenditures			
Salaries & wages	370,924	342,766	284,289
Paid interest	242,069	147,448	137,412
Interest adjustments & interest write-offs	10,481	8,447	21,326
Remunerations	7,520	6,574	6,232
Fuels, parts, tires & supplies	27,800	27,778	16,936
Fixed assets depreciation	6,031	5,180	4,795
Miscellaneous expenses	44,940	24,210	20,337
Postage, telegraph & other	1,017	1,155	1,280
Rent	2,235	1,706	1,413
Other expenses	19,775	16,253	6,041
TOTAL EXPENDITURES	732,792	581,517	500,061
NET PROFIT	225,127	317,992	216,082

Source: ACC 1980–82.

Institutional: The ACC is governed by the 1963 Law No. 12, and the ensuing by-laws which regulate its organisation and operations. According to the Law, the ACC is a semi-government body with financial and administrative autonomy. It has a Board of Directors on which the Government, the banking sector and farmers are represented. The Chairman of the Board is himself the Director-General of the corporation, assisted by a Deputy Director-General. Both executives are appointed by cabinet resolutions confirmed by royal decree. The Board decides upon overall policies concerning the ACC's operations and follows up on lending activities and financial status. The Director-General of the ACC, assisted by the Deputy Director-General, is fully in charge of the administration of the ACC and the implementation of its policies and regulations.

Work at head office is distributed between three main departments handling finance, loans and administration. Each department is headed by a senior executive entitled Assistant Deputy Director-General. Three other senior officials carry out the supporting functions of internal auditing, general supervision and foreign relations. Branch managers who report to the Deputy Director-General are responsible for all functions in their branches. Figure 9.1 shows the organisational structure of the ACC's administration. Out-of-the-ordinary personnel affairs are handled by the Director-General assisted by a special personnel committee in accordance with the provisions of the 1963 By-law No. 3 which, among other things, sets the salaries for ACC personnel.

Figure 9.1: Agricultural Credit Corporation: Organisational Structure

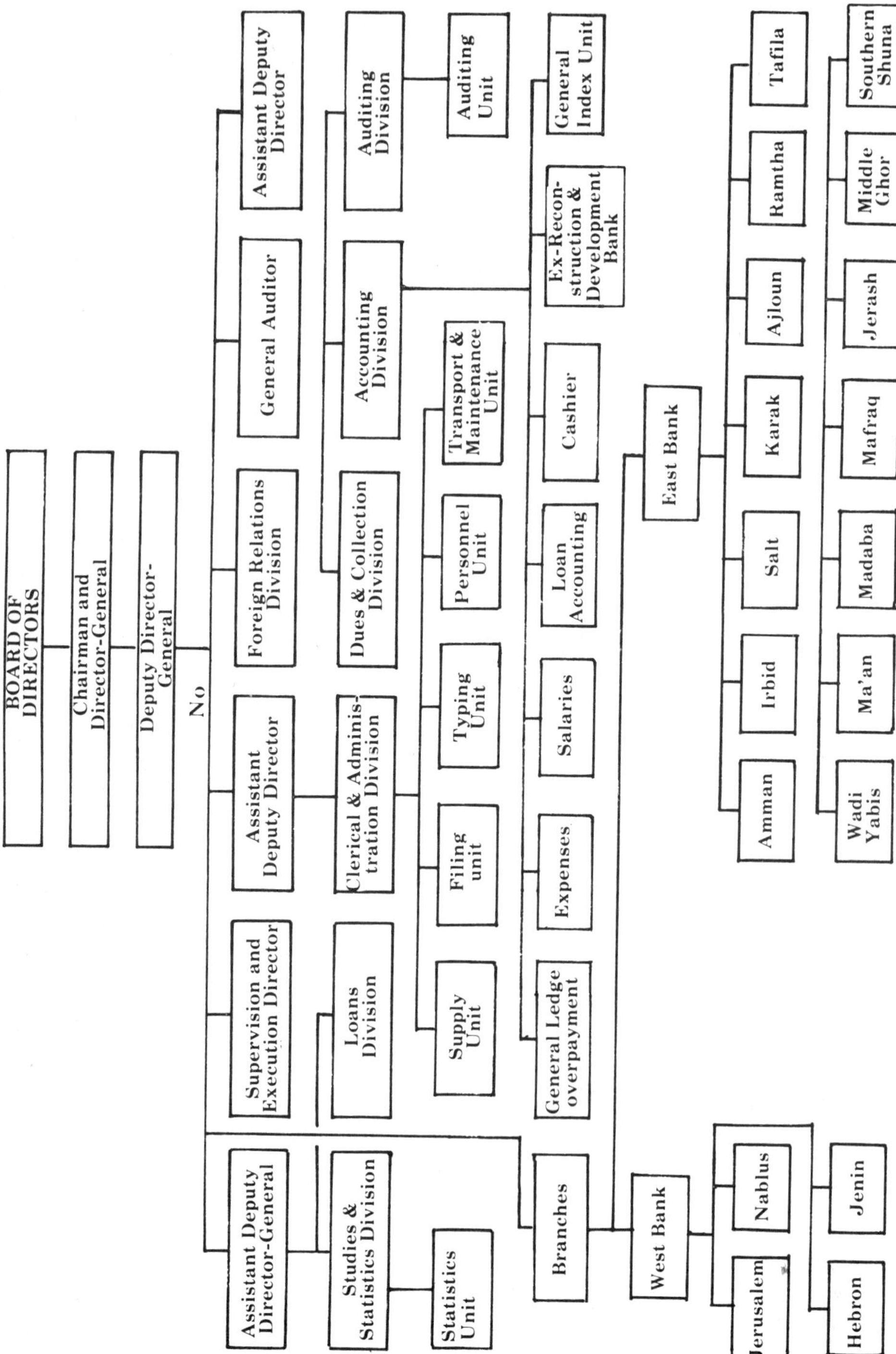

By virtue of its statutes, the ACC is a non-profit-making development agency, and this fact underlies and dictates its attitude towards farmers. Loans are given under close supervision to guarantee proper handling by farmers. Attempts are always made to simplify and facilitate lending rules and procedures. Collection policies are also flexible and take into consideration the unavoidable problems which farmers frequently encounter. These policies need to be reviewed with the objective of raising collection rates, which are currently unjustifiably low.

Regular working hours at the ACC are six hours daily, with one day (Friday) as a weekend day off. These are the normal working hours for government agencies and banks in Jordan.

Coordination between the ACC and other government agencies is provided through these agencies being represented on the ACC's Board of Directors and also by the ACC being represented on the boards of directors of other agencies or institutions. Most important is the ACC Director-General's membership of the Higher Agricultural Council (HAC) of Jordan, which establishes the overall agricultural policies in the country. The National Planning Council (NPC), the Ministry of Agriculture (MOA), the Land Registry Department (LRD) and the Central Bank of Jordan (CBJ) are represented on the ACC Board by senior officials. The ACC has to take into consideration national plans and policies relating to agriculture which are mainly set out by the HAC and the NPC. Where public laws and regulations require it, the ACC instructs its borrowers to obtain certificates or licences from the respective government agencies which issue them, such as the Ministry of Health (MOH), the MOA, the Ministry of Municipal Affairs and Rural Environment, and the Natural Resources Authority (NRA).

Any competent citizen of Jordan is eligible for credit from the ACC to finance a productive agricultural project. The ACC statutes place no upper limits on the size of any individual loan, provided it is for productive purposes and the amount does not exceed 75 per cent of the assessed value of the security provided by the farmer to guarantee the loan. Land and landed property are the most commonly used collateral. Although a very unusual practice, the pledge of two guarantors can be accepted by the ACC to guarantee loans of a limited amount.

Certain types of projects may not be financed by the ACC if the country or particular area already has more than enough of them. Licences, which have to be obtained from various government agencies to ensure eligibility, place certain restrictions on loans.

The ACC has been empowered to consolidate its right of redress in the event of default. It enjoys full exemption from taxes and customs duties as well as preferential treatment by government departments. Land or property mortgaged as security for any loan can be sold by the ACC, without recourse to official courts, by public auction. In using these powers the ACC moves slowly, cautiously and with much discretion, deliberately attempting to protect the rights of farmers as far as possible.

Informational: Borrowers usually make their own estimates of the farming activities they need to finance. They are requested to state the amount of loan required in the space provided on the application when they first apply for a loan at a branch office. Loan applications ask for introductory information which farmers have to fill in themselves. The borrower's previous loans, if any, are also listed on the application by the branch office, with details of current status and any existing arrears. Normally, dues from previous loans have to be fully repaid before a new loan can be extended.

If both the farmer and the project to be financed are considered eligible, the loan supervisor of the branch makes a visit to the farm site. Here he files, together with the farmer, a farm plan which includes detailed information concerning the borrower, his family, his farm, his assets and liabilities, as well as the farming activities he intends to carry out, with the itemised costs of each activity. The approximate costs are generally known to the ACC. The farm plan, together with an inspection report prepared by the loan supervisor on his farm visit, is attached to the loan application in one file carrying an identification number assigned by the branch to the borrower. The inspection report includes any specific information the loan supervisor might have collected, or observations which have a bearing on the loan to be made.

Farm business in Jordan is tax-exempted, so the financial resources of the farmer cannot always be obtained from formal records. Land ownership, on the other hand, is very well documented, so landed property is generally preferred as collateral.

A considerable amount of information is available on the aggregate size and types of loans extended by other specialised sources of credit in Jordan. It is more difficult to obtain information about the indebtedness of ACC borrowers to these agencies. As a precautionary measure, the ACC normally requires that the land or property offered to guarantee a loan be placed under first mortgage in its favour. In principle, the relationship between the ACC and other specialised credit sources is complementary rather than competitive.

Activities

Obtaining Finance: The ACC is entitled to enter into loan agreements with local or foreign sources of finance, subject to the approval of the Government. Foreign loans obtained by the ACC have all been guaranteed by the Government. The CBJ provides credit facilities against credit instruments up to JD 3.2 mn. The term for these loans depends upon the maturity dates of the documents but cannot exceed five years. The ACC is more likely to expand its foreign loans component along with expansion of its capital through government transfers to cover its future needs. Collections from principal and interest dues on farmers will provide the ACC with its basic operating funds.

The ACC is also entitled by its statutes to accept deposits from clients. Due mainly to the high rates of interest it has to pay on deposits relative to the reduced rates charged on its loans, this source of funds has not yet been tapped.

Granting Loans: The ACC extends loans to farmers in accordance with a set of rules and regulations adopted by the Board of Directors and applied by all branch offices. Some of these rules might occasionally be changed or amended to fit the circumstances of the country, or of a certain area.

Most ACC loans are medium-term, which may extend up to 10 years. Although allowed by the statutes, longer-term loans are very seldom made. The remaining ACC loans represent seasonal loans of up to one year to finance maintenance and operational farm costs. Medium-term loans are given to finance capital formation costs, which include land reclamation and development, tree planting, irrigation, installation of plastic houses, livestock and poultry farms, farm machinery, dairy farms, and so on. Loans for the purchase of farms are issued only under exceptional circumstances and mainly to prevent the fragmentation of small farms.

All competent citizens of Jordan are eligible for loans, provided the projects to be financed are agriculturally feasible on land owned or rented by the borrower. The borrower's ability to run the project, the sufficiency of the security offered to guarantee the loan and the past performance of a former borrower are additional criteria with regard to lending. According to the current lending rules and regulations of the ACC, the extension of a loan involves the procedures outlined below.

First, the loan application is submitted by the farmer duly filled in at the branch office under whose jurisdiction the project site lies. The branch office completes the application with information about any previous loans the applicant might have had from the ACC. Any amounts due from past loans must be repaid before a new loan can be considered. Land ownership and location are checked with the Department of Land Registration (DLR) to verify statements.

The branch office prepares a report with recommendations based upon a staff visit to the site, together with a farm plan which includes all necessary details about the farmer, his income, his guarantees, the project site and the project. A branch committee then reviews the application and makes its decision if the loan amount does not exceed JD 200. If it does exceed this amount, the committee then makes its recommendation to the Central Committee at the head office. Each branch has a branch committee headed by the branch manager with two other members representing the MOA and the farmers. The Central Committee is headed by the Deputy Director-General, with the assistant deputies of finance and loans and a farmers' representative as members. The recommendations of the branch committee are received by the Loans Department at the head office where they are evaluated, cost-checked and put up to the Central Committee for final consideration. This Committee makes lending resolutions which take effect upon approval by the Director-General. If the loan amount is over JD 30,000 the consent of the Board of Directors also has to be obtained. Normally, ACC loans do not cover the full cost of the project, in order to encourage the farmers' savings and commitment.

Second, approval of a loan is communicated to the relevant branch office which requests the LRD to mortage the property for the ACC's benefit and enters into a loan agreement with the borrower. The loan document (bond) specifies the amount, the purpose, the term, the repayment schedule and the interest rate of the loan. It also describes the properties provided as security and specifies the degree of mortgage placed upon them. Measures to be taken in the case of default are also outlined in this document. The average time taken to grant a loan is about three weeks.

Copies of the basic documents are kept in the branch office and at the head office under the same file number for each borrower.

Checking Credit-Worthiness: The credit-worthiness of a borrower is determined by various factors such as his reputation in the area, his previous loan record with the ACC, and his knowledge and experience of the farming activity financed by the loan. The adequacy of the guarantee is, in all cases, a major factor.

Setting Interest Rates: Interest rates are set by the Board of Directors. Currently, the rate charged on seasonal loans is 8 per cent, plus 1 per cent as penalty for default. Early payments are encouraged by a refund of 1 per cent. Medium- and long-term loans carry 7 per cent interest, unless the loan is for dryland farming with an amount of less than JD 5,000, when the rate drops to 6 per cent. Commercial banks charge an average rate of about 10 per cent on short-term loans up to one year. In setting the interest rates, the Board of Directors aims to provide credit to farmers at the lowest possible cost, allowing for the overall costs of running the ACC.

Method of Repayment: The term of a loan is determined on the basis of the productive life of the financed project. ACC loans are categorised according to type of activity with a standard term assigned to each type. Dates due for repayments are set to match the production season in each case as far as possible. Citrus tree-planting loans, for example, are given for a term of 10 years, with the first 5 five years as a grace period when only interest payments are required. Because full production of the trees starts only after the fifth year the loan is scheduled to be repaid in five equal instalments starting in the sixth year. The date due is set to be 30 November of each year, to match the production season of the trees.

Action in Cases of Default: Branch managers are responsible for follow-up on the collection of amounts due. Borrowers are normally notified ahead of time and reminded after dates due to make repayments. ACC collectors are always on the move for this purpose. Farmers, however, are apt not to make prompt repayments on or immediately after dates due and the ACC is not very strict in this respect. Partial payments are accepted easily and, where justifiable, rescheduling of instalments is not uncommon.

In cases of unjustified default the ACC is authorised to enforce the law, leading to the public sale of the borrower's mortgaged property. The procedure involved, however, is generally slow and lengthy and the actual transfer of a title is rarely carried through. Borrowers usually hurry to pay their debts or to arrange settlements before legal action reaches its conclusion. In general, the ACC resorts to legal action only as a last resort and tries to protect the farmer's rights in all circumstances.

Deposit-Taking: Although entitled by law to take deposits, the ACC had not used this right so far. The main reason may be the high rate of interest which has to be paid on deposits in a competitive market, compared with the relatively low rates charged by the ACC on its loans.

Payment Procedure: Branch managers receive renewable loan funds from the head office entitled 'loan advances' for farmers. A first payment is made to enable the farmer to start the work, and subsequent payments are made in accordance with the progress of work as reported by the branch staff after visiting the project site each time. Most loans are paid out in three instalments extending over a period which varies according to the nature of the project and the speed of work. Branch managers can pay cheques not exceeding JD 1,500 without reference to head office. Cheques for larger amounts have to be co-signed by authorised officials at head office. Payments are recorded on identical individual loan cards for each borrower kept at the branch office and at head office.

Repayment Procedures: Repayment of loans may be made by borrowers to field collectors, or at the relevant branch office, or at head office.

Amounts collected at the branch office are deposited in a special account at a local bank and transferred by cheque to head office every few days. Amounts collected at all branches, and collections made at head office, are deposited daily in a central bank account from which all payments, including loan advances to branches, are made.

At head office, the Financial Department classifies the collections into interest and principal counterparts and enters the relevant transactions into the accounting records. Copies are sent to the respective branch offices to enter the same data on their loan cards after being cross-checked.

The activities above are outlined in the following flow charts (Figures 9.2-9.5).

The Jordan Cooperative Organisation (JCO)

Introduction

While the ACC specialises in financing individually owned agricultural projects, the JCO aims to promote cooperation and the cooperative movement in all types of economic activities in Jordanian society.

The cooperative movement in Jordan was officially started in 1952, when the Law of Cooperatives was first enacted, setting off a process of organisation, gradual evolution and expansion. It was not until 1968 that the present JCO was established to take on the full responsibility of looking after all aspects of cooperation in the country. In carrying out its functions, the JCO provides various types of service necessary for the promotion of the cooperative movement. Most important among these is the extension of loans and administrative support to cooperative societies. The cooperatives served by the JCO are not confined to agriculture: they belong to various economic sectors and have a wide

Figure 9.2: Agricultural Credit Corporation: Lending Activities

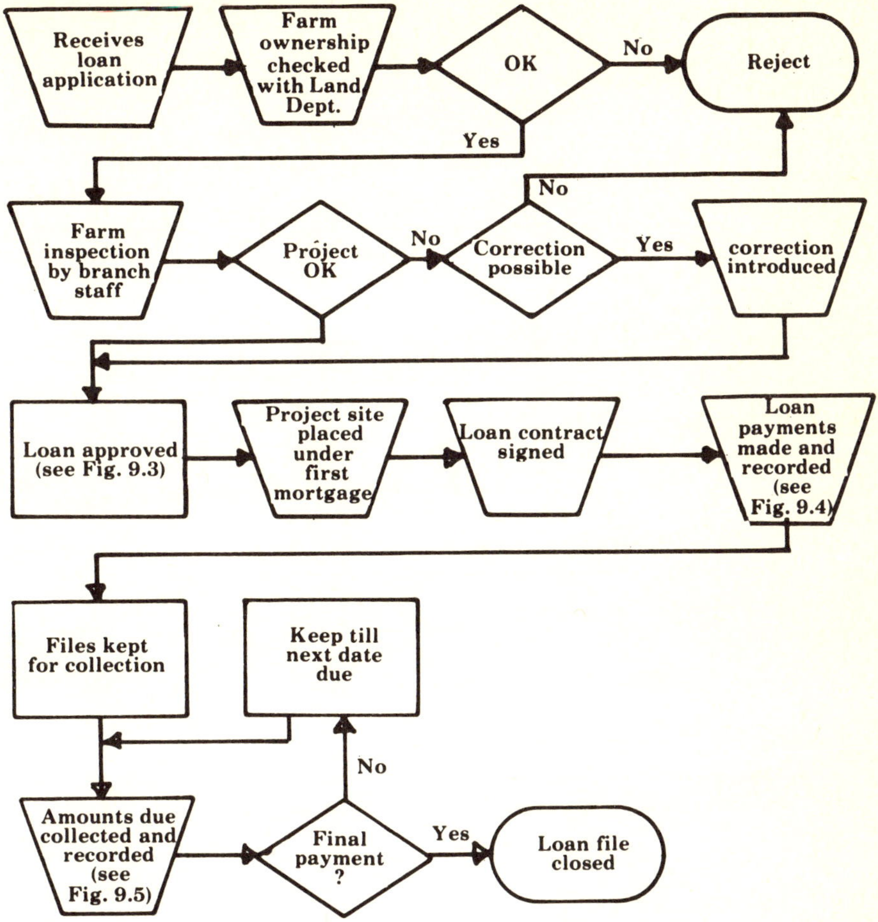

Figure 9.3: Agricultural Credit Corporation: Approval of Loans

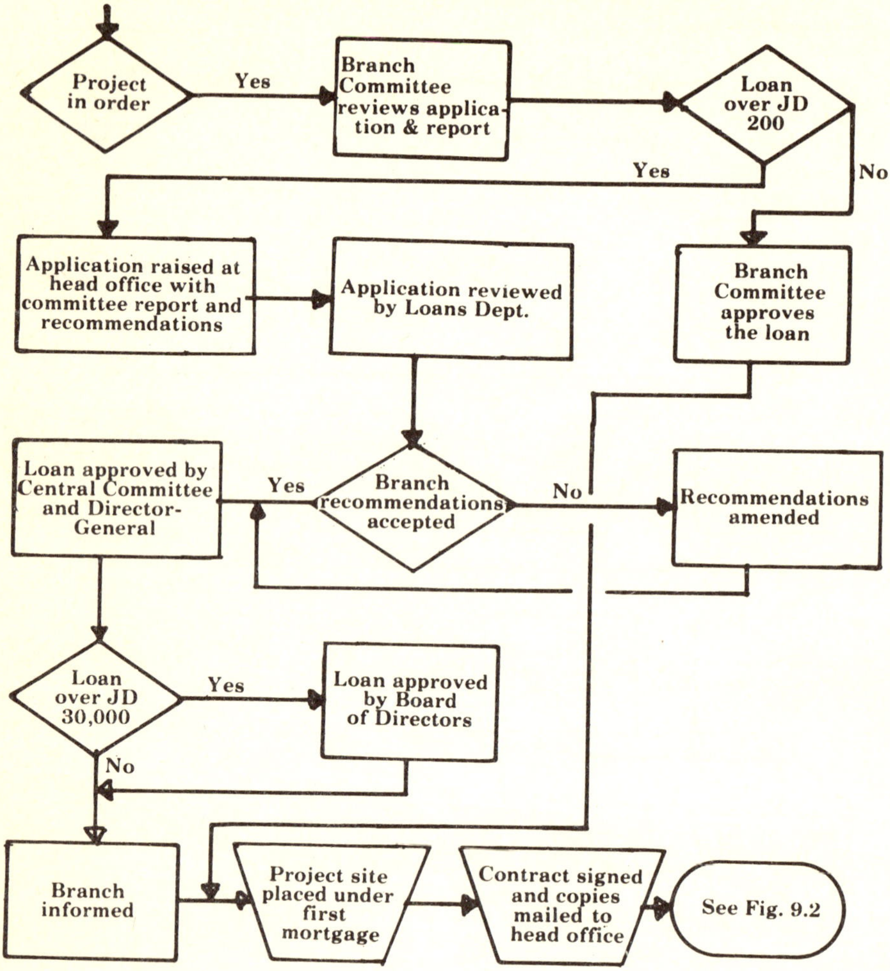

Figure 9.4: Agricultural Credit Corporation: Payment of Loans

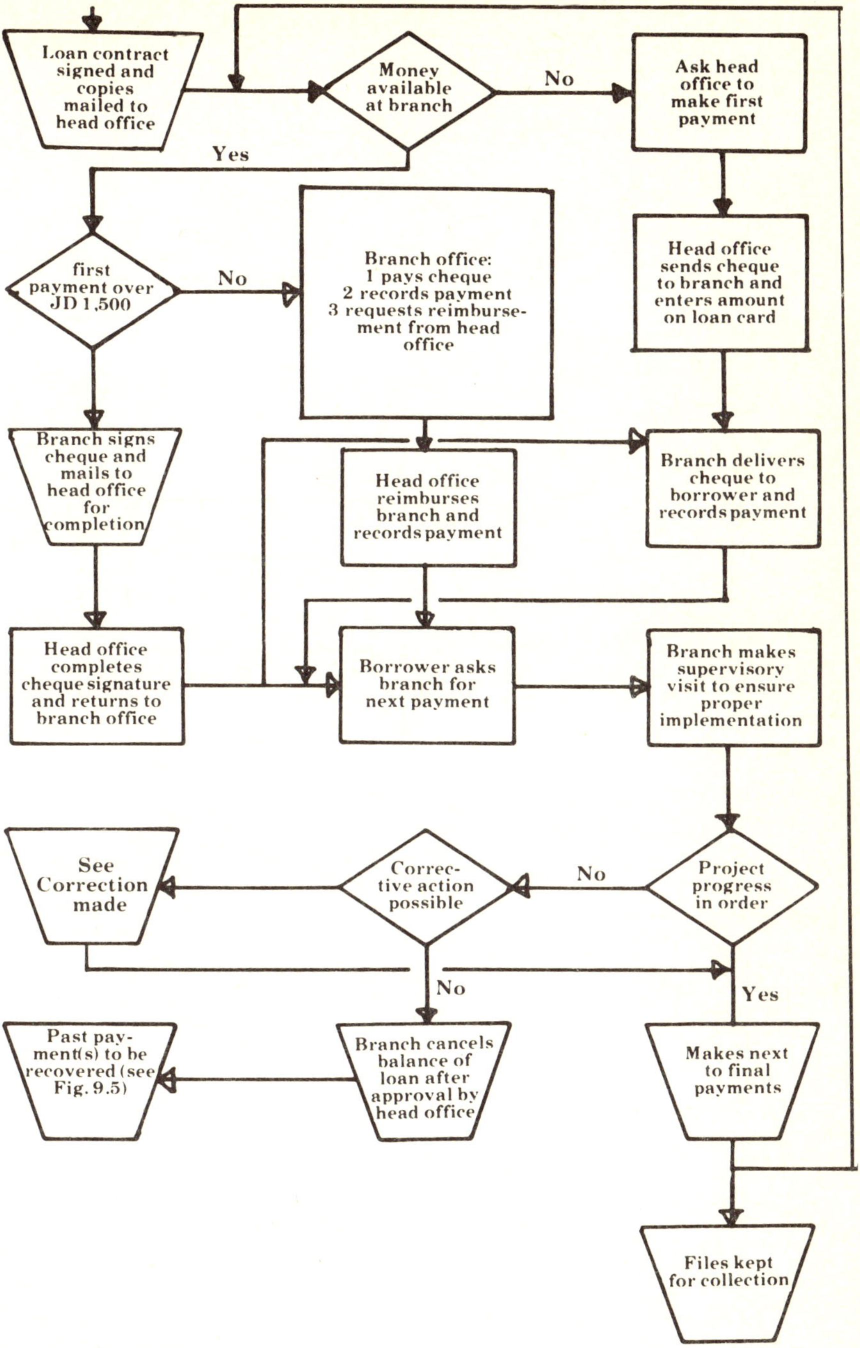

Figure 9.5: Agricultural Credit Corporation: Repayment of Loans

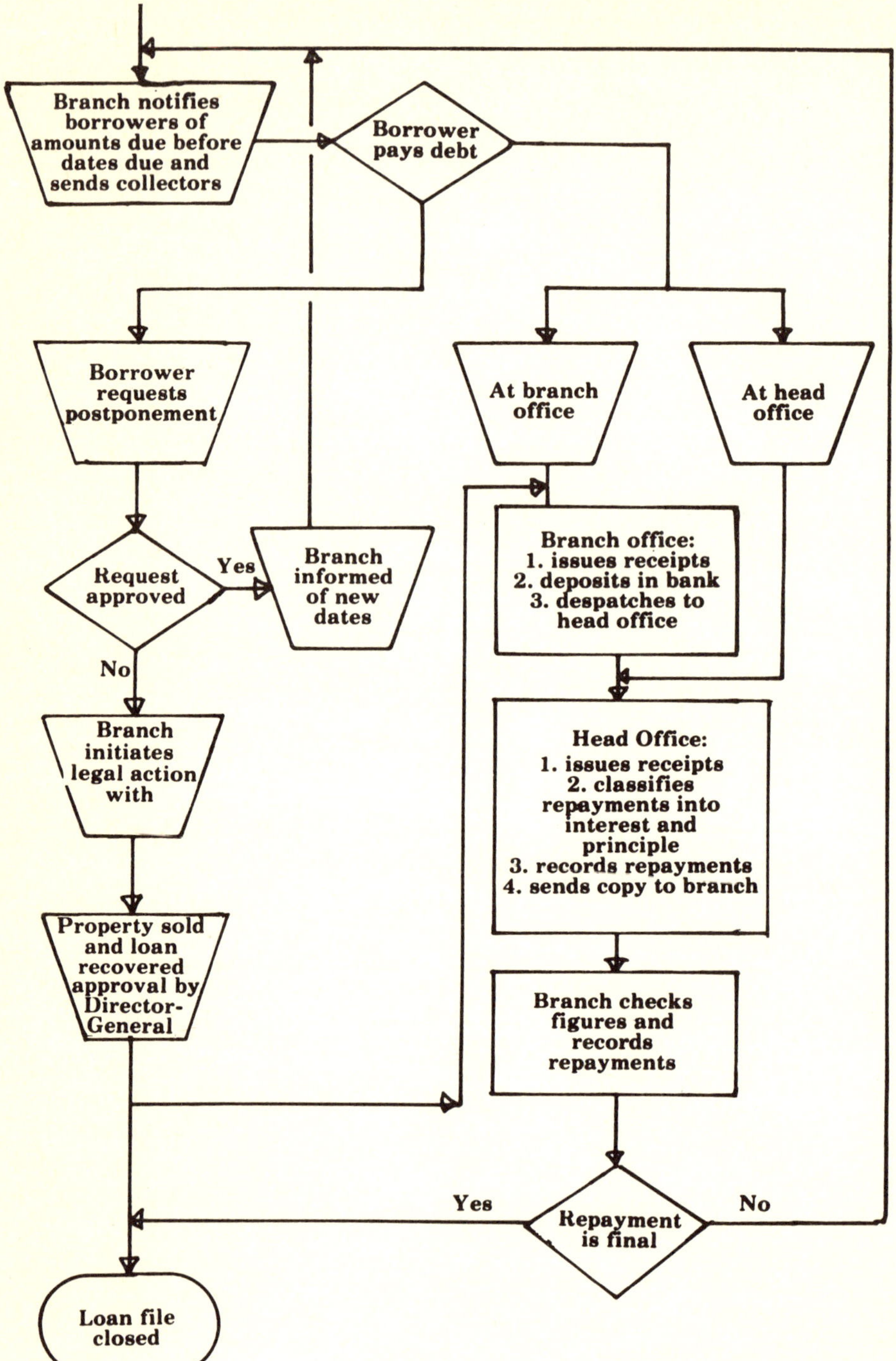

range of purposes. However, providing a service to rural and agricultural communities represents a major target of the JCO.

Agricultural cooperatives cover credit and saving, production, marketing, processing, and multi-purpose societies. Non-agricultural societies, on the other hand, include housing, consumption, credit and saving, multi-purpose, vocational, transport, and others. Out of a total number of 389 societies with a membership of 39,311 in 1981, agricultural societies accounted for 169 with a membership of 16,052. This means that agricultural societies represented about 44 per cent of the total number of cooperatives and 41 per cent of total membership. In 1981 loans extended by the JCO for agricultural purposes amounted to JD 4 mn out of a total loan volume of JD 4.68 mn (about 85 per cent) (JCO 1981). With most of its loans extended for agricultural purposes, the JCO occupies an important place among sources of agricultural credit in Jordan.

Cooperatives in Jordan have been growing at a moderate pace in both number and membership. The number of operational cooperative societies increased from 335 in both the East and West Banks of Jordan in 1961 to 389 in the East Bank only in 1981. Corresponding membership increased from 21,103 to 39,311 during the same period, reflecting an increase in the average membership from 63 to over 100 members (JCO 1973 and 1981). During this period, a considerable number of cooperatives were either dissolved due to failure or merged with others to form larger societies.

The total assets of the JCO grew from JD 0.59 mn in 1961 to JD 1.47 mn in 1971 and to JD 18.5 mn in 1981. Most of these assets (over 60 per cent in 1981) were in the form of outstanding loans in the hands of borrowers (JCO 1973 and 1981). In 1982 total assets approached JD 23 mn. Tables 9.3 and 9.4 show the balance sheet and the profit and loss statement of the JCO for 1981 and 1982.

Objectives

According to its statutes, the JCO aims to promote the cooperative movement as a means of improving the socio-economic standards of local communities and cooperative members and, in consequence, to increase their contribution to economic and social development in Jordan. In pursuance of these objectives the JCO performs the following functions:

1. Provides the public with services and information necessary to establish and register cooperative societies.
2. Provides services, information and technical assistance for existing cooperatives and their members, including management, accountancy and auditing of records.
3. Promotes cooperation through various channels of communication, but mainly through a Cooperative Institute, owned and managed by the JCO, where cooperative courses are taught.
4. Extends loans and banking services to cooperatives and their members through a Bank of Cooperatives owned and managed by the JCO at a reasonable low-interest cost.
5. Supplies cooperatives with all types of material inputs and provides marketing services for outputs at a reasonable low cost.
6. Represents the cooperative movement of Jordan and coordinates its relationship with associated agencies both at home and abroad.

Food security and land development are among the major national objectives which specialised credit agencies attempt to promote by financing agricultural activities. Like the ACC, the JCO is a non-profit-making organisation, although it tries to generate sufficient income to cover its administrative expenses. The Government makes annual contributions towards these expenses in addition to its contributions to the JCO's capital.

Environmental Characteristics

Physical: The JCO operates through its head office in Amman and its 11 cooperative directorates throughout the East Bank of the Kingdom.* Like the ACC, the head office of the JCO is easily accessible and is very close to the ACC's head office. Access to the JCO is facilitated by the directorates operating throughout the country.

* Three other directories used to operate in the West Bank of Jordan before the Israeli occupation in 1967.

In 1981 the JCO employed 306 people distributed between the directorates and the head office. Each directorate is staffed with the personnel necessary to carry out all normal JCO duties and functions. A computerised management information system has recently been installed and is expected to facilitate and improve operations.

Table 9.3: Jordan Cooperative Organisation: Consolidated Balance Sheet for the Years 1981 and 1982

Assets		1982			1981	
		(JD)	(fils)		(JD)	(fils)
Fixed assets		559,041	042		495,611	186
Fixed assets under establishment		37,026	848		—	
Long-term loans		1,582,702	721		1,044,326	993
Medium-term loans		3,572,641	580		4,357,618	129
Current assets						
Closing stock	1,206,981,171			728,854,257		
Investments	753,595,720			432,200,000		
Receivables	1,783,329,822			2,280,461,249		
Short-term loans	1,797,621,846			2,735,865,300		
Loans due	4,783,482,320			2,826,154,269		
West Bank loans & others	463,247,633			467,431,413		
Other receivables	980,007,610			615,962,707		
Overdraft account	745,198,185			—		
Cash in hand & in banks	4,507,504,244	17,020,968	551	2,550,887,525	12,637,816	720
TOTAL ASSETS		22,772,380	742		18,535,373	028

Liabilities	1982		1981	
	(JD)	(fils)	(JD)	(fils)
Capital	2,128,735	578	1,864,749	678
Reserves & accumulated surplus	549,000	414	357,275	884
Provision & funds	699,507	910	604,383	532
Long-term loans	5,043,878	132	5,009,260	648
Advances (Central Bank & Ministry of Finance)	7,094,144	204	4,473,268	407
Current liabilities				
Current deposit & savings accounts	6,540,509	833	5,381,211	193
Payables	246,702	037	383,006	375
Other credit accounts	298,141	043	220,357	181
Net surplus	171,761	591	241,860	130
TOTAL LIABILITIES	22,772,380	742	18,535,373	028

Table 9.4: Jordan Cooperative Organisation: Statement of Revenues and Expenditures for the Years 1981 and 1982 (JD)

Revenues	1982	1981
Revenue on trade	206,475	145,824
Revenue on credit	1,175,513	911,782
Government assistance	276,700	253,300
Other revenues	230,032	104,633
TOTAL REVENUES	1,888,720	1,415,539
Expenditures		
Wages, salaries & allowances	914,939	764,731
Financing costs	531,787	376,638
Administrative expenses	227,447	148,606
Depreciation	38,836	28,502
Other expenses	3,949	3,318
Net profit (transferred to the balance sheet)	171,762	93,744
TOTAL EXPENDITURES	1,888,720	1,415,539

Source: JCO 1981 and 1982.

In addition to its main functions of providing services to cooperatives, the JCO was designated in the 1981-85 Five-Year Plan to carry out certain projects of special importance to the cooperative movement. Among these are projects for the provision of farm machinery and improved seeds in rain-fed areas, the protection and improvement of grazing areas, and the promotion of consumer cooperatives as a channel of marketing agricultural products. The JCO has also established what are called 'parallel markets' in the capital city and two other towns. The parallel market represents a centre where goods (mainly foodstuffs) are sold at prices calculated to reflect the fair price which consumers have to pay under prevailing market conditions.*

The JCO provides numerous services to cooperatives and their members, including the supply of inputs and the marketing of outputs through specialised departments which aim to protect the interests of their clientele. The Bank of Cooperatives was established in 1977 to provide various banking services to the JCO, to cooperatives and their members and to the public at large, particularly in rural areas. Currently it operates through a central office located at the JCO head office in Amman and two branch offices in the Jordan Valley. The JCO plans to expand the number of its branches in the future. The Cooperative Institute of the JCO provides educational services aimed at promoting the cooperative movement and spreading awareness of cooperatives throughout the country. These duties give weight to the special role which the JCO is expected to play overall, as well as in agricultural development.

The JCO is well aware that cooperation is most needed in rural areas where agriculture is the major source of income and employment. Although it provides services to non-agricultural groups, farmers are the main group so far benefiting from the JCO's services. The attitude of the JCO to cooperatives and their members in general and to farmers in particular is one of total support, based upon its overall objectives.

* No formula for calculation is available, but parallel markets are supposed to pay competitive costs and charge competitive prices which include a reasonable profit margin.

Institutional: The JCO is currently governed by the 1971 Law of Cooperatives No. 20 and the ensuing by-laws which regulate its organisation and activities. According to the Law, the JCO is an independent legal body with administrative and financial autonomy. As a development agency, however, it is allowed certain privileges such as full exemption from taxes, custom duties, and fees normally paid by private businesses to the government. Like the ACC, the JCO enjoys preferential treatment from related government agencies, particularly the LRD, which at the JCO's request places mortgage seals on land and landed properties used by borrowers as loan guarantees, and also removes the seals. The JCO's loan documents are also considered enforceable evidence without further need for action by related government agencies.

By virtue of the said Law, the JCO is run by a Board of Directors on which the NPC, the MOA and the CBJ are represented. The Board also has eight members representing cooperative societies throughout the country. The Chairman of the Board is actually the Director-General of the JCO, who as chief executive is fully responsible for the management of the organisation and the implementation of the Board's policies. The Director-General is appointed, and the terms of his employment specified, by the Council of Ministers.

The highest level of authority in the JCO is vested by law in the General Assembly, where each registered society in the Kingdom is represented by one member. The Director-General is also the Chairman of the General Assembly. Board members representing the societies are not elected by the General Assembly but by gatherings held at regional level. The Director-General is assisted by a Deputy Director-General who acts for him in his absence.

The work-load of the JCO is distributed between various departments, divisions, branch directories and offices which employed 306 people in 1981 and 371 in 1982. Figure 9.6 is an organisational chart.

Due to the complexity and multi-faceted nature of the JCO's operations, the organisational structure of only those functions having to do with agricultural credit is discussed here. Two main functions are of special relevance in this respect: the credit and supply functions within the Departments of Cooperation and Projects, the Commercial Department and the Bank of Cooperatives. The supply division was established in 1975 to supply agricultural inputs to cooperative socieities which in turn resell these inputs to their members. It aims to provide farmers with as many of their inputs as possible, in sufficient quantities, at the desired time and for moderate prices. The division sells to cooperative societies either for cash or on three months' credit without interest. If the JCO is providing credit for the purchasing societies, the division provides the inputs and the Bank carries the loans.

The Bank of Cooperatives is the organ of the JCO through which loans are disbursed to cooperatives. It was established to provide finance for cooperative societies and to carry out general banking operations in rural areas. The Bank's capital is JD 5 mn, fully paid up. The JCO's Board of Directors shoulders the responsibility of organising the Bank's affairs and drawing up its general policies. The Board appoints the Director of the Bank, who works under the supervision of the Director-General. The Director of the Bank is in charge of carrying out the banking operations and coordinating the activities of the administrative, banking and loans divisisons of the Bank.

The JCO operates six hours a day, normally from 8.00 am to 2.00 pm, six days a week. These are the official working hours for most government offices and banks in Jordan.

All the JCO's operations have to be coordinated with related government agencies to keep in line with the national Development Plans. This coordination is facilitated by the representation of some of these agencies on the JCO's Board

Figure 9.6: Jordan Cooperative Organisation: Organisational Structure

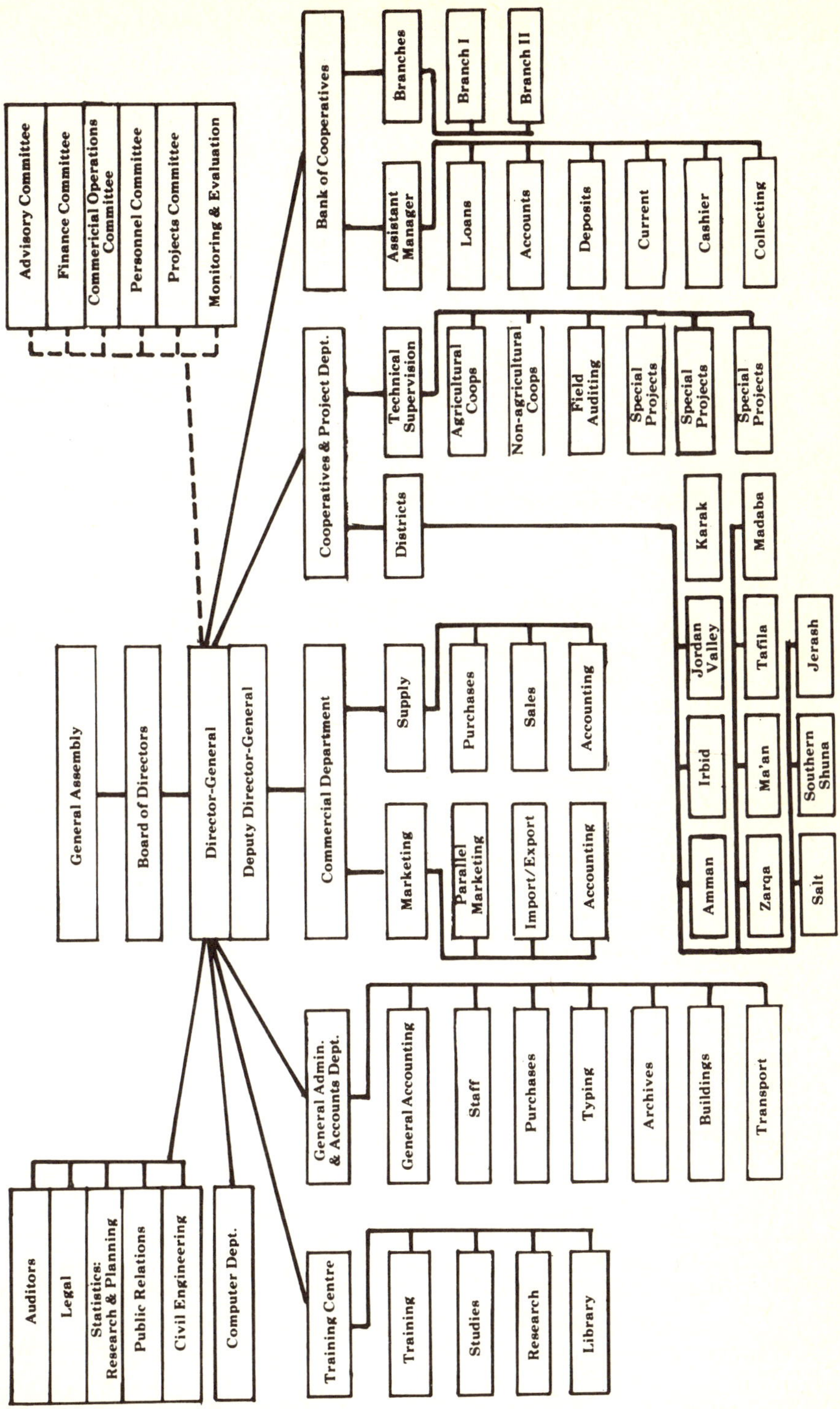

of Directors and also by the JCO Director-General's membership of the HAC. Borrowers are given referrals by the JCO to obtain licences from related government agencies whenever the projects to be financed require this. Although it has to cooperate and coordinate its activities with other agencies, the JCO is not subservient to any other institution.

All cooperative societies and their duly registered members are eligible for loans from the JCO. Farmers who are not members of societies also have access to seasonal loans under certain conditions relating to the type of project and to securities. In no case can a new loan be issued to a borrower who has not repaid loans due or arranged a settlement for this purpose with the JCO. Members of cooperatives have to obtain the approval of their societies before the JCO can consider their individual loan applications. Cooperatives must have an account at the Bank of Cooperatives and loans issued by the JCO are supervised loans aiming to finance specific activities. A cooperative society needs to have a minimum of 10 members and to subscribe to at least two shares of the JCO's capital upon registration and to one share annually therefore.* Before obtaining a loan, the cooperative must be fully paid up to the value of at least 10 shares (i.e. JD 100). The total outstanding loans of any society cannot exceed 5 per cent of the JCO's capital and reserves.† The guarantees provided, as well as the authorized capital of the society, place further limits on the size of its loans.

The JCO requests as security for its loans to any society a letter of guarantee signed by the administrative committee of the society, together with a bill signed by society members authorised to conclude financial transactions on behalf of their society. Where necessary, the JCO might require the mortgage of immovable property as additional security. In all cases the loan contract remains the basic reference document for both the society and the JCO.

The financial documents submitted by cooperatives as security represent legally binding evidence and are, for the JCO, enforceable without further legal action. The same applies to mortgaged properties. Collection of loans due can also be enforced under a special law regulating the collection of all funds owed to the government.

Informational: The JCO possesses a considerable amount of information about the financial position of each cooperative society operating in Jordan. Any farmer applying for a loan has to submit his application through his cooperative. The management of the cooperative knows about the farm requirements of each of its members. In most cases, the endorsement of the cooperative is considered to be adequate evidence. Sometimes, however, an inspection report based on a visit to the farm by JCO officials is requested to back up the application. Approved loans, moreover, are often disbursed in instalments to allow supervision of the loan's use. In all cases, follow-up visits are supposed to provide full information about each loan.

Information on other specialised credit sources is available to the JCO in aggregate form. It is more difficult to obtain information about individual farmers' debts to other creditors. Under these circumstances, loan guarantees are considered to be crucial for the extension of loans.

Activities

Obtaining Finance: The capital of the JCO consists of funds provided by the Government and subscriptions of member societies. The sources of JCO operating funds include its capital, reserves and allocations, retained earnings, external and internal loans and deposits of clients. Owned capital, reserves and retained

* If membership of the society is over 100 the annual subscription is raised to two shares.
† For cooperative unions and agricultural multi-purpose societies, this limit can reach 15 per cent.

earnings of the JCO constituted 17 per cent of total sources of funds at the end of 1981. Long-term loans constituted 27 per cent, Central Bank and Treasury advances 24 per cent and current liabilities (mostly deposits and current accounts) 32 per cent. The Government makes annual contributions towards the current expenditures of the JCO.

The JCO can obtain advances from the CBJ of up to JD 6 mn, repayable in five years at 3.5 per cent interest. It can also enter into long-term loan agreements with external sources with the approval and guarantee of the Government. Deposits and current accounts have grown to be an important source of funds in recent years. These are funds attracted by the Bank of Cooperatives. Cooperatives are encouraged to keep their deposits at this Bank and must maintain current accounts there to be eligible for loans from the JCO. By the end of 1982 total deposits and current accounts amounted to JD 6.54 mn, compared with JD 5.38 mn at the end of 1981. The Bank operates through its head office and two branches in the Jordan Valley. More branches are planned for the future.

Granting Loans: The JCO extends loans according to a set of rules and regulations agreed by the Board of Directors. There are three types of loans: seasonal loans, extending from one to fourteeen months; medium-term loans, extending from two to seven years; and long-term loans, extending beyond seven years with a grace period that matches the productivity of the financed project.

Seasonal loans are given to finance the operational expenses and the purchase of seasonal inputs. Medium-term loans are issued to finance capital expenditures on new or existing projects such as the establishment of poultry, sheep or dairy projects, land reclamation and improvement projects, acquisition of farm machinery and equipment, irrigation projects, and so on. Purchase of farm land can also be financed by medium-term loans under certain exceptional circumstances. Long-term loans, on the other hand, are extended for agro—industrial projects such as olive oil presses, dairy products plants, and similar projects. These loans are extended only for cooperatives unless otherwise approved by the Director-General of the JCO.

Societies and their members must have repaid loans due or arranged a settlement of them with the JCO in order to be eligible for credit.* Projects have to be feasible and loans fully secured. Loans to individual farmers are extended through cooperative societies by the Bank of Cooperatives. Seasonal loans may cover the full needs of the farmer, while medium-term loans may cover 80 to 95 per cent, depending on the size of the loan. Smaller loans generally receive a higher percentage of financing.

Lending decisions are made by the Finance Committee, which is chaired by the Director-General with a membership comprising a Board member representing the cooperatives, the Deputy Director-General and the Director of the Bank of Cooperatives. The Committee can approve loans within amounts allocated annually for each society by the Board of Directors on the basis of requests submitted by cooperative societies for such allocations.

In two directorates there are branch lending committees which are authorised to issue seasonal loans not exceeding JD 1,500 for irrigated land and JD 750 for rain-fed land.

Checking Credit-Worthiness: The credit-worthiness of any borrower is checked by reference to previous records which have to be default—free before a new loan can be issued. In all cases, a resolution by the management of the cooperative has to be made to back the loan application of the cooperative or any of its members.

Making Contact: Cooperative directorates make contact with farmers to promote

* Eligibility is restored if 70 per cent of loans due are paid by societies in irrigated areas or 60 per cent in rain-fed areas.

cooperative ideas and to encourage the formation of cooperative societies. TV and radio channels are occasionally utilised for this purpose.

Setting Interest Rates: Interest rates on JCO loans to borrowers are set by the Board of Directors and called 'service fees'. Current rates charged by the JCO range between 6 and 8 per cent: seasonal loans carry 8 per cent fees, medium-term in irrigated areas 7 per cent, medium-term in rain-fed areas 6.5 per cent, and long-term loans 6 per cent. Cooperatives charge their members a commission of 2 per cent. The Bank refunds 1 per cent in cases of full repayment of dues on or before dates due.

Interest paid by the Bank on deposits is normally determined by the CBJ. It is currently 2 per cent on demand deposits and 6-7.5 per cent on term deposits, depending on the term in each case as well as the size of the deposit.

Method of Repayment: The method of repayment is determined by the loans committee in the same loan approval decision. The nature of each project dictates this method of repayment as the JCO tries to match dates due with production times.

Action in Cases of Default: In the event of non-repayment of any part due, the whole amount of a loan becomes due. Borrowers are warned, and if they do not repay during the warning period, legal action is started to enforce collection. This is normally a slow process which requires announcements in the *Official Gazette* and/or local newspapers and ends in the public sale of mortgaged property or in the forced collection of loans. If the borrower is a cooperative society, the JCO normally knows about its financial status through its Audit Department, which is responsible for auditing the accounts of all member societies. As a rule, the JCO only resorts to legal action as a last resort. Normally postponement, extension and refinancing measures are followed wherever circumstances dictate. The balance of amounts due is automatically postponed when 70 per cent of these amounts is paid up by societies in irrigated areas, or when 60 per cent is paid up in rain-fed areas.

Deposit-Taking: As mentioned earlier, the Bank of Cooperatives accepts deposits from member societies, members of cooperatives, and the public at large. The Bank carries out all financial activities of JCO loans. It pays out loans to borrowers and receives repayments of loans.

Payment Procedure: Approved loans are communicated to relevant societies and JCO directorates by the loans division at head office. This division opens a file for the loan where all basic documents are kept and the Bank makes initial payment and waits for progress reports if required for subsequent payments.

Cheques are made out to cooperative societies, which in turn pay their members after deducting 2 per cent of the disbursed amount as commission. Cooperative societies must follow up on the collection of these loans from their members. Generally, payment of each loan is made in accordance with the lending resolution of the Finance Committee, which specifies the methods of payment and repayment, and assigns the responsibility for any required inspection visits or progress reports. The resolution may request the cooperative society, the cooperative directorate or the head office of the JCO to carry out these duties. All cheques have to be signed at the head office by two duly authorised officials of the JCO and the Bank.

Repayment Procedure: Cheques made out by any society in repayment of its loans are credited to the current account of the society at the Bank, while the repayment is classified and proper transactions entered into the Bank's records according to information provided by the society making the repayment. Societies can make repayments to the Bank of Cooperatives or to the JCO account at other banks.

The following flow charts (Figures 9.7-9.9) describe the lending and collecting procedures of the JCO. To avoid too much detail, loans in cash or materials are not differentiated, although the provision of agricultural inputs on credit is an important function of the JCO.

The Jordan Valley Farmers' Association (JVFA)

Introduction

The JVFA was established in 1974 with a view to grouping all farmers in the Jordan Valley into an association aimed at developing agriculture and improving the economic status of farmers. Every farmer who operates a farm unit in the Valley is a member by law and has to pay a subscription fee of JD 2 and an annual membership fee of JD 3 to the Association.

The Association is more or less a private agency, although it has government representatives on its Board of Directors and it gets financial support from the Government.

Objectives

To achieve its general objectives the JVFA specifically aims to provide loans and inputs to members and provide services common to farmers in the Valley such as machinery, transportation and marketing. It may establish plants for the production of inputs or for the processing of farm outputs. It is entitled to enter into agreements with local and foreign donors of loans with the approval of the Council of Ministers. It is required to cooperate and coordinate with government agencies such as the MOA, the ACC, the JCO, the Agricultural Marketing Organisation (AMO) and mostly with the Jordan Valley Authority (JVA). In purchasing and supplying farm inputs, the JVFA is required to operate competitively as far as possible. The Association is exempted from all taxes, fees and duties.

Environmental Characteristics

Physical: The JVFA started its operations in 1977 with a registered membership of 1,887 which increased to 4,563 in 1982 (JVFA 1980-82). The Jordan Valley is divided into 33 areas, each called a 'development area', and each comprising a minimum of 8,000 du and 200 farmers. These areas are serviced by seven branch offices, in addition to the head office which is centrally located at Deir Alla in the Valley, and a liaison office at Amman.

The Jordan Valley is characterised by its farming activities under permanent irrigation and by a semi-tropical climate which together create a type of intensive production adaptable to technological innovation. It has therefore been given special status in the development plans of Jordan as a regional development area. The JVFA is a farmers' organisation trying to provide better services at lower costs. However, it still needs to expand its activities considerably to win the loyalty and cooperation of farmers.

Institutional: The JVFA was established in accordance with 1974 Law No. 19 as an independent legal body. By the same law, however, it has to operate in cooperation with, and under the guidance of, certain government agencies, as mentioned earlier. Most of its finance is provided by or through the Government. It is therefore categorised here as a public specialised credit institution.

Member farmers in any given development area constitute the area general assembly, which elects a council to look after its affairs. Every 50 members are represented by 1 member on this council. Councils of the 33 development areas constitute the General Assembly of the JVFA, which elects 10 members to the Board of Directors from outside the Assembly. The Board of Directors is composed of the 10 elected members in addition to representatives of concerned government agencies comprising the ACC, the JCO, the MOA, the AMO and the

Figure 9.7: Jordan Cooperative Organisation: Lending Activities

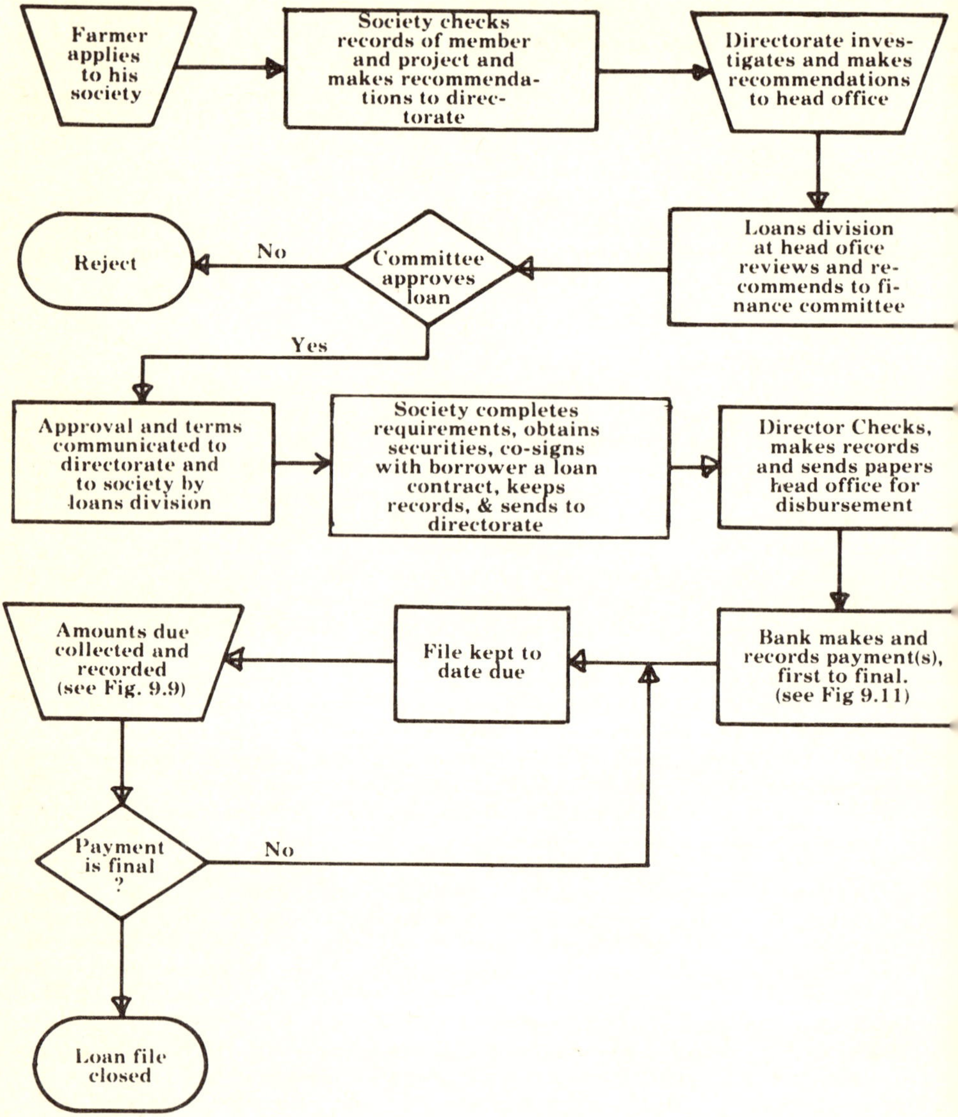

Figure 9.8: Jordan Cooperative Organisation: Payment of Loans

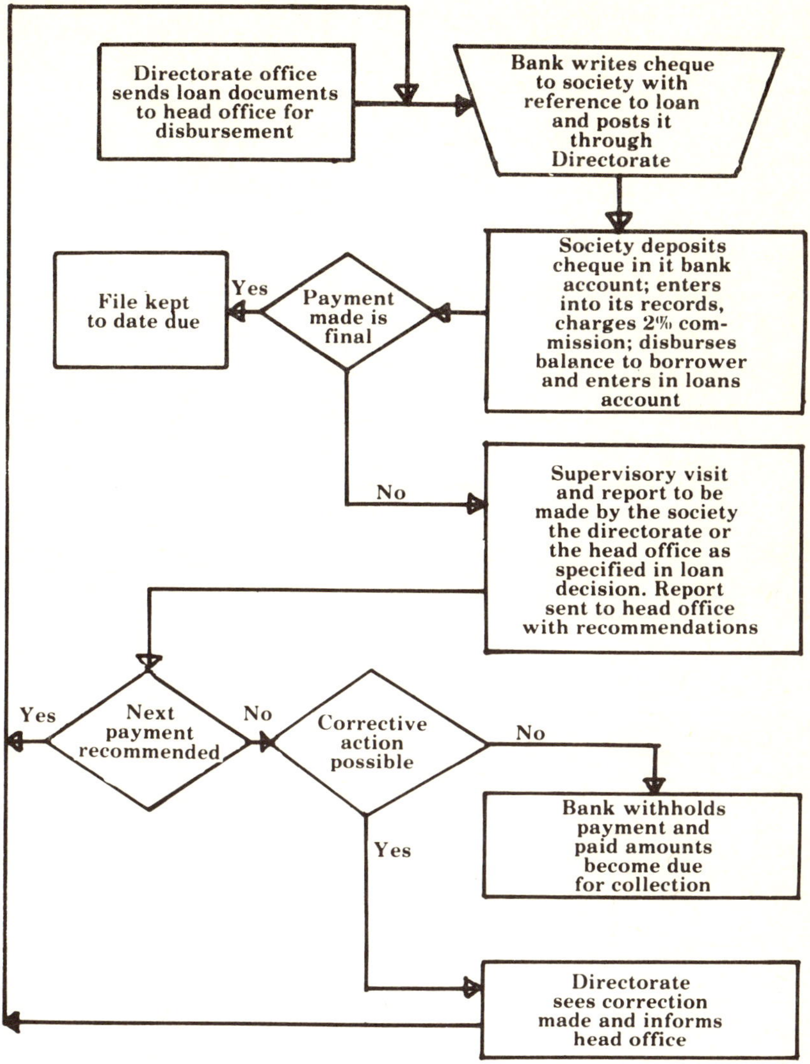

Figure 9.9: Jordan Cooperative Organisation: Repayment of Loans

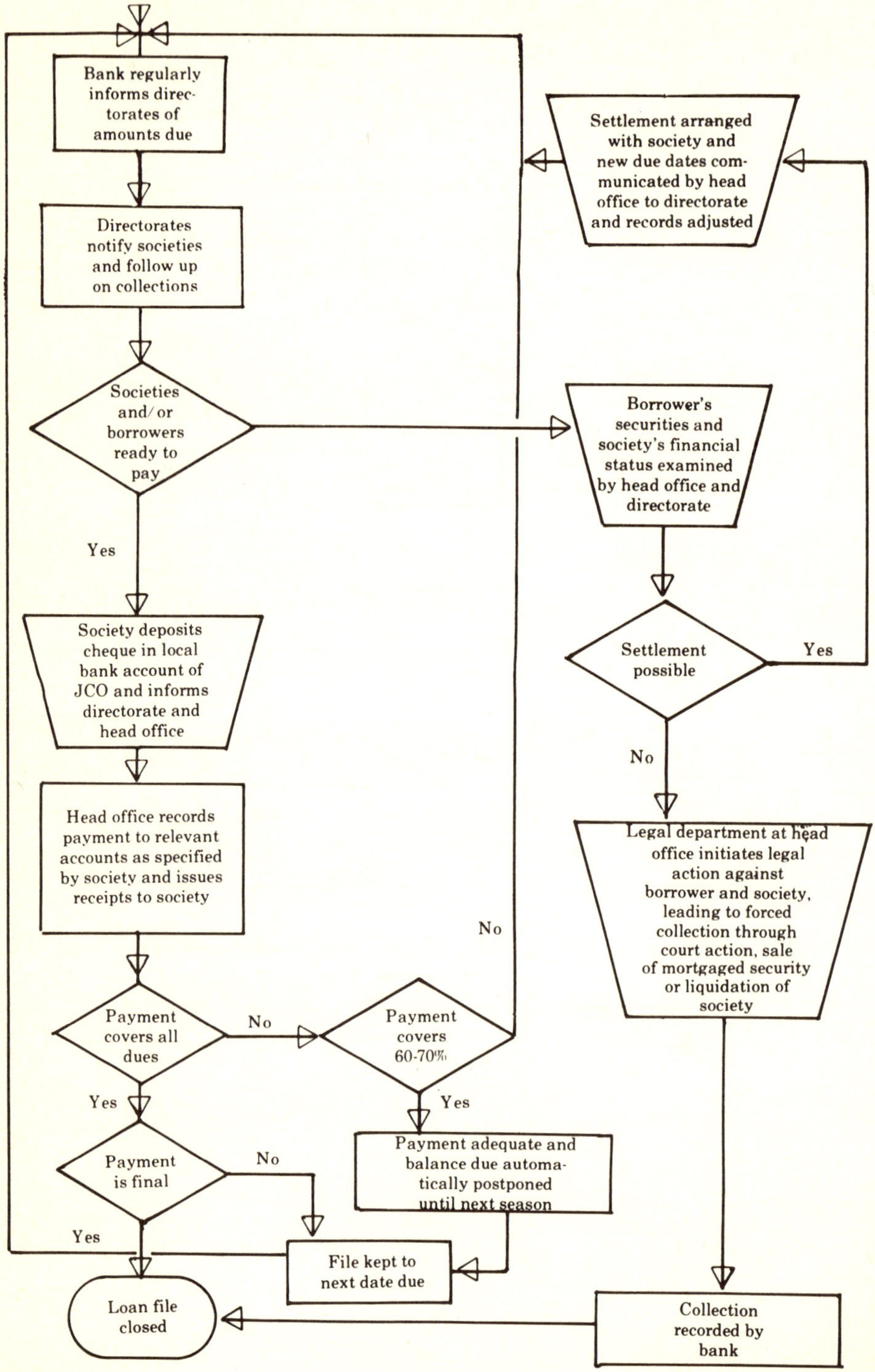

JVA. The Board elects a chairman, who, by law, represents the Association, carries out the policies of the Assembly and signs contracts for the Association. The Board also appoints a Director-General to look after the overall administration of the Association and to execute the policies of the Board.

The Director-General is the chief executive of the JVFA. He carries out his duties by authorisation from the Board and is responsible to the Board for the proper functioning of the Association. He is assisted by several department heads and branch managers looking after the departments and branch offices, as shown in the following organisational chart (Figure 9.10).

The JVFA operates six hours a day, six days a week; but due to the warm—often hot—weather of the Valley, staff start an hour earlier than in the uplands during summer.

By virtue of its statutes the JVFA is an independent entity but it is actually tied to government policies and agencies, and to some extent subservient to the JVA which, among other things, has to approve the by-laws of the JVFA as a condition for the ratification of these by-laws by the Cabinet.

There are no legal restrictions on the loans which the JVFA can extend within the scope of its objectives. Availability of funds and security requirements, however, represent the main limitations on loans. Unlike the ACC and the JCO, the JVFA has no juridical powers to facilitate legal action in cases of default. It must therefore use discretion and, where necessary, ordinary legal channels to collect its dues from farmers.

Informational: The JVFA has access to a considerble amount of information on individual members, their farm units and their farming operations. The Membership and Farm Service Department has a special file for each member which includes such detailed information. Information on the indebtedness of members to other sources of credit, however, is not easily accessible. Due to the close relationship between the JVFA and its member-farmers, there should be no lack of information of any type necessary for successful lending programmes. Each member has to fill in a membership card which includes basic information about himself, his family and his farm. This information is referred to in estimating his credit requirements.

Activities

Obtaining Finance: The authorised capital of the JVFA is not specified by law. Sources of capital, however, are specified to include net revenues, government grants, assistance from other sources, and any funds received by the Association and designated as capital by the Board of Directors. The Association started its operations with funds contributed by the Government, the Food and Agriculture Organisation of the United Nations (FAO), the United States Agency for International Development (USAID) and other material grants. In 1981 an amount of JD 28,000 (10 per cent of capital) was allocated as legal reserve. Most of the working capital of the JVFA is in the form of loans. Apart from its contributions towards the Association's expenses, the Government provided the JVFA with an interest-free medium-term loan (six years) of JD 576,000 in 1979 and 1980. The ACC extended a line of credit to the JVFA at 5 per cent amounting to JD 1 mn. USAID extended a long-term loan of about JD 450,000 at 2 per cent for the first 10 years, and 3 per cent for the final 20 years.

Granting Loans: Loans issued by the JVFA are currently confined to seasonal loans for the financing of farm inputs such as seeds, fertilisers, pesticides, insecticides, feedstuffs, labour and so on. The cost of renting farm machinery can also be loaned. Other purposes may be allowed by a Board resolution upon recommendation of the Director—General. Loans are extended for terms which match the productivity of the projects financed but do not normally exceed 12 months.

Figure 9.10: Jordan Valley Farmers' Association: Organisational Structure

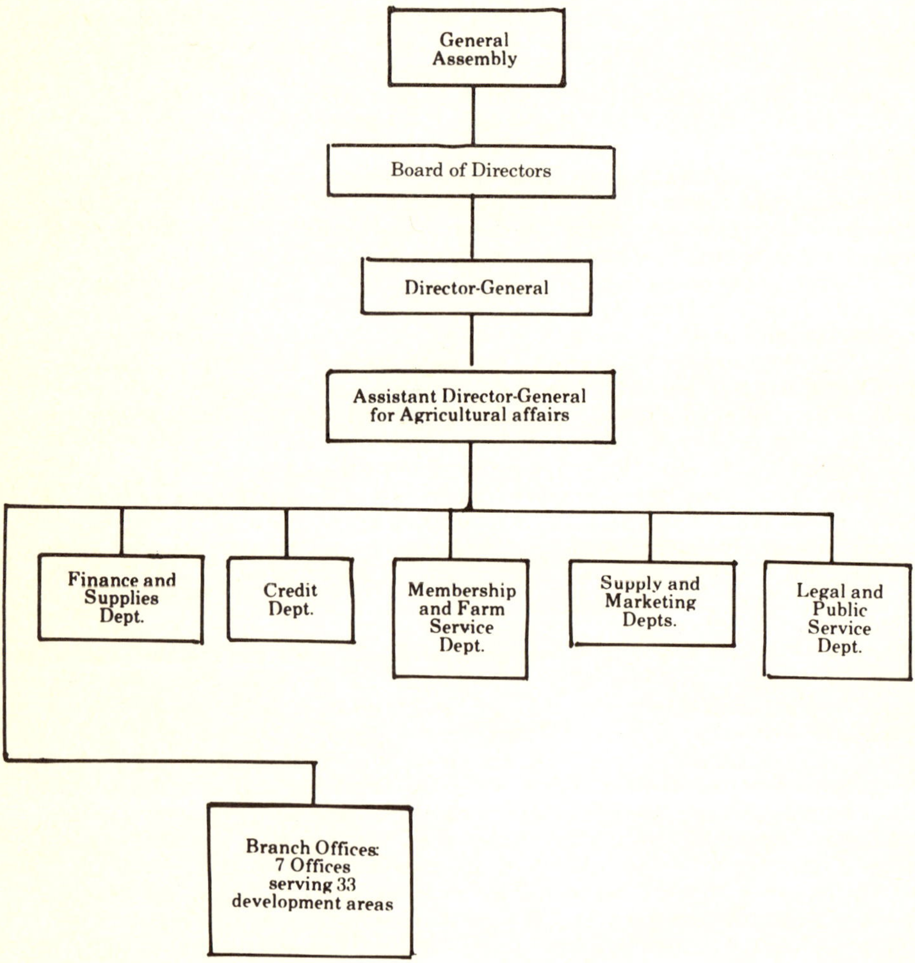

Only competent Jordanian farmers registered as members of the Association, with subscription fees and loans due (if any) fully paid up, are entitled to JVFA loans. They include owners, share-croppers, and lease-holders of farm units in the Jordan Valley. Cooperative societies operating in the Valley are also eligible, provided that there are no arrears in their loan accounts or in the subscription-fee accounts of their members.

The statutes of the Association do not impose any restrictions regarding the purposes or types of loans which can be extended. Due to a shortage of funds, and in order to avoid duplication with the ACC, the JVFA currently refrains from extending medium- or long-term loans.

Guarantees provided by borrowers to support their loans consist of one or more of the following main types:

1. Land and landed properties, to back loans whose value may not exceed 100 per cent of the assessed guarantee value if the land is in the Jordan Valley area, or 70 per cent if the land is located elsewhere.
2. Marketable stocks and bonds to back loans which may not exceed 80 per cent of their nominal or market value, whichever is the lesser.
3. Personal guarantees by owner-members of the JVFA for loans not exceeding JD 1,500 each and by two owner-members for loans above JD 1,500 up to JD 3,000. If the borrower is a share-cropper his guarantor should be the owner of the farm unit he operates. No guarantor may guarantee more than two borrowers.

Loan applications are filled in by farmers at branch offices where they are processed and supplemented by farm inspection reports based on visits by loan officers. Lending decisions are made by several authorities depending upon the size of each loan. Branch managers are authorised to approve loans not exceeding JD 750 each, upon recommendation by loan officers. The head of the Credit Department at head office may approve loans above JD 750 up to JD 1,500 upon recommendation by branch managers. The Director-General may approve loans above JD 1,500 up to JD 3,000 upon recommendation by the head of the Credit Department. Loans exceeding JD 3,000 and up to JD 5,000 are approved by the Central Loans Committee, chaired by the Director-General with the heads of the Credit and Farm Service Departments as members, together with three farmer representatives appointed by the Board for a term of two years. The head of the Credit Department acts as chairman of this committee in the absence of the Director-General. The Committee can hold meetings with a quorum of four members which should include the Director-General or the head of the Credit Department. The Director-General has the decisive vote in cases of split voting. Any loan exceeding JD 5,000 has to be decided upon by the Board of Directors with recommendations from the Central Committee.

In every case, loan decisions have to be based upon detailed and fully documented information concerning the borrower, his farming activity, purpose of the loan, inputs needed and time of their use, guarantees, financial status of the borrower and financial feasibility of the proposal. A detailed inspection report with recommendations by the loan officer are basic requirements.

Checking Credit-Worthiness: Credit-worthiness is checked by reference to the loan application, the inspection report and previous loan record (if any) of the borrower. Like other agencies, the JVFA relies more on the adequacy of guarantees than on the past record of borrowers.

Making Contact: Contact between the JVFA and farmers is easier and closer than in the cases of other credit sources, due to geographical proximity and the organisational structure of the JVFA which allows for close relations with farmers. The fact that the JVFA supplies farmers with inputs and provides a rental service for certain kinds of farm equipment makes day-to-day contact even easier. Contact is also facilitated by the central location of the JVFA's main office and the well-distributed branches in the Jordan Valley area.

Setting Interest Rates: The Board of Directors of the JVFA sets the rates of interest, called 'service charges' or 'service fees'. Currently, the Association charges 7 per cent on cash loans to be collected with the principal amount on the date due for each loan. An additional fee of 1 per cent is chargeable on arrears for the period of delinquency. A moderate margin is added to the cost of material inputs supplied to farmers and the resultant sum represents the loan amount repayable at maturity date.

Method of Repayment: The method of repayment is specified in the lending resolution by the same lending authority. The amounts to be repaid, the service fees on cash loans and the date of repayment are normally specified in the loan approval decision. The date due is set so as to match the production season, and hence the time of peak cash availability.

Action in Cases of Default: In cases of unjustified default, the JVFA resorts to ordinary law courts to enforce collection. Such action might result in the foreclosure and sale of properties and/or crops of borrowers. This approach, however, is not favoured by the JVFA and is taken only as a last resort. The JVFA may, and often does, seek the assistance of the JVA, which governs Valley affairs and provides farmers with irrigation water and other services, in the collection of such loans due. Even after legal action has been initiated, a settlement might be accepted by the JVFA.

Deposit-Taking: The JVFA does not take deposits from its borrowers or accept deposits from the public. Loans are disbursed in full under supervision of the staff concerned. Several banks have opened branches in the Jordan Valley area and compete strongly to attract deposits from the public. Given the low return on its loans, the JVFA is in no position to compete with commercial banks on interest payments on deposits.

Payment Procedure: When all lending requirements are completed, and upon approval of each loan, the borrower is informed and a loan contract is signed in the Credit Department at head office. A first payment, not exceeding 50 per cent of the cash loan, is disbursed by the Finance Department, and the balance is paid according to progress of work, or as recommended by supervision reports and approved by the head of the Credit Department. Input supplies are dispensed to borrowers by the Supply and Marketing Department on the basis of a written request from the Credit Department indicating the amounts of inputs and their credit value. The head of the Supply Department provides the borrower with a payment voucher against which materials can be delivered to the borrower by the JVFA warehouse in his development area. The warehouse officer provides the Credit Department and the Finance Department with copies of the receipt signed by the borrower.

Input requirements not available at the JVFA are financed by direct payment to the supplier of the amount allocated for this purpose. Payment is transacted through a voucher and a receipt document of the materials duly signed by the borrower. At least 50 per cent of the total amount of each loan should go towards the purchase of materials.

If a borrower misuses the first payment or if, for any reason, his membership of the JVFA ceases, the amount disbursed from the loan becomes due, and the delinquency charge mounts up until repayment is made.

Repayment Procedure: Loan repayments may be made at the head office of the JVFA, at branch offices or to collectors in the field. In all cases a receipt must be issued indicating the amount paid, the borrower's name and branch office, and the date of payment. A copy of every receipt is dispatched to the Finance Department, together with total amounts collected which are promptly deposited at the bank.

The Finance Department makes the proper accounting entries and maintains up-to-date loan records for all borrowers. The Department is required to issue notices to all borrowers indicating amounts due one month before the date due. The notices are delivered by JVFA collectors at branch offices. Lists of borrowers, their loans and amounts due are also passed to the Marketing Department, which assists in the collection of dues through its handling of sales revenues accruing to borrowers who sell their products through the Department.

In certain circumstances beyond the control of borrowers, repayments may be postponed by the Board of Directors.

Tables 9.5 shows the balance sheets of the JVFA for 1981 and 1982. Tables 9.6 exhibits revenue and expenditure for the same years. The flow charts (Figures 9.11 and 9.12) which follow summarise the overall lending and collecting activities of the JVFA; minor details are omitted, as they are to be found in the text above.

Table 9.5: Jordan Valley Farmers' Association: Consolidated Balance Sheet for the Years 1981 and 1982 (JD)

	1982	1981
Assets		
Cash in banks	149,903	306,390
Loans outstanding	1,490,626	710,464
Interest due	105,312	45,859
Subscription fees due	19,150	20,515
Delinquency fees	22,231	13,545
Inventories	408,929	150,070
Letters of credit	67,799	248,304
Other receivables	51,847	18,817
Fixed assets	139,920	135,376
Minus: depreciation	(32,105)	(16,618)
TOTAL ASSETS	2,423,612	1,632,722
Liabilities		
Accounts payable	90,200	23,915
Government loan	576,000	576,000
USAID loan	448,595	448,595
ACC loan	824,205	271,420
Capital	280,000	280,000
Legal reserves	28,000	28,000
Retained earnings	176,612	4,792
TOTAL LIABILITIES	2,423,612	1,632,722

Table 9.6: Jordan Valley Farmers' Association: Statement of Revenues and Expenditures for the Years 1981 and 1982 (JD)

	1982	1981
Revenues		
Ordinary revenues[1]	345,821	179,483
Extraordinary revenues[2]	125,635	86,631
TOTAL REVENUES	471,456	266,114
Expenditures		
Wages & salaries	179,765	135,822
Administrative expenses	119,872	80,583
Net profits	171,819	49,719
TOTAL EXPENDITURES	471,456	266,124

Notes: [1] Including interest, subscription fees and profit on sales of inputs.
[2] Including grants from the Government and foreign assistance in kind.

Conclusions

The specialised agricultural agencies presented in the preceding sections provide an essential service without which the Jordanian agricultural sector cannot function properly. The agricultural development which has taken place in Jordan during the last decade would not have been possible without the services of these agencies. The services themselves have been improved and upgraded according to requirements dictated by the varying circumstances and development stages.

Even so, specialised agricultural credit has not fully achieved the long-range targets for which it was established. Farmers still have to seek financial support from other, more expensive sources. Collection of loans due is still a difficult task and arrears are generally at high levels. Non-objective factors still sometimes play a role, in the granting of loans and the avoidance of prompt repayment. Fully objective methods of allocating funds to projects and borrowers have to be introduced to give farmers equal opportunities to obtain credit. Although most of the agencies' finances come from or through the Government, government supervision of their lending and collecting operations is almost negligible. Top management wields great power with very little delegation of authority. There is still much room for improving salaries and other work incentives for employees. They ought to be well compensated for working in public agencies which offer only limited opportunities for promotion and where transfers to positions in other government departments are virtually blocked.

Agricultural credit agencies have to be regarded and run as banking institutions and their management and staff have to be able to operate efficiently. Boards of directors need to have more active members, particularly those with banking background and experience.

A major issue which affects all specialised credit agencies negatively is the lack of coordination concerning the indebtedness of their borrowers. A farmer can easily obtain loans from all of these agencies at the same time, and possibly for the same activity. This is a drawback from which both farmers and credit

Figure 9.11: Jordan Valley Farmers' Association: Lending Activities and Payment of Loans

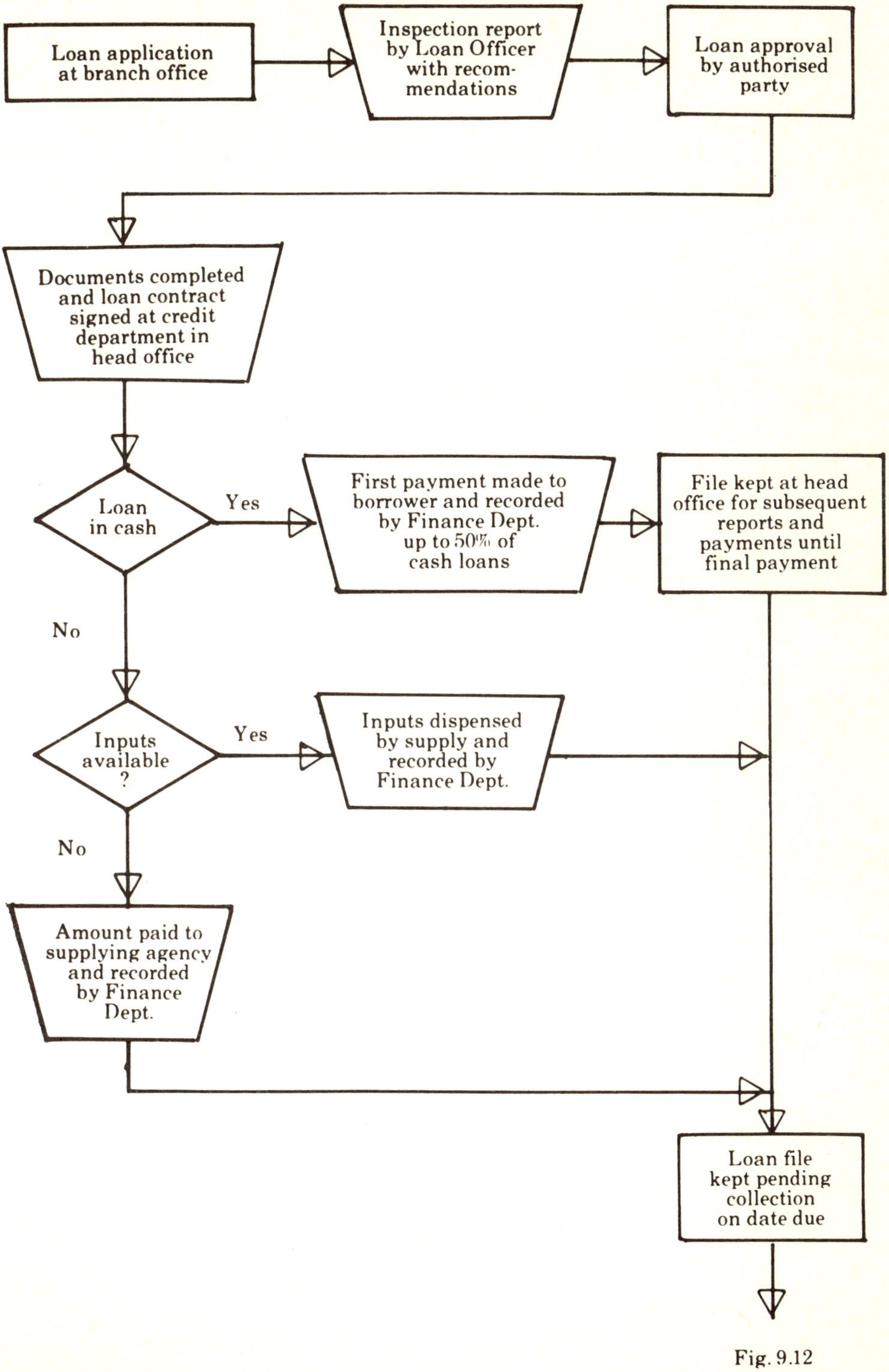

Figure 9.12: Jordan Valley Farmers' Association: Repayment of Loans

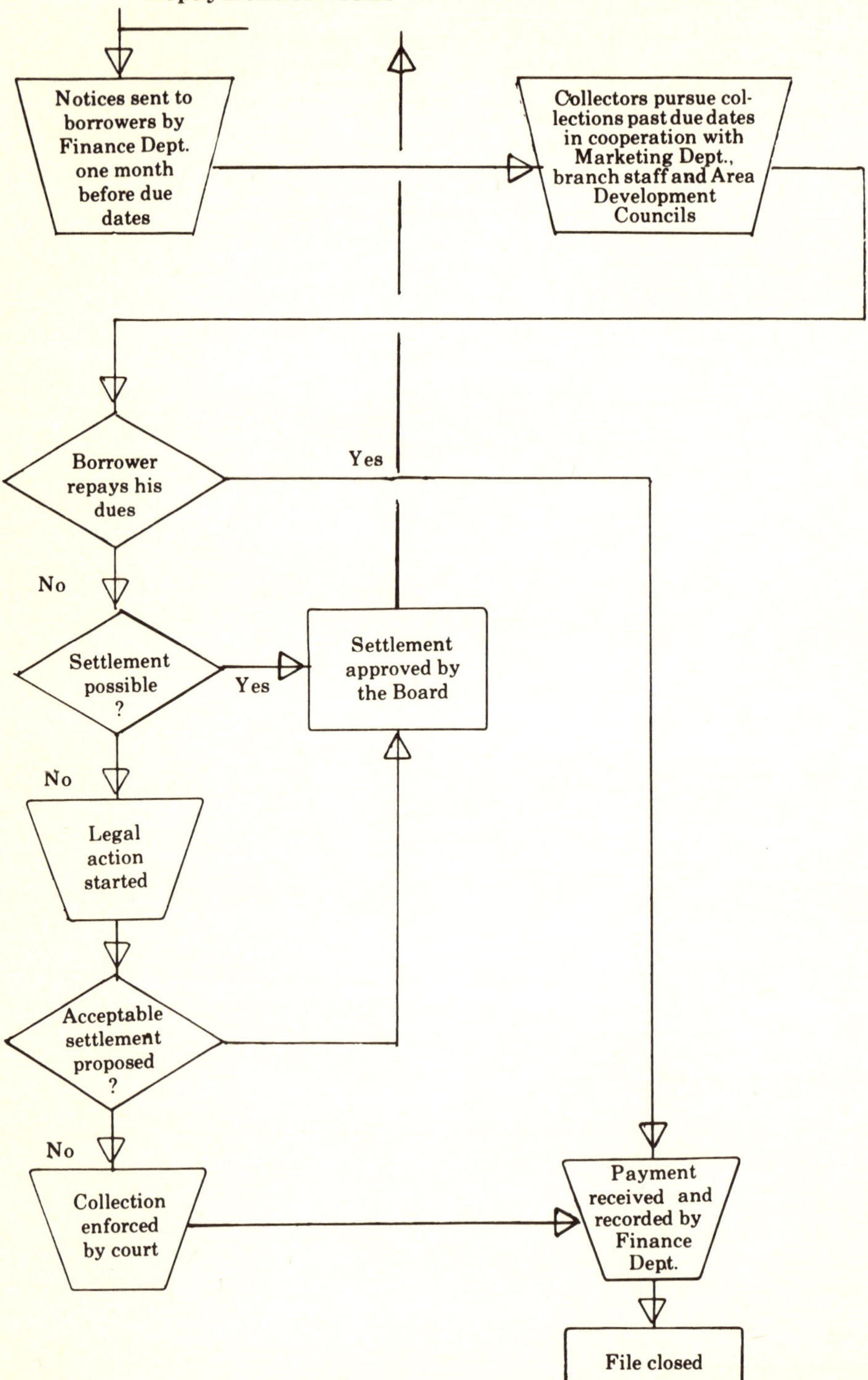

agencies may suffer. Coordination could take various forms, ranging from a complete merger of some or all of these agencies to the mere exchange of information on an explicit and regular basis.

Due to a strong demand for subsidised agricultural credit, interest rates must be reconsidered with the aim of increasing them progressively according to loan size until they reach market rates. The supply of funds for loans will always fall short of demand at the present rates of interest.

Coordination is also necessary between specialised agricultural credit agencies on the one hand and agencies involved in the marketing of agricultural products on the other. In general, market conditions and marketing aspects which are not usually emphasised in the loan-making process have to be examined on the macro-, micro- and farm-specific levels as major factors in making loan decisions. For this purpose, an efficient system of market research, data collection and analysis is indispensable.

III Non-Official Sources of Agricultural Credit in Jordan

There are several sources of agricultural credit other than those previously described as specialised agencies. Here, such credit is not based on formal rules and procedures stipulated by law but mostly on the basis of personal relations between lenders and borrowers. These sources include commercial banks, suppliers of inputs, purchasers of outputs, landlords, middlemen, families and money-lenders. An important feature of all these sources is that they are privately owned, profit-motivated businesses. Although no reliable information is available on the size of agricultural credit they extend to farmers, there is sufficient evidence that their role in overall agricultural finance is an important one. Farmers eligible for loans from specialised sources are not likely to seek loans from these higher-cost and short-term sources. Farmers borrowing from them, therefore, are mainly those who do not qualify for loans from formal sources and/or those whose projects or purposes are not compatible with formal requirements.

In the following pages each of these sources is discussed briefly, detailed information not being available.*

Commercial Banks

During the last 10 years, commercial banks have experienced a phenomenal expansion in Jordan. The following figures (CBJ b. 1979 and 1983) summarise the major indicators of this expansion.

	1974	mid-1983
Owned Capital	5.6	81
Reserves	3.9	52
Total deposits	112.0	1,288
Outstanding agricultural loans	16.0	28

The number of operating commercial banks has increased from 9 banks having 37 branches in 1968 (Hurani 1979: 32) to 19 banks with 169 branches in 1982 (Ramadan 1982: 20). The most relevant figures for the purposes of this study are those concerning the volume of credit facilities extended by commercial banks to the agricultural sector, which increased from an outstanding balance of JD 16 mn

* Information on this section is mainly based on discussions with farmers and lenders as well as personal knowledge and experience of the author. See Taqieddin 1979.

in 1974 to JD 28 mn in mid-1983. This reflects a substantial contribution to agricultural finance, exceeding the corresponding figures for both the ACC and JCO. It is not possible to obtain details from commercial banks on the distribution of their credit facilities by occupation of beneficiaries or purposes and terms of credit; yet it is evident that these facilities are obtained for purposes other than farm production, by non-farming borrowers such as merchants dealing with the import and export of agricultural inputs and outputs. Most of these facilities are short-term credits since commercial banks are not attracted by any kind of investment which extends beyond a one-year term. They are mainly letters of credit for importers and exporters as well as current accounts on which clients can draw up to a specified limit reviewed annually or every six months. Commercial banks are stricter in collecting their dues and less hesitant to take legal action for this purpose. Private businesses selling inputs, equipment, machinery and services to farmers can afford to pay the higher cost of commercial credit, due to the higher profitability of their businesses. The cost of farm machinery and equipment is more often than not paid by specialised credit agencies direct to selling companies in one payment, on behalf of the farmers. Companies, therefore, prefer to sell through credit agencies. Normally, they would be willing to sell to farmers on credit and/or accept payment by instalments.

To the extent that farmers finance their purchases from bank-financed companies and businesses through loans from specialised credit agencies, there is a double accounting of agricultural credit. The CBJ has been trying to persuade commercial banks to increase their participation in medium- and long-term finance of agricultural production at lower than commercial rates. Certain steps might be needed to motivate and oblige banks to move more vigorously in the direction of this policy. The organisation of marketing channels and other measures aimed to reduce risks in agricultural business, however, are prerequisites for their effectiveness.

Suppliers of Inputs

Farmers cannot obtain all the credit they need from formal agricultural credit sources. Lending rules and regulations of these sources prohibit lending for certain purposes, such as consumption. Guarantee requirements, distance to be travelled and slow bureaucratic procedures are additional reasons. Many farmers prefer to deal with merchants and suppliers according to religious law, as interest is prohibited in Islam. In some cases, farmers buy on credit from suppliers the same inputs for which they have borrowed from specialised credit agencies, in an attempt to save cash for other uses.

Most of the credit extended by merchants is seasonal, for the supply of seeds, feed fertilisers, insecticides, one-day old chicks, and so on. Some suppliers sell equipment on medium-term credit. The credit price is generally higher than the cash price and varies according to types of inputs, personal relations and circumstances. Suppliers are generally cautious and selective in their dealings with individual farmers and security requirements are not therefore very strict. Suppliers' knowledge of the financial status and credit-worthiness of their clients is a decisive factor as to how much credit to give in each case. Sometimes farmers are required to sign bills as collateral or to provide an acceptable guarantor. Suppliers aim to establish permanent relationships with their farmer customers and, if prospects are not encouraging, they can insist on cash payment.

Many suppliers operate in various market places, so that this type of business is highly competitive. In cases of default, social pressure and legal action are used as means of collection, receipts, bills and other documents being used as evidence.

Purchasers of Outputs

This group includes merchants who purchase agricultural products from farmers to resell them to retailers or to export them to neighbouring countries. Relationships between farmers and these merchants are established on a mutual interest basis. Merchants provide seasonal credit in cash to their clients in return for securing agricultural products needed for local or export markets at going market prices. A farmer, on the other hand, needs cash to finance his living costs, including food, clothing, medical care, transportation, education, and so on, during the production period. Such costs are not normally financed by specialised credit sources.

Merchants may themselves be exporters of agricultural products, in which case they benefit from good-quality agricultural products which their clients supply at normal prices, but which can be sold on the export market at high profit. Some of these merchants are themselves suppliers of farm inputs as well.

Here again personal relationships are important and based on the merchants' long experience and thorough knowledge of the financial standing and credit-worthiness of their farmer clients. Generally, no interest charges are involved for credit from this source.

Landlords

Landlords may provide credit direct to tenants or share-croppers working on their farms, or they can help them obtain credit from specialised sources by providing securities or acting as guarantors. Relations here are more personal and based on agreements made between the parties concerned. Credit provided includes cash for consumption purposes during the production period. Normally, landlords collect their money at harvest time. In bad seasons, landlords have to postpone collection and also have to continue financing some of the operational costs. Hence, landlords try to minimise their financial involvement as far as possible.

Middlemen

Credit extended by this group includes the provision of cash to farmers during the production period to cover some living and operational expenses. In return, farmers are informally obliged to sell their products through the lending middleman, who charges a commission for selling their products. The middleman is supposed to act in the best interest of his clients, which he usually does, to maintain good relations with them. The middleman recovers his money from the sale price of the agricultural produce.

Some middlemen are themselves merchants providing inputs to farmers or exporting outputs. In such cases farmers are open to exploitation by the merchants through overpricing of inputs sold on credit and underpricing of outputs purchased to settle debts.

Families

Family sources of credit still exist but on a very limited scale, and are generally confined to close relations. Family members will more often provide land to be mortgaged by official credit sources as a loan guarantee for another member of the family who does not own such land or guarantee. The land might be owned in common by the family or individually owned by some member(s). Cash is also provided in certain cases where formal credit is not possible. Family ties and relations govern the type of arrangements through which such credit is provided. Normally it does not involve interest or cash costs.

Money-Lenders

Individual money-lending based on excessive interest rates has virtually disappeared from the agricultural credit market in Jordan. Specialised credit agencies and banks in general have greatly reduced the business opportunities of this group. Credit thus extended to farmers can only be negligible.

Conclusions

In summary, the non-specialised agricultural credit sources, including commercial banks, still play an important role in financing agriculture, but no statistics or accurate information are available on the extent of its magnitude. Credit activities from these sources are not regulated and are mainly based on personal, individual relationships.

References and Other Works Consulted

ACC (Agricultural Credit Corporation), *Annual Reports, 1960* and *1980-82*, Amman (in Arabic).

— Law No. 12 for the Year 1962 and By-Laws (in Arabic)

CBJ (Central Bank of Jordan), a. *Annual Reports, 1975-1982*, Amman (in Arabic).

— b. *Monthly Statistical Bulletins, 1979* and *1983*, Amman (in Arabic).

Hurani, A., 1979, 'Banking Institutions in Jordan', Municipalities and Villages Development Bank, Amman, 1979 (in Arabic).

JCO (Jordan Cooperative Organisation), *Annual Reports 1973* and *1980-1982*, Amman (in Arabic).

— Law No. 20 for the Year 1971 and By-Laws (in Arabic).

JVFA (Jordan Valley Farmers' Association), unpublished Report for 1980-82, Deir Alla (in Arabic).

— Law No. 19 for the Year 1974 and Regulations (in Arabic).

Mazur, Michael P., 1979, *Economic Growth and Development in Jordan*, Westview Press, Boulder, Colo.

Mitchell, Mark, 1983, 'A policy-oriented system analysis of the agricultural sector', Wye College, University of London (draft).

National Planning Council, *Five-Year Plan, 1981-85*, Amman.

Ramadan, Z., 1982, 'Management of Banking Enterprises', Business Department, University of Jordan, Amman (in Arabic).

Taqieddin, N., 1979, 'The role of middlemen in agricultural credit and marketing in Jordan', Country Study for Food and Agriculture Organisation, Rome.

10 THE MARKETING SYSTEM

Suleiman Arabiat

General Background

Introduction

The importance of agricultural marketing increased during the 1960s, when the production of vegetables in the Ghor expanded and commercial farming was introduced, resulting in a surplus for export to foreign and neighbouring Arab countries. Agriculture has made remarkable progress in a short period. Some of the agricultural sub-sectors have witnessed a transition from small units producing for the local market and family consumption into modern large-scale production units responding to foreign demand and local processors. The increase in production and flow of large quantities of agricultural commodities from the farms to the market has put pressure on the limited marketing facilities. A high population growth rate combined with urban expansion have pushed towards the establishment of a marketing system.

Government concern in agricultural marketing began in 1964 with the initiation of a government project supported by international funds which contributed to the construction of the central wholesale market. The Government's role was limited to administration and implementing laws, while marketing functions and services were left to the private sector. Despite government investment in the infrastructure of the market, the role of middlemen, including wholesalers, exporters and merchants, has been influential and obvious and has thus become a controversial matter.

The objectives of this chapter are to identify and describe the marketing system, marketing channels, central markets and marketing services. A major objective has been to examine environments and activities related to the marketing of tomatoes, cucumbers and bananas.

Production of Vegetables and Fruit

Vegetables and fruit are the major crops for local consumption and export to neighbouring Arab countries. The average production of vegetables was 543.5 thousand t for 1979/80. Tomatoes, eggplants, cucumbers, cauliflowers, squashes, cabbages and watermelons represented 42.1, 15.5, 9.6, 7.2, 5.2, 3.4 and 2.8 per cent of the total production, respectively. The average production of fruit was 128.3 thousand t for 1979/80. Grapes, oranges, lemons and apples represented 30, 10.5, 7.4 and 1.4 per cent of total production, respectively.

The average growth rate of vegetable and fruit production presented different percentages during 1970/80. Tomatoes occupied first place, while eggplants, cucumbers, cauliflowers and cabbages showed a significant increase in importance. Squashes, muskmelons and potatoes increased slightly, while watermelons decreased in importance. The average growth rate of the total production of fruit was 1.18 per cent during 1970/80, while it was 2.47 per cent for grape production and 3.17 per cent for other fruit. Oranges showed a decrease of 5.15 per cent and apples a 0.35 per cent decrease.

Vegetables are grown during two seasons, summer and winter in irrigated

Table 10.1: Average Distribution of Vegetables according to Growing Season, 1979/80 (per cent)

Crop	Winter	Summer (irrigated)	Summer (rain-fed)	Total
Tomatoes	74.96	22.73	2.32	100
Eggplants	95.16	4.84	—	100
Squashes	86.38	11.86	1.86	100
Cucumbers	57.56	42.44	—	100
Cauliflowers	100.00	—	—	100
Cabbages	100.00	—	—	100
Watermelons	20.96	1.80	77.25	100
TOTAL	74.29	20.52	5.19	100

Source: Department of Agricultural Economics, MOA, unpublished data.

Table 10.2: Volume of Vegetables and Fruit Available for Consumption 1976/80, and Projected for 1985 and 1990 (tonnes)

Crop	1976	1977	1978	1979	1980	1985	1990
Vegetables	263,485	281,012	224,341	299,385	342,486	429,306	700,354
Fruit	247,661	181,303	180,545	189,437	184,148	264,444	374,376

Source: Department of Agricultural Economics, MOA, unpublished data.

areas, and summer in rain-fed areas. Table 10.1 shows the distribution of vegetable production during the two growing seasons. Production during the irrigated winter season extends from June to September. The production of summer vegetables in rain-fed areas extends from July to November. Fruit production differs according to variety. While bananas are available throughout the year, citrus production extends from November to May. In the rain-fed areas, fruit production extends from April to December.

Demand for Vegetables and Fruit

The demand for vegetables has changed with time due both to population growth and to increases in *per capita* income. Table 10.2 shows the total amount of vegetables and fruit available for consumption from 1976 to 1980 and projected for 1985 and 1990. Vegetables available for consumption were 224,341 t in 1978

Table 10.3: Volume of Exported Vegetables and Fruit, 1976-80 (thousand tonnes)

Crop	1976	1977	1978	1979	1980
Vegetables	132.7	150.65	207.83	296.58	301.96
Fruit	171.04	170.54	161.18	180.75	188.02

Source: Department of Agricultural Economics, MOA, unpublished data.

and 342,486 t in 1980, and are expected to reach 700,354 t in 1990. Fruit available for consumption ranged from 180,545 t in 1978 to 247,661 t in 1976, and is projected to be 374,376 t in 1990.

Table 10.4: Value of Exported Vegetables and Fruit, 1976-80, at Current Prices (million JD)

Crop	1976	1977	1978	1979	1980
Vegetables	4.51	4.19	7.91	8.95	11.22
Fruit	7.89	9.17	6.72	5.09	6.42

Source: Department of Agricultural Economics, MOA, unpublished data.

Imports and Exports of Vegetables and Fruit

External trade constitutes a major component of the national economy because it provides the population with consumer and capital goods and contributes to GDP. The value of exported goods increased from JD 9.3 mn in 1970 to 120 m in 1980. The value of imports increased from JD 65.9 mn in 1970 to 720 mn in 1980. This means that the deficit in the balance of trade increased from JD 53.7 mn in 1970 to 560 mn in 1980.

Table 10.5: Volume of Imported Vegetables and Fruit, 1976-80 (thousand tonnes)

Crop	1976	1977	1978	1979	1980
Vegetables	20.30	14.16	30.25	42.31	64.15
Fruit	153.06	148.94	120.76	128.44	129.25

Source: Department of Agricultural Economics, MOA, unpublished data.

Jordan is a leading exporter of vegetables. Table 10.3 shows the amount of vegetables and fruit exported from 1976 to 1980. The amount of vegetables increased from 132.7 thousand t in 1976 to 301.96 thousand t in 1980. Fruit exports increased from 171.04 thousand t in 1976 to 188.02 thousand t in 1980.

The value of vegetables exported increased from JD 4.51 mn in 1976 to 11.22 mn in 1980. The value of fruit exports decreased from JD 7.89 mn in 1976 to 6.42 mn in 1980. Table 10.4 shows the value of fruit and vegetables exported from 1976 to 1980.

Table 10.6: Value of Imported Vegetables and Fruit, 1976-80, at Current Prices (million JD)

Crop	1976	1977	1978	1979	1980
Vegetables	2.89	3.23	2.54	3.16	3.37
Fruit	3.17	1.25	1.96	3.50	3.12

Source: Department of Agricultural Economics, MOA, unpublished data.

Imports of vegetables increased from 20.3 thousand t in 1976 to 64.15 thousand t in 1980. Fruit imports decreased from 153.06 thousand t in 1976 to 129.25 thousand t in 1980. Table 10.5 shows the amount of vegetables and fruit imported from 1976 to 1980.

The value of imported vegetables increased from JD 2.89 mn in 1976 to 3.37 mn in 1980. The value of fruit imports decreased from JD 3.17 mn in 1976 to 3.12 mn in 1980. Table 10.6 shows the value of imported vegetables and fruit from 1976 to 1980. The increase in value was not stable throughout this period.

Figure 10.1: Marketing Channels for Major Fruits and Vegetables

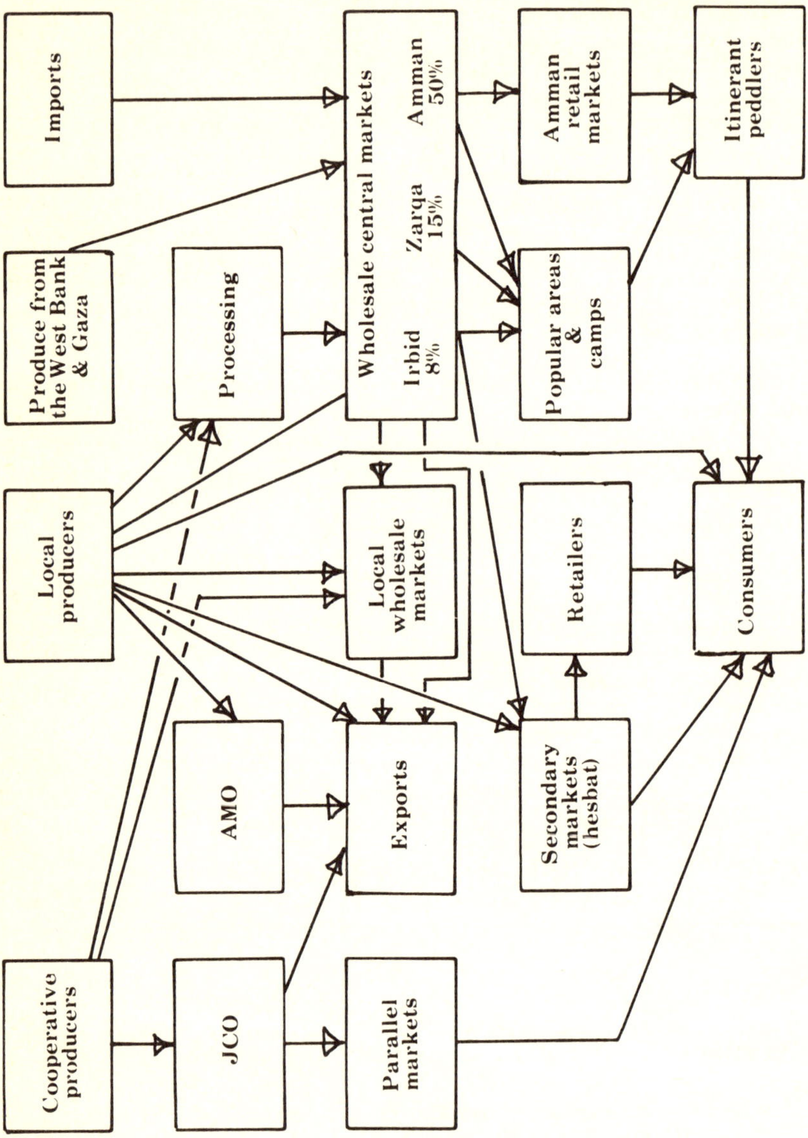

Source: AOAD 1982: 74

The Marketing System

Marketing Channels
The sequence of intermediaries and markets through which goods pass *en route* from producer to consumer is known as a marketing channel. It comprises local and cooperative producers, wholesalers, commission agents, exporters, retailers and consumers. Any attempt to understand the marketing system requires total knowledge of the marketing channels of agricultural commodities. Figure 10.1 shows the marketing channels for the major home-produced and imported fruit and vegetables.

Domestic products are supplied by the private sector as well as by a limited number of cooperative farmers. Farmers sell their produce either through middlemen or the cooperative society, or the Agricultural Marketing Organisation (AMO) and exporters. Some produce is sold directly to processing plants or wholesalers. Buying and selling mainly take place at the central markets. Imported goods are sold through them, and they handle 73 per cent of production.

Local Wholesale and Assembly Markets
Local wholesale and assembly markets have been established in the production area in the Ghors. Before the Central Market of Arda was built, two assembly markets were available in Karama and Sawalha. Normally the farmers bring their produce to these markets and it is auctioned through the middlemen. The assembly markets have not been subject to government supervision and farmers do not pay fees to municipalities, while middlemen earn 5 per cent of the total value of the goods sold.

The newly established Central Market of Arda is administered by the Jordan Valley Authority (JVA). Two other central markets exist in Wadi Eliahes and Southern Shuna, including 20 stores in each.

Central Wholesale Markets
The high rates of population growth and rural-urban migration have called for a flow of goods from productive areas to those where consumers are living. There are three major wholesale central markets:
Amman Central Wholesale Market (ACWM): The market was built in 1966. It is administered by the Municipality, while middlemen conduct the buying and selling of goods. The Municipality charges a fee equivalent to 4 per cent of the value of produce from the seller and the buyer (2 per cent each). The middlemen earn up to 5 per cent of the value of the goods sold.

This is the largest market, with 50 per cent of the total amount of fruit and vegetables being handled by 56 licensed middlemen. Twenty middlemen sell 70 per cent of the total goods exchanged in the market.

Middlemen provide farmers with input supplies and loans which are repaid at the end of the season. Farmers have to sell through the middlemen who provided them with loans and input supplies.

The ACWM serves the people of Amman and those living in surrounding towns and villages. Retailers and exporters trade in this market.
Irbid Wholesale Market: Irbid Wholesale Market was established in 1968. It has 21 stores, one of which is occupied by the AMO; the remainder are rented by wholesalers and exporters. Five wholesalers control 70 per cent of the total value. Selling is done by auction. The amount sold here constitutes 8 per cent of the total amount sold in the central markets.

Zarqa Wholesale Market: Zarqa Wholesale Market was established in 1972. It has 18 stores, one of which is occupied by the AMO. Amounts sold in this market are

12 per cent of the total exchanged in the central markets. Three middlemen organise most of the sales and collect 4-6 per cent commission of the total value. Middlemen offer loans in cash and kind to farmers.

Secondary Wholesale Markets
Secondary wholesale markets (called *hesbat*) are located in a number of towns. Limited amounts of goods are exchanged here. The produce comes from the central markets and the surrounding farms. Selling is by auction. In most cases, municipalities do not collect fees; middlemen, who perform the auctions and provide the place, earn commissions of 4 per cent of the total value.

Retail Markets
Retail markets are scattered throughout the populated areas and commercial districts of Jordan. Some retail markets are shared, where many stores sell goods in the centre of town. The quality of goods in these markets is not high. There are specialised shops in towns for selling fruit and vegetables, but in the villages these goods are sold at general shops.

Parallel Markets
The Government established a number of retail markets in Amman in 1973. The 30 stores were rented to retailers on condition their prices conformed with government prices. In 1976 the Government transferred the stores to the Jordan Cooperative Organisation (JCO). The new name 'parallel markets'* started at that time. The JCO arranges contracts with cooperative farmers and producers and buys their produce. The goods are graded and then sold in plastic bags. In addition to shops, mobile vehicles which sell directly to consumers are also utilised in the populated areas.

Marketing Agents

Middlemen in Central Wholesale Markets
Middlemen in the central markets perform other functions as well as their main one of commission agent. They are also wholesalers or exporters. The JCO and the AMO are represented in the central markets.
Commission Agents: The main function of these middlemen is to facilitate buying and selling and thus get a commission which ranges between 3 and 5 per cent of the total value. They do not own the produce but sell it on behalf of the farmers.
Wholesalers: Wholesalers comprise the majority of middlemen in the central markets. They buy and sell for themselves and for farmers, and collect 4-6 per cent commission from farmers. They offer loans in cash and kind to farmers.
Exporters: Wholesalers may be exporters at the same time. Exporters deal with farmers direct, or buy from the Assembly Market and Arda Central Market.
The JCO: The JCO acts as an agent at Amman and Arda Central Markets. It buys from farmers to fulfil contracts or sells on behalf of farmers.
The AMO: The AMO has stores in the central markets. It buys from farmers direct or by contract to export their produce. It imports certain goods through agreements with organisations outside Jordan.

Middlemen in Retail Markets
Retailers in Combined Markets: Retailers in shared markets, where many stores are located, buy goods from the central markets and sell to consumers. These are mainly in the refugee camps.

* The term 'parallel market' as used here is specific to Jordan.

Retailers in Populated Areas: Scattered in different towns and villages, retailers buy from the local and central markets or from producers. In high-income areas, some retailers sell top-quality goods at high prices. Low-quality produce is exchanged in areas of low income.
Itinerant Retailers: Itinerant retailers sell their produce from pick-ups in the slums and populated areas. They buy from local and central markets. They sell at lower prices than other retailers.
Retailers on Side Roads: Sales along main roads started up in places near farms where farmers sold their own produce. This developed to include the highways near main towns and Amman. These retailers sell at high prices, and are not subject to municipality or Ministry of Supply (MOS) control. They do not sell by weight but by container, in order to escape announced retail prices.

Marketing Services and Margins

Marketing Services
Marketing services include the collection of produce, its movement to the areas of demand and its transformation to the form required by consumers. Marketing activities include transport, storage, grading, processing, financing and market information.
Transport: Shipment of goods between the farm and market is one of the major services and it is part of each marketing stage. Transport of crops from the farm to the wholesale market is carried out by small vehicle or truck. Tractors pulling trollies are used to transport produce from the farm to the local central (wholesale) markets. Produce for export is transported by large, refrigerated vehicles. Table 10.7 shows the number and value of refrigerated vehicles from 1975 to 1980. The total number of vehicles was 1,086 and their total value JD 26.9 mn. The drop in the number of vehicles noted between 1979 and 1980 was due to a decrease in exports.

Table 10.7: Number and Value of Vehicles Imported, 1975–80

Year	No. of vehicles	Value (JD mn)
1975	132	2.670
1976	185	4.207
1977	149	3.804
1978	222	5.651
1979	270	7.523
1980	128	3.015
TOTAL	1,086	26.870

Source: Department of Agricultural Economics, MOA, unpublished data.

Storage: Storage is essential to conserve agricultural commodities, particularly those which are perishable. There are insufficient refrigerated stores of the right size for fruit, vegetables, meat, eggs and food products. The available capacity of refrigerated stores is 26,826 m^3. Most of the storage facilities, other than wheat silos, are owned privately.
Sorting, Grading and Packing: There are three stations for sorting, grading and packing vegetables in Jordan. None of these was working for a time; one station for grading citrus is now functioning. Containers are made locally to meet 40 per

cent of the demand and the rest are imported. The problem with containers is that they are not standard in size and are made of many different materials.

Grading is not practised except by some progressive farmers and exporters who do manual sorting and grading of goods for export.

Processing: Surplus production and increased vegetable growing has led to a demand for the establishment of a company for processing fruit and vegetables. Tomato paste has mainly been produced. In 1981 another processing plant for tomato paste and juice was built at Arda with a capacity of 350-400 t per day; it started production in 1983. Another plant for the same purpose was constructed in Southern Ghors with a capacity of 360 t per day, and is expected to go into production in late 1984.

Market Information: Market information is meant to be the amount of production, market demand and prices. Despite modern means of communication, however, this information is still limited. Middlemen and exporters make connections through their agents in the Gulf area and neighbouring countries. Prices and supply forecasting are not available either for producers or for exporters. Retail prices are the only information available.

Financing: Marketing activities require financing to maintain their functions. There are three sources of funds, including the AMO, the JCO and the Jordan Valley Farmers' Association (JVFA). The commercial banks and middlemen provide loans to farmers. The main purpose of these loans is to finance production, only a minor portion going to marketing carried out by farmers, wholesalers and retailers.

Marketing Margins

The marketing margin is defined as the difference between the price received by the farmer and the price paid by the consumer. These margins are devised from individual margins for those who did the buying and selling. Marketing margins are important in identifying marketing efficiency for a particular commodity. It has been found that marketing margins constitute a large proportion of retail prices, especially for perishable goods. To determine the magnitude of marketing margins, estimates were made for tomatoes, cucumbers and bananas.

Tomatoes: Tomatoes are perishable and not suitable for storing for a long period.

Table 10.8: Marketing Margin for Tomatoes (per kg)

Item	Value (fils)	% of retail price
Farm price	68.93	41.35
Container	20.00	12.00
Transport	5.00	3.00
Commission (5%)	5.05	3.02
Municipality fees (2%)	2.02	1.21
Wholesale price	101.00	60.58
Municipality fees (2%)	2.02	1.21
Transport	10.00	6.00
Storage	10.00	6.00
Wastage (20%)	20.20	12.12
Bags	4.00	2.40
Profit margin	19.49	11.69
Retail price	166.71	100.00

Source: AOAD 1982: 112.

The percentage of wastage and loss amounts to 20 per cent. An estimate of the tomato-marketing margin was calculated for 1 kg in 1980, to include the cost of the container, transport, storage, municipality fees and the middleman's commission. Table 10.8 shows the marketing margin for tomatoes per kg. The farmer's share of the consumer price is about 41 per cent; the highest cost goes to wastage, followed by the container.

Table 10.9: Marketing Margin for Cucumbers, 1980 (per kg)

Item	Value (fils)	% of retail price
Farm price	174.01	67.80
Container	20.00	7.79
Transport	5.00	1.95
Commission (5%)	10.70	4.16
Municipality fees (2%)	4.25	1.67
Wholesale price	214.00	83.37
Municipality fees (2%)	4.28	1.67
Transport	10.00	3.90
Storager	10.00	3.90
Wastage	5.13	2.00
Bags	4.00	1.56
Profit margin	9.27	3.60
Retail price	256.68	100.00

Source: AOAD 1982: 113.

Cucumbers: The wastage and loss for cucumbers is about 2 per cent. The items which comprise the marketing margin are the same as for tomatoes. The cost of the container is the highest element. The farmer's share of the consumer price has been about 68 per cent while in tomatoes it has been 41 per cent. Table 10.9 shows the marketing margin for cucumbers per kg in 1980.

Bananas: The elements of the marketing margin for bananas differ from those for other fruit. The wholesaler, who is the owner of the ripening plant, buys direct

Table 10.10: Marketing Margin for Bananas, 1980 (per kg)

Item	Value (fils)	% of retail price
Farm price	156.26	60.80
Loading	4.00	1.55
Transport	5.00	1.95
Price paid by wholesaler	165.26	64.30
Municipality fees (2%)	4.24	1.65
Storage & ripening	9.00	3.50
Wastage (5.8%)	12.20	4.79
Profit margin	12.20	8.25
Price paid by retailer to wholesaler	212.00	82.49
Municipality fees (2%)	4.24	1.65
Commission (2%)	4.24	1.65
Transport	2.50	0.97
Wastage (6%)	12.72	4.95
Bags	6.20	2.41
Profit margin	15.10	5.88
Retail price	257.00	100.00

Source: AOAD 1982: 110.

from the producer. Table 10.10 shows the items in the marketing margin for bananas. It was found that the farmer's share of the consumer price is 60.8 per cent and the profit margin for the retailer and wholesaler together is 14.13 per cent of the consumer price.

Grain Marketing

This section will discuss the production, consumption, import, export and marketing of wheat.

Wheat Production
Wheat is one of Jordan's major crops. Land allocated to it during the period

Table 10.11: Wheat Area and Production, 1974-80

Year	Area (000 du)	Production (000 t)
1974	2372.5	243.9
1975	1183.1	50.1
1976	1396.4	65.5
1977	1264.6	62.4
1978	1345.6	53.3
1979	989.9	16.2
1980	1331.9	133.5
Average	1412.0	89.3

Source: DOS a. 1974–80.

1974-80 averaged 1.412 mn du. This area varied greatly from year to year, ranging from 0.989 mn du to 2.372 mn du. Production was extremely variable, with an average of 89.3 thousand t, but ranging from 16.2 to 243.9 thousand t (see Table 10.11).

Demand for Wheat
Demand for wheat has increased due to population growth and increased incomes. In 1974 demand was 360.5 thousand t; this had grown to 435.3 thousand t in 1980.

Imports and Exports
Imported wheat comprises a high proportion of all imported agricultural commodities, in both quantity and value. This is due to a shortage in production. In 1975, 44.8 thousand t of wheat and 59.7 thousand t of flour were imported; these

Table 10.12: Amount of Value of Imported Wheat and Flour, 1975-1980

Year	Amount of wheat (000 t)	Value ($US mn)	Amount of flour (000 t)	Value ($US mn)
1975	44.8	8.8	59.7	11.4
1976	132.5	28.7	68.3	15.1
1977	139.0	20.30	107.6	20.6
1978	173.3	28.8	85.1	16.8
1979	201.2	45.2	78.5	18.6
1980	162.9	46.7	118.1	28.3

Source: DOS a. and b. 1975–80.

Figure 10.2: Marketing Channels for Wheat and Flour

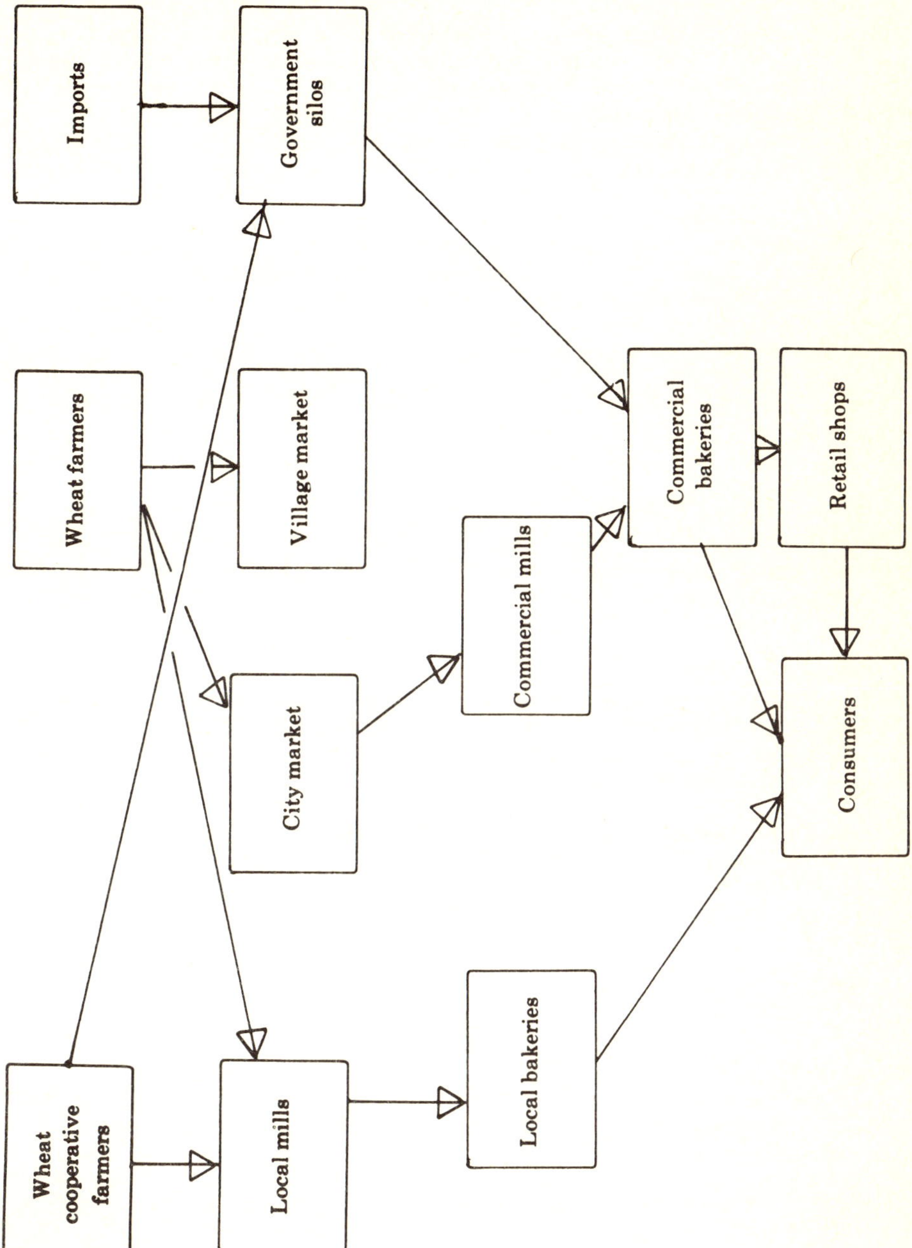

imports increased to 162.9 thousand t and 118.1 thousand t respectively in 1980. The value of imported wheat and flour was $US 20.2 mn in 1975 and rose to 75.0 mn in 1980 (see Table 10.12).

Marketing Channels

Until recently, the private sector controlled the wheat trade. This situation has now changed, and the public sector has assumed the main role. Government involvement became necessary as a result of the high prices of wheat and other imported food commodities.

The marketing of domestic wheat is shared by local merchants who purchase it from the farmers. The marketing system includes village and town wholesale merchants, as well as the city merchants. The village wholesale merchants are small businessmen who handle small quantities of wheat. They buy it either on their own or as intermediary agents (middlemen) for the town merchants. The latter are located in the districts of Amman, Irbid, Balqa, Karak and Ma'ân. They purchase either from village merchants or directly from the farmers, and store the wheat to resell to the Government, to consumers or to the flour mills. City wholesale merchants are located in Amman. Their wheat trade has declined because of government intervention. At present, the Government is the sole wheat importer. In order to understand the wheat marketing system, it is useful to look at its marketing channels. Figure 10.2 exhibits these channels for wheat and flour.

Sheep Marketing

Sheep, goats, cows and camels are a major source of red meat. This section describes the production, consumption, import and marketing channels of sheep and red meat.

Sheep Production

Sheep are the main source of red meat in Jordan. The number of sheep was 644 thousand heads in 1970, and increased to 852 thousand in 1980.

Goats come next in importance. The number was 349 thousand in 1970, increasing to 445 thousand heads in 1980.

Cattle did not show any significant improvement. Domestic cattle decreased from 32 thousand heads in 1970 to 26 thousand in 1980. Red meat production increased from 8.4 thousand t in 1970 to 13.2 thousand t in 1980.

Demand for Red Meat

Average *per capita* consumption was 9.7 kg per year during the period 1970-80, ranging from a minimum of 6.1 kg to a maximum of 12.5 kg. Total consumption of red meat was at a minimum level of 3.15 thousand t per year and reached a maximum of 27.6 thousand t during the same period.

Imports and Exports

Jordan does not produce sufficient red meat to meet local consumption demands. It therefore imports meat and live animals from different sources. The self-sufficiency ratio is about 48 per cent and the average amount imported was 9.2 thousand t per year during the period 1970-80. The maximum amount imported was 17.1 thousand t per year. Exports of live sheep amounted to 2.2 thousand heads in 1978 and increased to 172.9 thousand heads in 1980.

Marketing Channels

The market for red meat is not organised. The Government imports refrigerated

Figure 10.3: Marketing Channels for Sheep and Red Meat

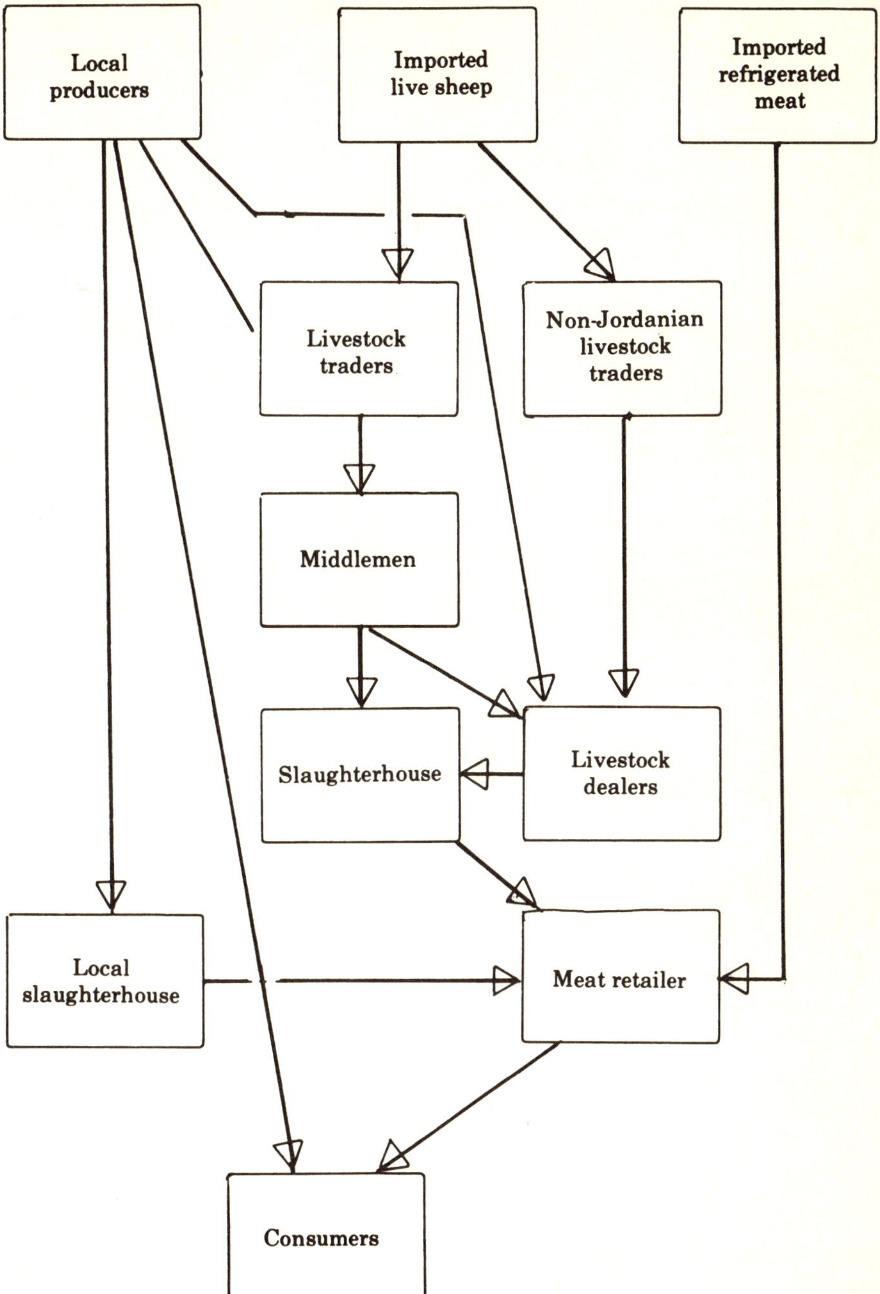

meat and live animals and distributes the meat to retail shops. Local producers have three channels through which to sell their animals: to livestock traders, through middlemen or to livestock dealers. The only central market for selling sheep is in Amman, and the Municipality there has made arrangements to move the market outside the city limits. The trade in other towns takes place in the traditional manner. Figure 10.3 shows the marketing channels for sheep and red meat.

Poultry Marketing

Poultry includes broilers and laying hens. Both industries have witnessed a remarkable quantitative and qualitative growth in recent years. This section will be devoted to the production, consumption, imports, exports and marketing channels of broilers.

Broiler Production
The broiler industry grew considerably during the period 1970-80. The number of chicks increased from 1.5 thousand in 1970 to 5.6 thousand in 1980. Chicken farms also increased in number: from 395 in 1970 to 965 in 1980. The production of broiler meat increased from 7.7 thousand t in 1970 to 29.1 thousand t in 1980.

Demand for Broiler Meat
The increase in broiler production was accompanied by an increase in *per capita* consumption and the self-sufficiency ratio of broiler meat. *Per capita* consumption was 5.8 kg in 1970; this increased to 15.7 kg in 1980. Total consumption grew from 8.7 thousand t in 1970 to 15.7 thousand t in 1980.

Imports and Exports
Despite the growth in the production of broilers, Jordan imports some of its poultry requirements in the form of frozen chickens. Imports fluctuated between 0.4 thousand t and 10.4 thousand t per year during the period 1970-80.

Marketing Channels
Broilers are 50-60 days old when sold. They are marketed mostly through wholesale dealers who are often the regional agents of feed companies and hatcheries. City wholesalers may purchase broilers from the producers or wholesale dealers and then distribute them to retail poultry shops. In these shops, the chicken are displayed alive in cages and sold to consumers on their live weight basis; they are then slaughtered and cleaned at the request of the customers. Some of the wholesale dealers take the broilers to the slaughter-house in either Amman or Irbid; after being processed they are sold to supermarkets and groceries. Figure 10.4 displays the marketing channels for broilers.

Institutions and Organisations Concerned with Agricultural Marketing

Numerous government and autonomous institutions, as well as commercial and consulting companies, are involved in the agricultural sector. Those organisations and institutions which are active in marketing are essential to agricultural development.

The agricultural sector suffers from duplication and a conflict of interests of those institutions working in it. This is due to a lack of coordination and the absence of agricultural policy. Agricultural marketing and marketing policy are not related to any one institution which maintains an integrated series of marketing activities. The institutions which influence marketing are as follows:

Figure 10.4: Marketing Channels for Broilers

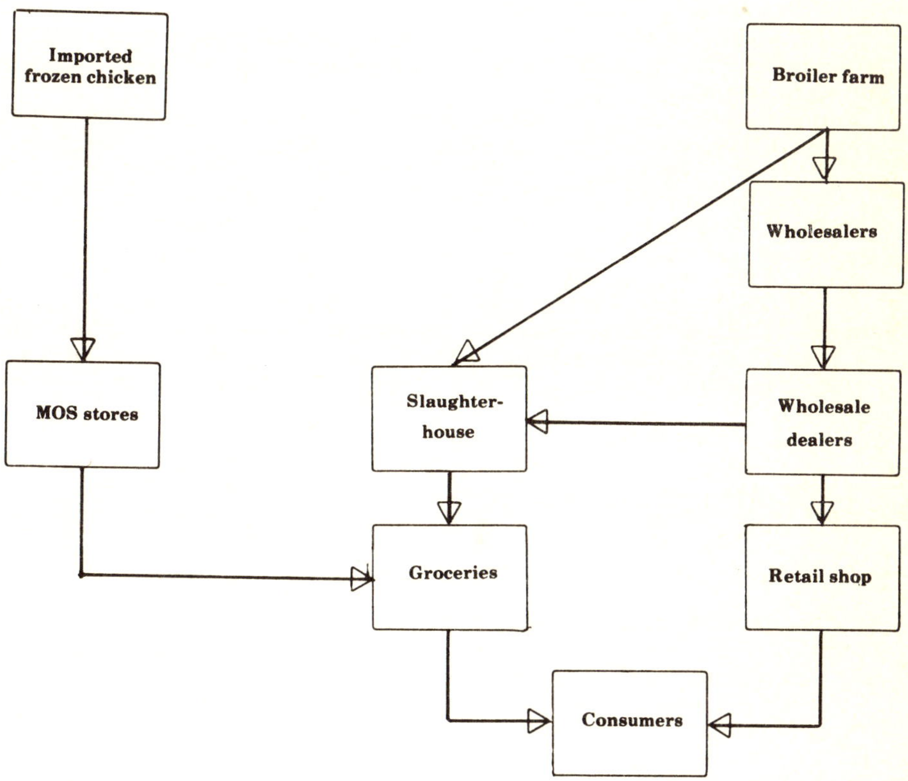

The Agricultural Marketing Organisation

There have been many attempts to organise the marketing of agricultural products. A private corporation was established by farmers, exporters and the Government, but it was transferred into a parastatal body after failing to meet its obligations. The AMO was established in 1970, its objectives being to increase marketing efficiency, develop exports, construct agri-based industries and make forward contracts with farmers.

The AMO has established a grading station at Wadi Yabis and a processing plant in Amman. Currently the Organisation imports and exports various commodities, especially to Syria and Iraq. The AMO is not the only body involved in import/export activities and a variety of individual exporters exists.

The Jordan Cooperative Organisation

The JCO was established in 1968. Its laws state that the organisation will build up a number of cooperative societies to provide supplies to its members and to market their products.

Marketing societies have not been particularly successful in marketing vegetables. The cooperatives' experience in marketing is limited. Competition from middlemen has been serious and the financial and administrative resources of cooperatives constrained. In 1980 the JCO reactivated its marketing activities, whose success has been limited by financial and management constraints.

The Jordan Valley Authority

The JVA does not undertake marketing activities according to the law, but its statutes have allowed it to execute projects which assist agriculture. The Authority has built three marketing centres in the Jordan Valley and for several seasons has bought farmers' produce, part of which is processed in its own plant.

The Jordan Valley Farmers' Association

This Association was established in 1974 with the objectives of formulating agricultural policies and participating in the implementation of development projects in the Valley. It plans to build processing plants, make containers, market, produce and provide farmers with inputs and loans. The JVFA has not proposed any marketing programme.

The Ministry of Agriculture (MOA)

The MOA is the principal institution responsible for formulating agricultural policy, implementing the cropping pattern, providing veterinary services and conducting research. The marketing side is restricted to issuing a monthly plan for imports and exports. The Ministry has the right to establish central markets and grading stations outside the limits of the municipalities.

In 1972 the Department of Agricultural Economics and Planning was established. Its obligations are to supervise the domestic markets, imports and exports. A monthly plan and the collection of price statistics are the functions of this Department.

Municipalities

In 1966 a law was issued to organise the wholesale central markets: the contract system was cancelled and the municipalities became responsible for the establishment of central markets for vegetables and fruit in towns. The municipalities organise market operations, and carry out inspections to ensure that the law to guarantee total competition between middlemen is being implemented.

The ACWM has a Board chaired by the Mayor. The Municipality charges a 4 per cent fee on the total value of the products sold. In Irbid Central Market, the

Municipality appoints a Director who is responsible for enforcing the law. A private firm on contract offers services to farmers and collects a 4 per cent fee on the total sales. In Zarqa, the Municipality administers the Central Market and has a Director assisted by inspectors and two employees to collect 4 per cent of the total value of agricultural goods sold. Municipalities play no part in the secondary markets (*hesbat*), except for supervising the implementation of the law.

Jordanian Company for the Marketing and Processing of Agricultural Products (JCMPAP)

In 1982 the Higher Agricultural Council (HAC) discussed the organisation and development of agricultural marketing in Jordan. It concluded that the marketing system suffered from a lack of services, the domination of middlemen and the absence of grading. The Council recommended the establishment of the JCMPAP, with a capital of JD 10 mn. The public sector owns 51 per cent of the shares and the private sector 49 per cent. The objectives of the Company are to trade, process and market agricultural products as well as to administer the central markets. It is expected that the Company will start work in 1984.

References and Other Works Consulted

AOAD (Arab Organisation for Agricultural Development), 1982, *Supply and Demand of Fruit and Vegetables*, Vol. 2, Hashemite Kingdom of Jordan, Khartoum.

DOS (Department of Statistics), a. *Agriculture Sample Surveys*, 1974-80, Amman.

—— b. *Agricultural Statistics Yearbooks, 1975-80*, Amman.

Steitieh, M. Akram *et al.*, 1978, *A Manual for the Vegetable Crops Grown in the East Jordan Valley*, University of Jordan, Amman.

PART IV

TECHNOLOGY TRANSFER AND FEEDBACK

11 AGRICULTURAL RESEARCH AND EXTENSION

Subhi Qasem

Introduction

Agricultural research is a technical and scientific activity through which information and technology are generated. Extension is the activity through which the generated technology is communicated, demonstrated and applied in farmers' fields. The final objective of both research and extension is to optimise productivity of resources and inputs in the process of producing food, fibre and other agricultural commodities. Research and extension are but two ingredients in a package of inputs in the overall system of agricultural production. It is in this context that we analyse research and extension in the agricultural system in Jordan.

Historical Background

Agricultural research and extension in the formal sense are both relatively new to Jordan. It was in 1952 that the first agricultural experiment station, Deir Alla Station was established. Deir Alla Station is still in operation, and is located almost in the middle of the Jordan Valley. The setting up of the Station was a joint venture between the Jordanian Government and the British Overseas Development Administration (ODA). The first formal experimental work there was started on cereals, at that time the predominant crops in the farming system of the Jordan Valley, as well as nationwide. In 1954 extension activities were started up through a technical assistance programme from the United States Agency for International Development (USAID). Extension activities at that time were the dissemination of transferred technology and information on: new stone- and pome-fruit varieties introduced from Europe (mainly from France); (b) the use of pesticides, introduced by various departments of the Ministry of Agriculture (MOA), and private trading companies from the US and the UK; (c) new vegetable varieties, introduced in the same way as pesticides, but mostly from the US and France; (d) the improvement of productivity of local and introduced varieties of cereals; and

(e) the improvement of cultivation of both local and introduced cultivars of citrus fruit species.

In 1958 research work was formally brought into a Department of Agricultural Research within the MOA, with support from another USAID project. In the same year extension work was also upgraded to warrant a Department within the MOA, again with the support of a USAID project. In 1969 the extension and research activities of the two Departments were reorganised under a single Department of Agricultural Research and Extension (DARE), which still survives as such within the MOA.

In 1973 the University of Jordan established the Faculty of Agriculture (FOA), with the objective of spearheading an integrated programme of teaching and research in the service of agricultural development. FOA's research role in 1974, and gained momentum in the years that followed.

Agricultural Research Institutions

As mentioned above, there are only two formal institutions involved in the implementation of research in Jordan, the FOA and the Agricultural Research Section within the DARE.

Type and Location of Physical Facilities Allocated to Research

The facilities of the FOA are allocated to the two functions of teaching and research and they comprise: (a) a main building; (b) specialised production units for dairy cattle, sheep and poultry; (c) glass houses and plastic houses; (d) the University Farm in the Jordan Valley (see Table 11.1 for details of areas and functions); and (e) access to several MOA nurseries or small stations

The facilities of the DARE comprise: (a) a main building on the campus of the University of Jordan (soon to move to offices in Amman); (b) Deir Alla Experiment Station in the Jordan Valley; and (c) access to several stations or nurseries run by various departments of the MOA (see Table 11.1 for details of areas and functions, and Map 11.1 for location of stations).

Evaluation of Type and Geographical Coverage of Facilities: The MOA has lost two main rain-fed agricultural experiment stations. The first became the campus of the University of Jordan at Jubeiha in 1962, and the second became one of the two campuses of the University of Yarmouk, near Irbid, in 1976.

Over the years, stations and nurseries run by the MOA throughout the country have been used less and less for experimental work. The stations at present serve many purposes, including seed production, fruit-tree nurseries, fruit-tree trials and forest-tree nurseries. There is not a single experiment station devoted to research and demonstration activities for rain-fed agriculture.

The Jordan Valley irrigated lands are served by two stations , one under the MOA (Deir Alla) and the second under the University of Jordan (University Farm). There are two smaller sub-stations in the northern part of the Jordan Valley but they have been increasingly used as citrus nurseries rather than for research.

Map 11.1: Agricultural Stations in Jordan, 1984

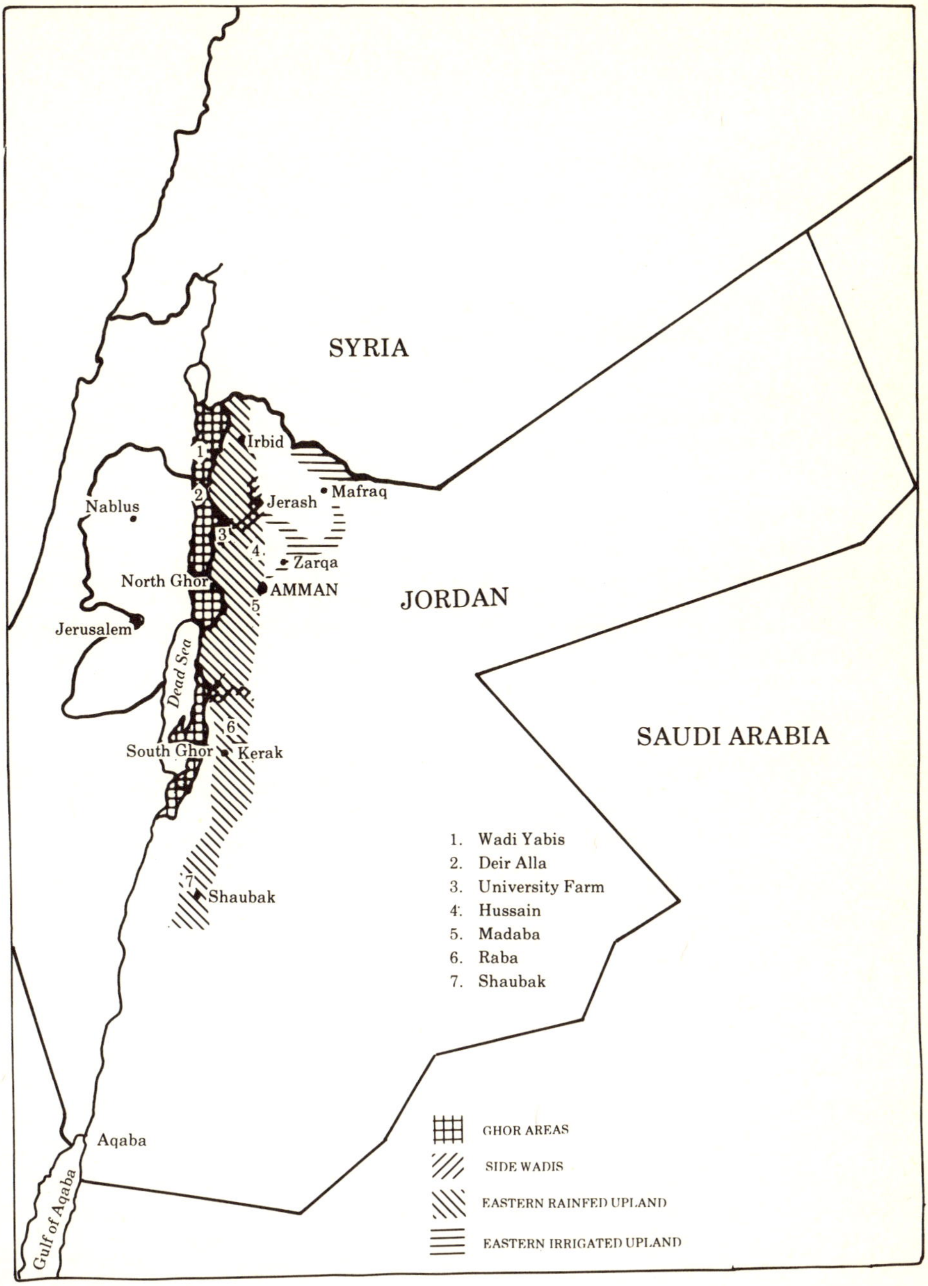

Table 11.1: Physical Facilities Available for Agricultural Research in Jordan

Faculty of Agriculture, University of Jordan

1. Main building: located on the University campus at Jubeiha, a suburb of Amman. The total area is 12,000 m², comprising the following:

Type	*Number*	*Comments*
Staff offices	52	About 14 m² each
Lecture halls	2	Seat 180–220
Classrooms	5	Seat 40–60
Seminar rooms	5	—
Teaching laboratories	5	Soils, plant protection, plant production, animal production, and nutrition
Research laboratories	20	—
Insect museum	1	—
Reading room	1	Small reference library
Dairy processing plant	1	—
Stores	2	—

2. Six glass houses with main house for teaching and research.
3. Poultry farm for teaching and research.
4. Sheep farm for teaching and research.
5. Machinery workshop and machinery units for teaching.
6. Experimental plots for cereals and legumes plus some demonstration orchards for stone and pome fruit trees.
7. University Farm located in the centre of the Jordan Valley. The Farm has experimental land devoted to irrigated vegetables in the open and under plastic, demonstration fruit tree orchards, a dairy farm and small-scale poultry facilities.

Department of Scientific Research and Extension

1. Main building, located on campus of the University of Jordan. Built in 1960 at what was known as Jubeiha Agricultural School. The land was given to the University of Jordan but the building remained in the use of the MOA. The total area is about 1,200 m² comprising the following:

Type	*Number*
Staff offices	22
Research laboratories	4
Reading room	1
Meeting hall	1
Store	1

2. Land allocated for trials at several Agricultural Stations located in various parts of country as follows:

Station	*Main activities*
Deir Alla Agricultural Experiment Station (centre of Jordan Valley)	Experiments on: irrigated wheat, vegetables in field, vegetables under plastic, forages. Plus nursery for production of forest and fruit-tree seedlings.

Shaubak Agricultural Station (southern uplands)	Experimental fruit-tree orchards, legume experimental plots in cooperation with ICARDA, and nursery for production of fruit-tree seedlings (mainly stone and pome fruits).
Hussain Agricultural Station (near Amman)	Fruit-tree plots, nursery for production of seedlings of fruit trees (mainly olives and stone fruits).
Raba Agricultural Station (southern uplands near Karak)	Experimental plots for cereals (wheat and barley) and for legumes. All land is rain-fed.
Madaba Seed Protection Station (south of Amman)	Multiplication of cereal seeds mainly wheat and experimental plots for cereals. All land is rainfed.
Wadi Yabis Station (northern Jordan Valley)	Experimental plots for citrus trees, plots for irrigated vegetables and cereals.

A summary of geographical and commodity coverage of the experiment stations is as follows:

Commodity and/or region	State of coverage
Citrus research (irrigated)	Very weak[1]
Olive research (rain-fed)	Very weak
Pome and stone fruits (rain-fed)	Weak[2]
Cereals (rain-fed and irrigated)	Acceptable[3]
Vegetables (irrigated)	Acceptable
Legumes (rain-fed)	Acceptable
Poultry	Very weak
Sheep & goats	Weak
Dairy cattle	Very weak
Irrigated areas (in general)	Weak to acceptable
Rain-fed areas (in general)	Weak to very weak

Notes: [1] *Very weak:* No organised experimental orchards and/or facilities consistent with research needs.

[2] *Weak:* Some experimental orchards and/or facilities which constitute part of either an experiment station or the multi-purpose stations of the MOA. Research mainly on the introduction of new germ-plasm and some pest control. These usually cover more than one major cropping region of the country.

[3] *Acceptable:* Land and facilities in experiment stations and/in multi-purpose stations which cover major areas of cultivation. Research is strong in some areas; but in others, there are gaps.

Fluctuating Levels of National Support for Agricultural Research

The budgets of neither the MOA nor the University of Jordan have line items for expenditures allocated to research. Therefore, to determine total expenditure a search had to be made through budgets and research projects together with interviews. Estimates of total recurrent expenditures are presented in Table 11.2. Total allocations are characterised as follows.

Table 11.2: Expenditure on Agricultural Research in Jordan, 1973-82 (Totals for Faculty of Agriculture and Department of Research and Extension in the Ministry of Agriculture (thousand JD)

Year	FOA Salaries	FOA Operational costs	FOA Total	MOA Salaries	MOA Operational costs	MOA Total	Grand total (rounded figures)
1973	0	0	0	80.0	35.8	115.8	116
1974	6.6	7.1	13.7	83.3	35.8	119.1	133
1975	16.9	12.2	29.1	120.1	48.4	168.5	198
1976	35.4	20.3	55.7	132.5	52.1	184.6	240
1977	62.9	35.5	98.4	133.4	54.2	187.6	286
1978	95.4	42.3	137.7	125.4	53.4	178.8	317
1979	94.4	46.4	140.8	94.5	54.5	149.0	290
1980	112.4	54.3	166.7	124.4	45.4	169.8	337
1981	78.8	36.8	115.6	127.8	54.4	182.2	298
1982	68.6	38.8	107.4	138.6	70.4	209.4	316

Commodity	Value (JD mn)	% of commodity value from AGDP	Allocations of research to commodity (JD 000)	% of commodity allocations to total
Field crops	7	15	110	33
Vegetables	15	33	128	38
Fruit trees	10	22	58	18
Animal production	14	30	38	11
TOTAL	46	100	334	100

Source: The data were calculated by extraction from budget items of the MOA and FOA. Figures include operational funds provided by technical assistance programmes from FAO, USAID, IDRC and the British Overseas Development Ministry (ODM).

First, allocations for research are low by international standards and are less than 1 per cent of total value of AGDP, as shown by the following, where all sums are in current prices:

Year	1976	1977	1978	1979	1980	1981	1982
Value of AGDP (JD mn)	37.3	41.7	58.6	43.6	64.6	76.6	85.3
Total expenditure for research (JD 000)	240	286	317	290	337	298	316
% of research expenditure to AGDP	0.6	0.7	0.5	0.6	0.5	0.4	0.4

The allocations are comparable to those in some developing countries which have low percentages, but are much lower than the levels of those in developed countries, which reach 2-4 per cent of AGDP.*

Second, allocations have fluctuated from one year to another and have been decreasing in constant value (prices) since 1977. Total expenditure was JD 286,000 in 1977, went up to JD 337,000 in 1980, and then down to JD 315,000 in 1982. Calculated at constant 1977 prices, allocations to research decreased by 25 per cent from 1977 to 1982 (calculated at 8 per cent inflation each year from 1977 to 1982).

Third, the increase in allocations in real terms has been more for staff salaries than for operating funds. Hence, as years have passed, the funds available to carry out research both in the laboratory and in the field have been decreasing.

Distribution of Allocations for Major Commodities

The allocations for our major commodities have been calculated to determine the importance given to each for the period 1976-80. Results were as follows, with all figures being averages for the period 1976-80:

Commodity	Value in (JD mn)	% of commodity value from AGDP	Allocations of research to commodity (JD 000)	% of commodity allocations to total
Field crops	7	15	110	33
Vegetables	15	33	128	38
Fruit trees	10	22	58	18
Animal production	14	30	38	11
TOTAL	46	100	334	100

Distortions in allocations are clearly apparent in the above figures. The average value of animal wealth made up 30 per cent of the total value of agricultural production but received only 11 per cent, while cereals received 30 per cent of total allocations for agricultural research but constituted only 15 per cent of the total value of agricultural production.

Manpower in Research

The number of workers involved in agricultural research fluctuated from one year to another. In 1974 the total number was 27, increasing to 72 in 1980, but falling to 54 in 1982. Table 11.3 provides details of numbers and qualifications. There are two reasons for such fluctuations. First, research workers in the MOA are treated the same as any other workers in the Ministry and hence may be being transferred from one department within the Ministry to another by ministerial decision. Second, the full-time equivalent of man-years devoted to research by staff of the FOA depends on how many are involved in teaching. In 1981 and 1982 staff became increasingly involved in teaching, and the trend of future years is in this direction.

The qualifications of researchers in the MOA are on the whole of BSc level,

* Allocations for agricultural research as percentage of AGDP for 1980 were: 1.15 in Brazil, 0.70 in Nigeria, 0.65 in Mexico, 0.41 in Pakistan, 1.08 in Kenya and 0.48 in Bangladesh (Oram and Bindlish 1981).

Table 11.3: Numbers (full-time equivalent) and Qualifications of Manpower Working in Research in Jordan, 1974-82

Year	Level of qualification	FOA	DARE	Total
1974	BSc	1	19	20
	MSc	1	3	4
	PhD	1	2	3
1976	BSc	2	24	26
	MSc	3	5	8
	PhD	6	2	8
1978	BSc	4	24	28
	MSc	8	5	13
	PhD	16	2	18
1980	BSc	4	24	28
	MSc	12	8	20
	PhD	22	2	24
1982	BSc	1	22	24
	MSc	6	8	14
	PhD	12	4	16

Sources: Extracted and compiled by author from:
Faculty of Agriculture 1974-82.
DARE 1974-82.

Table 11.4: Number of workers (full-time equivalent) either qualified in the field of specialisation (MSc or PhD) or simply working in that field (BSc), 1976 and 1980

	1976				1980			
	PhD	MSc	BSc	Total	PhD	MSc	BSc	Total
Cereals	1	2	5	8	2	4	4	10
Legumes	0	0	1	1	1	1	2	4
Vegetables	1	1	4	6	2	2	4	8
Fruit trees (deciduous)	0	1	3	4	2	1	3	6
Soils & irrigation	2	1	6	9	6	3	6	15
Animal nutrition	1	0	1	2	2	4	2	5
Citrus	0	0	0	0	0	0	0	0
Olives	0	0	0	0	0	0	1	1
Poultry	0	0	0	0	1	1	1	3
Plant protection	2	3	5	10	6	5	4	15
Agricultural economics	1	0	1	2	2	2	1	5
TOTAL	8	8	26	42	24	20	28	72

Sources: Extracted and compiled by author from:
Documents of research projects submitted by researchers to Faculty of Agriculture, University of Jordan, Amman.
DARE 1976-80.

which is considered too low for the identification, formulation and implementation of research work. Students at BSc level receive little if any training in research techniques and/or research implementation.

Another factor is the attrition rate in research workers. A characteristic of manpower in research is the discontinuity of service, and this is especially true of those working in the MOA. Although the number of research workers may remain the same from year to year in the DARE, the same individual may or may not continue from one year to the next. Those with 10 or more years of continuous service in research in the MOA do not exceed 10 per cent of the total manpower in any one year.

There are also gaps in fields of specialisation of manpower in agricultural research and there is no apparent plan for manpower development. This, plus the fact that staff in research change jobs, has led to several gaps in the fields of specialisation of researchers. At present, for example, there are no individuals qualified in citrus or olives. Persons qualified in animal production and working in research are very few indeed and are not congruent with the needs of the country. A summary of manpower working either in major fields or on commodities is given in Table 11.4.

Structure of Research Institutions

The research function is organised in a Division within the DARE. The Division has research units which will carry out research in the separate areas of field crops, vegetables, tobacco, fruit trees, plant protection, animal production, and soils and fertilisers.

The FOA is organised into six departments: Plant Production, Plant Protection, Animal Production, Nutrition, Soils and Irrigation, and Agricultural Economics. In the FOA, research is carried out in every department, but for certain programmes, like wheat improvement, activities of the programme cut across departments.

Research Activities

There are two major categories of research activity. First, major research programmes which are commodity-oriented (multi-disciplinary), for example, wheat improvement and food legume improvement. These programmes are usually few and funded jointly by national institutions and foreign donor or technical assistance agencies. Second, there are individual research projects which are most uni-disciplinary and are usually funded by the core budget or from national resources. (See Table 11.5 for details.) The following is an account of the steps through which activities go, together with comments on them.

Identification of Research Projects

Research projects do not emanate from a national research plan with defined priorities. They originate from one of the following sources:

a. Personal judgement of the researcher
b. Suggestion by a technical assistance project
c. Decision by a group of researchers
d. Suggestion by researchers, approval by planners with minimal scrutiny and inclusion in the Five-Year Development Plan.

Farmer's inputs to problem identification are either minimal or nil.

Table 11.5: Summary of Major Research Areas Carried out by the Faculty of Agriculture and/or Department of Research and Extension, 1978-82

Commodity	Nature of research activity	Implemented by
Wheat	Introduction of new varieties, seeding rate, date of planting, weed control, fertilisation rates & time of application, regional and international nurseries	Most of the work is part of a cooperative programme between FOA, DARE and ICARDA
Barley	Introduction of new varieties, seeding, rate, date of planting, international nurseries	DARE and FOA
Chick-peas & lentils	Introduction of new varieties, international, chick-pea nursery, date of planting, seedings rate, fertilisation rate	Mostly by FOA in cooperation with ICARDA
Tobacco	Introduction of new varieties, growing methods, fertilisation rate, breeding new varieties	DARE
Vegetables	Introduction of new varieties, epidemiology and control of major diseases and insects, fertilisation rate, date of planting, water requirements under irrigation and cultural practices. Most work is on tomatoes , eggplants, cucumbers, squashes, cantaloupes & beans.	Varieties: DARE Pest control: DARE, FOA Fertilisation: DARE, FOA Water requirements: FOA
Olives	Mechanical harvesting, methods of propagation by vegetative growth, fertilisation, epidemiology and control of important insects.	DARE and FOA
Peaches	Introduction of new varieties, epidemiology and control of powdery mildew	FOA
Grapes	Introduction of new varieties, hormone studies, control of powedery mildew and insects	DARE and FOA
Citrus & apples	Behaviour of some introduced varieties	DARE
Pest control	Integrated pest control studies on viruses of tomatoes, cucurbits, also on broomrape (Orbanche) of tomatoes, nematode on trees & vegetables	Mostly by FOA; some by DARE
Pest survey	Survey & identification of nematodes, weeds, diseases & insects invading crops, especially in Jordan Valley	Mostly by FOA; some by DARE
Plastic agriculture	Vegetable varieties (tomatoes and cucumbers) for plastic production, water requirement & irrigation regimes, cultural practices & and pest control methods. Also size, orientation plastic quality & dimensions of plastic houses in Jordan Valley	Mostly by FOA DARE: vegetable varieties
Sheep	Lamb fattening studies & use of crop residues in feeding	FOA
Poultry	Feed rations, behaviour of some breeds of egg layers & production practices	FOA

Table 11.5 continued

Agricultural economics	Economics of production of major crops, economics of some input in crop production & production/consumption studies of some food commodities	FOA
Fertilisers	Solubility of major nutrients in soils of Jordan	FOA

Review of Formulated Projects
Documented projects are subjected to committee review in the FOA. Projects of the DARE are rarely formulated or documented, and hence are rarely reviewed. Experimental design is rarely reviewed before implementation.

Implementation of Research Projects
Implementation in the field is usually carried out by technicians without direct supervision from senior researchers. This is particularly true of projects implemented by the DARE. Implementation is slowed down, hampered or interrupted by poor allocations of operating funds, especially for transportation.

Monitoring of Research Activities
No monitoring method is used except in projects supported by external sources of funding. In the latter case, biannual or annual progress reports are submitted.

Priority Determination
Research projects are either compiled in a list or presented individually. No criteria are used to determine priorities consistent with national objectives.

Formulation of Projects
Formulation documents showing materials, methods, plan of work and inputs required are rare, except in projects funded by external agencies or sources outside the institution. Documentation does exist for FOA projects. In the case of the DARE, projects pass directly from title identification to implementation.

Evaluation of Research Projects
Projects are not usually subjected to any evaluation. Some projects, especially crop breeding, go on and on without any checks. Annual reviews take place only where projects are supported by external sources of funding. Research projects are rarely terminated because of a review process, but rather because funds are not available, or the researcher has moved, or simply on the basis of the researcher's personal judgement.

Research Output and its Fate
Research results are reported either in annual reports, as in the case of the DARE, or published in scientific journals, as in the case for publishable work carried out by staff of the FOA. Results of all researchers supported by funds from outside the core budget are given in the form of a final report (FOA). English is used when research is published in scientific journals, while Arabic is used in the case of annual reports of the DARE.

Development of Research Results
The DARE publishes occasional extension pamphlets or bulletins. These are aimed at extension agents and farmers. The information included in the

published material derives from research results implemented either in Jordan or abroad. The quality of these publications is poor, their distribution is limited, and they are rarely updated to keep up with new information or to replenish stock. Seldom do farmers refer to these materials when seeking answers to problems. Furthermore, the topics covered are not necessarily those relevant to farmers.

Both the DARE and the FOA hold occasional workshops, or field days, to which concerned farmers are invited. Research results are presented in the workshops in simplified form. The workshops held by the FOA are the most successful; they cover major commodities like poultry production, sheep production and wheat improvement. Healthy interaction between researchers and farmers has been reported from such workshops. On the whole, workshops and field days are not held systematically and do not cover all the major concerns of farmers.

Materials and technology generated by research within the country do not have strong channels to facilitate their distribution to farmers. The avenues provided by the MOA are limited to wheat seeds, which are produced annually in quantities that cover less than 10 per cent of seeds needed. The varieties reproduced are old introductions of the DARE. Newer and better varieties did not take their rightful place in the limited seed production programme. The same applies to the fruit tree and olive seedling distribution programme. Seedlings distributed by the MOA are not always true to type, are not competitive in quality and type with imported ones, or even with those distributed by the private sector, and cover only a small percentage of farmers' needs. This is true of seedlings for grapes, stone fruit and citrus. In the case of olives, seedlings cover a larger proportion of farmers' needs but in some cases they are not true to type.

Research Links

Links between the Two Implementing Institutions

There are no formal links between the FOA and the DARE. It is only in cooperative research programmes like wheat improvement that workers from both institutions participate in the identification, formulation and implementation of such programmes. In the case of wheat improvement, cooperation is brought about through conditions set by the agencies supporting the programme—the International Centre for Agricultural Research in the Dry Areas (ICARDA), the International Development Research Centre (IDRC), the Ford Foundation and the Centro Internacional de Mejoramiento de Maiz y Trigo (CIMMYT).

Any other links are in the form of personal initiatives and efforts. On the whole linkages are weak. Despite previous efforts to coordinate programmes implemented by both institutions, the relationship remains far from satisfactory. As a result, one finds duplication of research efforts, non-uniformity in approach and methodology used in the implementation of activities, and difficulties in reaching a consensus on research output which should be promoted to farmers.

Links with other Agencies

The following is a summary of the links of the two institutions with other agencies both inside and outside Jordan. The links are evaluated as occasional, weak, acceptable or strong.

Links with	FOA	DARE	Remarks
Seed trade companies inside Jordan	Acceptable	Acceptable	Provide new seeds for testing
Chemical trade companies	Strong	Acceptable	Provide new chemicals for testing
ICARDA	Strong	Strong	Provides germ-plasm for testing, and training
ACSAD*	Weak	Acceptable	Provides germ-plasm for testing
USAID	Occasional	Strong	Provides funds for research support to DARE 1981-84
IDRC	Strong	Occasional	Provides funds for legume research and research documentation to FAO
Other research institutions in the region	Weak	Weak	Limited exchange of information
AOAD†	Acceptable	Acceptable	Provide funds for training
Other departments within the MOA	Occasional	Acceptable	Provide facilities for research; participate with DARE in extension information
Other institutions working in other agricultural services and development	Occasional but strong	Occasional but weak	Provide input for research needs

* Arab Centre for the Study of Arid Zones and Dry Lands.
† Arab Organisation for Agricultural Development.

Links with Extension

The FOA has no extension personnel and does not have any formal links with extension workers in the MOA. Both research and extension functions are the responsibility of the DARE. The relationship between extension and research within the DARE will be discussed below.

Links with Farmers

The staff of the FOA have no formal links with farmers, but contact them informally during workshops and field trips. The workshops are held once every year or two. Field trips depend on the individual researcher and vary in frequency from once a week to once a season.

Formal Extension Institutions

There is only one formal institution in charge of extension activities. This is the Extension Division (section) within the DARE of the MOA.

Manpower

There are about 180 agricultural engineers reported by the DARE as being involved in extension activities. This manpower falls into two categories. The first category consists of personnel working at the DARE headquarters in Amman and at the Deir Alla Agricultural Experiment Station. These make up the core staff of the DARE and carry out both research and extension functions. Although numbers change from year to year, they average 80. How much time each worker spends on each function is specified neither in the job description nor in the employment terms and conditions. The following is an estimate of time spent in each function for various specialities based on interviews with the workers.

Field of Work	% of the DARE core workers' time allocated to:	
	Research	Extension
Field crops	80	20
Animal production	40	60
Horticulture	80	20
Dry farming	20	80
Tobacco	50	50
Fertilisers and soils	40	60
Plant protection	40	60
Extension specialists	0	100

The second group consists of those with the title of extension agent, who number about 110, and are located in towns and cities of all governorates of Jordan.

Quality of Manpower

Almost all extension agents have BSc-level training in one of the agricultural fields; the remaining few hold only an intermediate diploma in agriculture.

Subject matter specialists* within extension manpower in the regions (outside Amman and the Jordan Valley) are not part of the organisation. The vast majority of extension agents are sent directly to work in the field upon joining the service. Such people have usually freshly graduated from universities outside Jordan and they do not go through any special training programmes. In past years (the early 1970s) the MOA ordered 6-12 month training courses for fresh graduates through a programme organised jointly by the United Nations Development Programme (UNDP) and the Food and Agriculture Organisation (FAO), but it was discontinued. In reviewing the way extension agents are distributed by region or town, there seems to be no apparent matching of needs between an area and the speciality of the extension agents working in it.

Geographical Coverage

Although the number of extension agents changes from one year to another in the same location, all major areas of agriculture are covered, but not without distortions. The Jordan Valley, for example, which produces 60 per cent of the total value of national agricultural production, is assigned about 30 extension agents, or about 25 per cent of the total force. Amman and Irbid have the next highest number of extension agents, ranging from 20 to 25 each. Extension agents avoid working in what they consider to be hardship areas or towns, like the Jordan Valley, Mafraq (in the east), Tafila (further south) and Ma'an (also in the south). The policy of the MOA is to have good coverage of all towns and agricultural areas, but living conditions in certain places and lack of incentives for those working in remote or hardship areas make it unattractive for them to continue service there.

Even more important than coverage is mobility, or the availability of transportation for extension agents in their outreach activities. All extension agents suffer from a shortage of transportation or even a lack of it in certain cases.

Activities

The 1982 report of the Extension Division within the DARE lists the following types and magnitude of activities:

Type of activity	Frequency	No. of farmers reached
Extension visits	1,479	2,692
Seminars on special topics	19	1,050
Demonstrations	11	202
Film shows	5	118
Field days	2	120
Meetings	18	166
Farmers' visits to extension agents	—	1,979
TOTAL		5,227

Notes: 1. This does not include the Jordan Valley and four other remote towns.
2. Figures do not show whether different farmers attended the same or a different activity.

* Subject matter specialists, like plant protection, vegetable production and wheat production specialists, are not considered part of the extension service. They work in other departments of the MOA or in the head of the DARE in Amman, and are called upon by extension agents when needed.

The number of full-time farmers or people working in farming is not accurately known for recent years. However, the 1981-85 Five-Year Plan puts the number of the labour force in agriculture in 1980 at 63,000. The percentage reached through reported activities of the DARE amounts to less than 10 per cent of the total agricultural labour force.

Impact

There are no published studies on the impact of formal extension on the promotion of information or material inputs in agricultural production. However, interviews carried out with 24 leading farmers and 10 leading agriculturalists in the field of agricultural development revealed that the role of formal extension is very weak. In fact, opinions gathered showed that the impact of formal extension is non-existent in certain areas and does not exceed 20 per cent in the majority of technical service inputs (see Table 11.6).

Linkages

Personnel working in research and extension at the DARE headquarters in Amman are part of the departmental organisation of the Ministry. However, personnel and extension agents in the towns and regions come under the direct supervision of Agricultural Departments or Directorates. In Irbid, for example, there is a Directorate of Agriculture which represents all departments of agriculture within the MOA, and extension agents work under the Director of the Irbid Directorate; such is also the case with the other regional Directorates located in the governorates of Jordan. This organisation is part of an MOA plan to decentralise (regionalise) its administration and services. However, extension agents feel cut off from the required sources and references of expertise available in Amman. Some specialists are working within regional directorates, but they are fully occupied with administrative work and cannot devote much time to coaching extension agents. On the whole, extension agents are not motivated to do more fieldwork since their promotion is mostly based on the accumulation of years of service rather than on achievements.

As to links between extension agents and farmers, overall this is evaluated as weak. Contact is limited to less than 10 per cent of the farming labour force. Furthermore, extension agents do not have the mobility to visit farmers, and the majority of farmers do not have the necessary transportation facilities to come to extension agents' offices. Usually visits are not to seek technical information, but rather to obtain answers to logistic and administrative questions.

Extension in Parastatal Institutions

Two parastatal institutions are involved in providing technical services to farmers in Jordan. These are the Jordan Cooperative Organisation (JCO) and the Jordan Valley Farmers' Union (JVFU). The JCO is responsible for the organisation, administration and support of the cooperative movement. In 1982 there were 408 cooperative societies throughout Jordan, of which 181 were agricultural while the others were mainly for housing, mutual benefits and multi-purpose. The JCO has developed centralised support services to member societies in production inputs, marketing and credit. It also offers some member societies machinery for rent and cereal seed-cleaning services.

Table 11.6: Evaluation of Role or Impact of Public Agricultural Research and Extension Institutions and/or Private-Sector Agricultural Companies in Technology (Material and Information) Generation, Technology Transfer from outside Jordan and Technology Dissemination and Application in the Agricultural Sector in Jordan (date of evaluation: 1983)

Type of technology input used in agricultural production	Public institutions			Private-sector companies		
	Generation	Transfer or adaptation	Dissemination & application	Generation	Transfer	Dissemination & application
Tomato varieties	0	1	1	0	4	4
Cucumber varieties	0	1	1	0	4	4
Parthenocarpic cucumber varieties	0	0	0	0	5	5
Potato varieties	0	3	2	0	2	3
Eggplants	0	1	1	0	4	4
Other vegetable varieties	0	1	1	0	4	4
Citrus varieties	0	3	2	0	2	3
Wheat varieties	2	3	2	Farmers' role 2—3		
Barley varieties	1	2	1	Farmers' role 3—4		
Olive varieties	0	3	3	0	2	2
Stone-fruit varieties	0	2	2	0	3	3
Grape varieties	0	4	4	0	1	1
Home-fruit varieties	0	2	2	0	3	3
Fungicides	0	1	1	0	4	4
Insecticides	0	1	1	0	4	4
Herbicides	0	1	1	0	4	4
Leaf nutrients	0	0	0	0	5	5
Fertilisers	0	1	1	1 (m)	4	4
Tools, small	0	1	1	0	4	4
Machinery	0	1	1	0	4	4
Plastic houses	0	1	1	0	4	4
Drip irrigation systems	0	1	1	2[1] (m)	2	4
Fertiliser applicator with drip system	0	0	0	4[1] (m)	1	5
Sprayers (pesticides)	0	1	1	2[1]	3	4
Packing materials & containers	0	1	1	4[1]	1	4
Soil fumigation injectors	0	1	2	0	4	3
Soil fumigation (how)	2	2	1	1	3	4
Soil fumigants (what)	2	2	1	1	3	4
Soil fumigants (when)	3	2	1	1	3	4
Fertilisers used (how)	1	2	1	0	3	4
Fertiliser techniques (what)	1	2	1	0	3	4
Fertiliser techniques (when)	1	2	1	0	3	4
Disease control techniques (what, when, how)	2	2	2	1	3	3
Insect control techniques (what, when, how)	2	2	2	1	3	3
Weed control techniques	2	2	2	1	3	3
Cultural practices	2	2	3	1	3	2
Seed-bed preparation (wheat)	4	1	3	0(F)	0	0
Harvesting methods (vegetables) (F)	0	1	1	1	1	1
Harvesting methods (olives) (F)	1	1	2	1	2	1
Harvesting methods (wheat) (F)	0	0	0	0	4	5
Grading (F)	1	1	1	1	1	1
Marketing information	1	1	1	1	2	2

Scale used: 0 to 5. 0 = no role. 5 = sole or strongest role.
m: manufacturing.
F: Farmers play major role.

The JVFU works mainly in the Jordan Valley and has been undergoing reorganisation in order to strengthen its services. Such services include extension activities and production input sales. In 1983 the JVFU employed 12 agricultural engineers who performed all regulatory work, including extension services.

Non-Formal Extension: Private Sector

Private Companies

There are 60 private companies (two are public share-holding) which deal with imports, manufacturing, development, production, wholesale, retail and/or distribution of inputs of agricultural production. For details of inputs and amounts, see Table 11.7.

Table 11.7: Private-Sector Companies Involved in Imports, Sales, Distribution, Production and/or Manufacturing of Agricultural Production Inputs in Jordan (date of study, 1983)

Agricultural production input	No. of companies involved	Remarks
Pesticides	18	All importation and distribution.
Vegetable seeds	12	Only one produce seeds of local leaf vegetables.
Fertilisers	9	
Tools, implements & small equipment	8	Equipment includes sprayers, bee keeping needs. Some sprayer tanks are made locally.
Drip irrigation systems	8	Six are involved in manufacturing; all supply and install drip systems.
Plastic houses	5	Two are involved in manufacturing.
Fruit-tree seedlings	8	Four produce some seedlings locally.
Heavy spraying equipment	2	
Plastic packing containers	4	All are involved in manufacturing.
Poultry feed manufacturing	3	Feed formula and pelleting.
Vegetable seedlings	2	
Day-old chicks	13	Incubator capacity: 4 mn eggs.
Veterinary products	9	Four manufacturers.
Consultancies	3	
Distribution of one or more of various inputs	24	No imports, only distribution throughout country.

Sources: Extracted and compiled by author from:
MOA.
Chamber of Commerce and Chamber of Industry.

Manpower in Companies

Private-sector companies employ a total of 116 agricultural engineers* who are qualified at BSc, MSc, or PhD level (98 BSc, 12 MSc and 6 PhD). (See Table 11.8.) Staff in these companies are paid higher salaries than their colleagues in public-sector institutions and enjoy greater incentives and fringe benefits. On the whole, they are considered to be better motivated and more active than those in the public sector. Furthermore, these engineers are provided with company transport and hence are much more mobile than their colleagues in public institutions. Many of these individuals are former public-sector employees attracted by private companies or private work. Some of them had wide experience in either research or extension work before they joined the private sector.

Services Rendered by Companies

The primary objective of all these companies is to trade in agricultural production inputs at a profit; they also offer a range of technical services. Agricultural engineers* visit farmers in the field as well as receiving them on company premises. They are referred to by both their companies and farmers as 'extension agents' or experts. The range of services they render include the following:

1. Farm visits to promote products, inspect premises and to discuss problems
2. Distribution of seed samples of new varieties
3. Distribution of pesticide samples
4. Soil sampling and soil testing
5. Provision of written and oral information
6. Organisation of field days
7. Organisation of public lectures
8. Drip irrigation system designing
9. Plastic house construction
10. Implementation of and/or supervision of pesticide spraying in citrus orchards

Geographical Coverage and Mobility

Activities of private-sector 'extension agents' are concentrated on the main agricultural production sub-sectors. The intensity of such activities is ranked as follows:

1. Plastic house agriculture in irrigated areas, mainly in the Jordan Valley
2. Irrigated vegetables and citrus mainly in the Jordan Valley
3. Poultry production farms, broiler and egg production throughout Jordan, but concentrated in Balqa, Zarqa and Irbid Governorates
4. Fruit-tree production, mostly stone fruits and grapes, in Balqa, Jerash and Irbid
5. Summer rain-fed vegetables in Balqa, Jerash and Irbid

Rain-fed areas, which usually require few inputs from trading companies, receive little attention. It is only when an input is promoted, like weed-killers for wheat, that these 'extension agents' have contacts with farmers in such areas.

Links between Informal 'Extension Agents' and Farmers

Private-sector companies and the 'extension agents' who represent them appear attractive to farmers because they meet much of their needs by:

* In Jordan, all holders of the BSc degree and above in agricultural science are known as agricultural engineers.

1. Providing inputs to farmers on credit
2. Usually having the latest inputs for agricultural production, e.g. new varieties, new pesticides, new fertiliser formulae and new drip irrigation systems
3. Responding quickly to farmers' questions and/or requirements
4. Being closer to farmers' problems through more frequent visits
5. Providing farmers with technical services. Some are supplied on the spot, such as the identification of pests; others are carried out elsewhere, such as soil testing and more difficult pest problems
6. Providing know-how and manpower required to design, construct and maintain such inputs as machinery, sprayers, drip irrigation systems and plastic houses
7. Delivering inputs to where farmers want them—most important

Impact of Informal 'Extension' on Agricultural Production

Table 11.6 above summarised the results of interviews carried out to measure the role and impact of informal extension provided by private-sector companies. It is quite evident from the table that impact is greatest in the transfer, dissemination and application of technology related to material inputs.

Table 11.8: Staffing, Type and Number of Private-Sector Companies Involved in Imports, Distribution, Production and/or Manufacturing of Agricultural Production Inputs in Jordan

Major commodities	No. of companies[1]	Total no. of agricultural engineers working	No. of technicians	Total no. of branches
Seed pesticides, tools and fertilisers	20	40	25	39
Drip irrigation, plastic houses, packing materials & mulching sheets	11	31	15	18
Fruit-tree seedlings	6	4	2	8
Poultry feeds, chicks & veterinary products	20	35	26	28
Technical consultancies	3	6	—	3
TOTAL	60	116	68	96

Notes: [1] Two are public shareholding companies.

Sources: Extracted and compiled by author from:
MOA.
Chamber of Commerce and Chamber of Industry.

Sources of Technology and Information Used by Farmers in Jordan

There are three main sources from which farmers obtain information, know-how, technology and material inputs needed in their agricultural production activities. These are as follows:

Transfer of Technology, Materials and Information to Farmers from Abroad, mainly through Private-Sector Companies (see Table 11.6 above)
Private agricultural companies, through their trade relations, have access to products and inputs developed by more than 80 companies throughout the world. Jordanian companies (within Jordanian laws and regulations) are free to trade with any company anywhere in the world. (Exception to this—under Arab boycott regulations—are companies in Israel or those which have special trade or manufacturing rights in Israel.)

The main companies with which Jordanian companies trade are in the US, France, the Federal Republic of Germany, the UK, Japan, Italy, the Netherlands, Spain and Canada. They deal with foreign companies as sole agents for Jordan and some other Arab countries, or on a free-trade basis. The range of products which Jordanian companies trade in, and hence 'transfer', include seeds of all vegetables grown in Jordan, seedlings of stone-fruit trees, seedlings of grapes, seeds or seedlings of citrus species, potato tubers for cultivation, sweetcorn, breeds of dairy cattle, parents or grandparents of day-old chicks of both broilers and layers, fertilisers, nutrileaf fertilisers, fungicides, insecticides, herbicides, plant hormones, veterinary products, tools, sprayers, tractors, fertiliser applicators, ploughs, mowers, cereal combine harvesters, drip irrigation systems, poultry feeders and drinkers, poultry cages, poultry feeds and components of feeds, and, last but not least, platic houses. Some of these inputs are manufactured in Jordan, either under licence from the parent company abroad or independently. Manufactured or locally produced inputs include some sprayers, day-old chicks for both broilers and layers, drip irrigation systems, some veterinary products, poultry feed, fertiliser applicators and plastic houses. (See Table 11.6 above.)

Once these inputs are imported, they become directly available on sale to farmers. There are few restrictions on the direct use of such inputs by farmers. Needless to say, the activities of the companies selling these inputs link Jordanian farmers to the most advanced technologies of developed and industrial countries.

It is worth noting here that individual farmers, government agencies, farmers' cooperatives and farmers' unions are all entitled (within the laws and regulations of Jordan) to import any agricultural production input. The role of these bodies in importing is minimal, however, since most of them find it easier to obtain what they need or seek through private companies.

Information and Materials Provided to Farmers by Formal Research and Extension Institutions
These are provided in one or more of the following forms:

1. Crop varieties developed by national research: Research output in this field has been limited to a few varieties of wheat and barley, which cover less than 10 per cent of the seeds used by farmers.
2. Varieties adapted or selected through research efforts: These varieties are provided to researchers either by private companies or by regional and international agricultural research institutions. They include vegetable seeds,

cereals, legumes, fruit trees and pasture crops. In the majority of cases, farmers will already be using such varieties while researchers are verifying their adaptation or performance under Jordanian conditions.

3. Type and methods of pesticides used in pest control: Researchers carry out trials on what, how much, when and how to use various pesticides to control pests under Jordanian conditions. Here again, private companies reach farmers and supply them (through sales and sometimes by the use of free samples) with new pesticides which researchers may or may not be testing.
4. Information leaflets, field days, films, farmers' meetings, radio and TV programmes: Research and extension workers provide farmers with information through these media. Most of this information is supposed to be verified through adaptive research carried out in Jordan. Information provided in this way includes varieties of vegetables, cereals or fruit trees, cultivation practices and pest control practices. The percentage of these sources to overall sources available to farmers is shown in Table 11.6.

Information and Materials Provided by Farmers Themselves

This is a dynamic and important source of information. Although it is difficult to quantify, farmers interviewed laid great stress on information and materials verified by other farmers. The original source of such information could be any one of the following: private companies; government agencies; or farmer's personal experience.

Throughout the years, and especially in the 1970s, farming has attracted three categories of farmer who have been pioneers in the introduction, adoption and skilful use of new technologies:

1. Ex-government employees (mostly ex-MOA employees) with good field experience
2. Entrepreneurs from other sectors of the economy (business and industrial leaders) with the financial capacity to introduce and apply new technologies
3. Fresh graduates of higher agricultural institutions who have opted to work in agricultural production rather than seek employment in government agencies or in private-sector companies

In addition to these three groups, there are those successful 'old-timers' who have been able to stay in farming armed with valuable experience and a good income. Such farmers have also been instrumental in the introduction and application of new technology.

The activities of these four groups, and the success they have achieved in the adoption of new technologies, have undisputedly set an example for other farmers to follow. The new fields in which these groups have been influential are citrus production, poultry production (both broilers and layers), plastic house agriculture and drip irrigation systems.

Levels of Technology and the Extent of their Application in Jordan

We shall address two questions under this heading: first, what is the level of technology used by farmers in Jordan compared to that in more developed countries; and second, to what extent are improved technology inputs and modern technical know-how used by farmers in Jordan? The answer's provided below were based on the following:

1. Author's judgement and experience
2. Results of interviews with 12 experienced farmers
3. Results of interviews with four 'extension agents' working in private companies
4. Known information and statistics on quantities of certain production inputs used in Jordan, such as pesticides, fertilisers, improved seeds and machinery
5. Published information on the subject

Before we attempt to provide answers to the above questions, the following remarks may help to clarify the prevailing situation in Jordan *vis-à-vis* technology level and its usage. First, one of the most well-known features of technology usage in Jordan is its variability. Technological inputs and technical knowledge may be available in one area of another, but that does not mean that they are used universally or even commonly by farmers. In certain production fields, such as cultivation in plastic houses, some producers fumigate and fertilise the soil, use the best-known improved germ plasm, apply insecticides, follow modern cultural practices, use drip irrigation and attain high production levels. Meanwhile, other producers growing tomatoes or cucumbers in the open field may utilise good varieties, but their use of any other improved technology is quite limited. One of the better-studied cases of technology adoption is that of wheat. Packages of technology which lead to high yields are available, but their use by farmers is very limited. The major factors affecting non-adoption of an available technology were found to be price policies, and the social and economic environment of wheat farming (Gotch 1976, El-Hourani 1975).

A second aspect of variability is the level of production per unit of land attained by different farmers. Statistics on yields are usually reported in averages. When yields attained by individual farmers are examined, there is considerable variation. In wheat, for example, some lands give zero yield in certain seasons because of drought, while others yield quantities as low as 50-100 kg per ha. In other cases, where wheat is planted under high rainfall conditions, yields reach an average of 2,500 kg per ha. In 1981 the average yield of all rain-fed areas was 480 kg per ha, which is less than 40 per cent of world yield averages. What is true of wheat is also true of many other crops. Why such great variation in yields exists needs closer examination. It must be emphasised here that the technical constraints of production are not common to all regions and to all farmers; they also vary greatly from one crop to another. An evaluation of the technology used and the extent of usage follows.

Evaluation of Level of Technology Used and Extent of Application

Scale used for measuring level of technology: A-E

A = More than 90 per cent of farmers:

1. Use improved seeds
2. Follow effective pesticide control schedule
3. Follow efficient irrigation schedule
4. Use fertilisers under best known information
5. Use modern cultural practices
6. Obtain high yields

B = 70-89 per cent of farmers follow points A. 1-6 above
C = 50-69 per cent of farmers follow points A. 1-6 above
D = 30-49 per cent of farmers follow points A. 1-6 above
E = Less than 30 per cent of farmers follow points A. 1-6 above.
Note: Variations within each class (from A-E) should be kept in mind.

Field of production	Scale	Greatest variation in:
Citrus	B	Pest control
Cucumbers under plastic	B	Pest control
Tomatoes under plastic	B	Pest control & fertilisation
Tomatoes under mulch & drip irrigation	B	Pest control & fertilisation
sector through		
Open field tomatoes under irrigation	D	Pest control, fertilisation & method of irrigation
Open field eggplants	C	Pest control, fertilisation & method of irrigation
Open field vegetables under irrigation	D	Pest control, fertilisation & method of irrigation
Wheat under irrigation	D	Varieties, weed control & fertilisation
Wheat under rain-fed conditions	E	Varieties, weed control, fertilisation & cultural practices
Barley under rain-fed conditions	E	Varieties, weed control, fertilisation & cultural practices
Vegetables under rain-fed production system	D	Varieties, cultural practices & pest control
Tobacco	C	Pest control
Rearing broilers	B	Pest control & management
Rearing layers	B	Pest control & management
Olive production	D	Varieties, fertilisation & cultural practices
Stone fruit	C	Varieties, fertilisation & pest control
Grapes	D	Fertilisation & pest control
Legumes	E	Varieties & cultural practices
Sheep production	D	Breeds, management, quality of feed & pest control
Dairy cattle	C	Pest control & management

Technical Constraints on Agricultural Production in Jordan

We shall attempt to address here those constraints which can be solved through research, extension development and other technical services. Other constraints which result from policy, and the economic and social environment in which farmers operate, are to be kept in mind, as they influence the support of technical services as well as the extent of technology usage.

Over the years, a certain level of technology, know-how and technical services has been made available to Jordanian farmers from the three sources discussed above. However, there are many gaps in the technical services provided, both in quality and in quantity, to meet all needs. There are also many production and marketing problems which meet either with unsustained and ineffective efforts to solve them, or no effort at all. Technical constraints as used here cover the following:

1. The absence of an institutional base for some technical services
2. The inadequate capacity of existing public institutions to provide systematic, regular and effective technical services for farmers, such as soil testing, water quality determination and usage, pest diagnosis and marketing information
3. The inadequate capacity of existing public institutions to provide sustained efforts to solve technical production problems
4. Gaps in research, development and extension programmes

In evaluating the present status of technological services provided both by public sector institutions and by private companies, scales of 0-5 will be used for quantity and quality as follows:

Sale of quantity or coverage: 0-5

0: absent
1. covers less than 20 per cent of farmers' needs
2. covers less than 40 per cent of farmers' needs
3. covers less than 60 per cent of farmers' needs
4: covers less than 80 per cent of farmers' needs
5: covers more than 80 per cent of farmers' needs

Scale of quality (refers to availability of manpower, facilities and operating funds): 0-5

0: absent
1: very weak
2: weak
3: acceptable
4: good
5: strong

Evaluation of Technical Services Available in Jordan in Support of Agricultural Production

Area of service	Quantity or coverage	Quality	Remarks
Macro soil survey	3	4	Jordan Valley
Micro soil survey	1	3	Jordan Valley mostly
Soil maps for productivity potential (macro)	2	3	Jordan Valley
Soil maps for productivity potential (micro)	0	0	
Fertility soil maps	0	0	
Salinity & alkalinity soil maps	0	0	
Systematic monitoring of soil erosion	1	2	
Soil testing services for proper application of fertilisers	1	1	Public sector
Leaf analysis services of crops to correct deficiencies of micro nutrients	0	0	
Monitoring of water resources for quality (irrigation water)	1	1	Public sector
Classes of water & their recommended usage	1	1	Public sector
Effective methods of water application	2	3	Public sector
Soil moisture conservation methods in rain-fed areas	2	2	Public sector
Breeding germ-plasm for wheat & barley	1	2	Public sector
Adaptation research for introduced varieties or lines of wheat & barley	4	4	Public sector
Availability of seeds of improved varieties of wheat & barley	1	3	Public sector
Breeding germ-plasm of all crops other than cereals	0	0	
Adaptation research on introduced varieties or lines of legumes	3	4	
Availability of improved legume seeds	1	1	
Introduction of new varieties of vegetables	4	4	Private sector
Adaptation research on introduced varieties of vegetables	3	2	Both public & Private sectors
Availability of vegetable seeds	4	4	All private sector through importation
Introduction of new citrus germ-plasm	1	1	
Introduction of new stone-fruit germ-plasm	2	3	Mostly private sector through importation
Introduction of new grape germ-plasm	2	3	Both public & private sectors
Introduction of new root stocks of fruit trees	1	1	

Introduction of new olive germ-plasm	1	1	
Adaptation research of new fruit-tree germ-plasm	1	1	
Seed testing services for purity and diseases	0	0	
Seed testing services for germination	2	2	Public sector, and only for imported seeds
Introduction of new pesticides and other chemicals	4	4	Mostly private sector; both public & private sectors
Adaptation research for use and suitability of new pesticides to conditions in Jordan	3	3	Both public & private sector
Availability of pesticides to farmers	5	4	All private sector
Development of spray schedules for pesticides, management and rationalisation of pesticide use	2	1	Public sector
Introduction of new fertilisers	3	3	All private sector
Optimisation of fertiliser use for various crops & different soils	2	3	Public sector
Survey of pests, including their distribution	4	4	Public sector
Monitoring pest outbreaks, seasonal changes & pest forecast	1	2	Public sector
Development of chemical control methods of fungal diseases of crops	3	3	Public sector
Development of chemical control methods of virus diseases of crops	3	3	Public sector
Development of chemical control methods of nematode diseases of crops	3	3	Public sector
Development of chemical control methods for weeds in crops	3	3	Public sector
Development of chemical methods for insects and mites on crops	3	3	Public sector
Development of integrated control methods for pests on crops	1	1	Public sector
Diagnostic services for crop pests	1	1	Public sector
Dissemination of pest control information	2	2	Both public & private sector
Diagnostic service for animal pests	3	3	Public sector
Development of integrated control methods for animal pests	1	1	Public sector
Availability of veterinary products for control of animal pests	4	4	Private sector; some vaccines by public sector
Development of cultural practices for cereals	3	3	Public sector

Development of cultural practices for vegetables & fruit trees	2	2	Public sector
Dissemination of information on cultural practices	1	1	Public sector
Dissemination of information on use of chemicals to control pests	2	4	Concentrated in Jordan Valley
Marketing information services for farmers	0	0	
Development of harvest packing, storage & marketing methods of agricultural commodities	1	1	Marketing methods are weakest if not non-existent
Development of processing	1	1	Public sector
Feeding requirements of animals, including poultry	3	2	Public sector
Management of poultry farms	2	2	Public sector
Socio-economic studies on farm management, input-output in farming, & attitudes of farmers to government policies	2	2	Public sector
Feasibility studies for public and private sector, agricultural development and business ventures	1	1	Public sector

Operations of a Typical Private Sector Trading Company

How does a typical private sector trading company operate in providing agricultural inputs and knowledge to farmers in Jordan? This question will be discussed in order to illustrate the role of private companies in what we may call 'direct transfer of technology'. This report is based on interviews and a study of eight leading companies working in Jordan.

The Environment

This section discusses the major features of the environment under which private companies operate in Jordan. There are a few agricultural production inputs which are partly produced in Jordan. (See Table 11.6 for details.) All imported agricultural inputs are exempt from customs duties, except spare parts for machinery and a few bee-keeping items. Companies are free to import inputs from any country in the world (except those which are black-listed by the Office of the Arab Boycott of Israel).

Government agencies produce and distribute some varieties of cereals, seedlings of a few stone-fruit varieties, citrus, olives, grapes and pome fruits. The seeds and seedlings produced cover less than 10 per cent of farmers' demands (except for olives and grapes). There is no government price control on inputs sold to farmers. Theoretically, there is a ceiling of 30 per cent profit on pesticides but it is neither enforced by the Government nor needed by companies. Profits made on company activities are subject to income tax, which may reach up to 40 per cent of net profits for private share-holding companies.

When seeds of crop varieties or pesticides are imported for the first time for sale to farmers in Jordan they have to go through a registration procedure administered by the MOA. The prodecure for seeds is as follows: The company provides samples from imported lots to the MOA for a germination test. Permission for sale and distribution is granted if seed lots pass this test. No registration of varieties or their characteristics is required. Potatoes are the only exception, where tuber samples of new varieties must be passed to the DARE for one year's trial and another year of demonstrations, the import licence being granted in the third year.

Registration is required for new pesticides and involves the following steps:

1. A certificate must be provided showing that the pesticide is registered and used in the exporting country for the same purpose.
2. The DARE must be provided with trial samples (about 4 l); the company can market the new product without waiting for the DARE trial results.
3. Special labels for packages on sale must be provided.

Physical Facilities and Manpower

Manpower (position/function)	No.	Comments
Director	1	Administration & business
Salesman	2-3	Not qualified technically
Store keeper	1	Experienced inventory keeping
Book-keeper	2-3	Experienced in accounting
Extension agent	2-3	Qualified to BSc level or above in one of the agricultural sciences; promotes products and disseminates technical information to customers

Commodities or Inputs Involved in an Operation

Vegetables seeds, pesticides, fertilisers, tools and implements, including sprayers.

Working Capital

Average: JD 200,000. Range of capital: JD 100,000-500,000, with banking facilities equal to capital.

Operations

Type of activity	*Comments*
Assess seasonal or annual demand for inputs in circulation proven to be superior and adapted to farmers' needs.	Study total needs of farmers. Estimate company share of total market.
Survey new products and identify those potentially suited to Jordan's needs for seeds and chemicals.	All ordered inputs in circulation have been through registration procedures required by the MOA.
Determine quantity to be ordered, obtain import licence, open letter of credit at bank and place order.	Import licence should be approved by the MOA to ensure that registration for relevant inputs has been completed.

Physical Assets

Type	No.	Comments
Main exhibition & sales store	1	In Amman
Branch sales store	1	Next to Central Market
Stores	1-2	
Offices	2-3	Next to main store
Cars	2-3	1 pick-up type

Type of activity	*Comments*
Order samples of new products. Seeds and pesticides identified as samples are allowed to enter country within a certain limited amount. Receive ordered inputs and prepare them for sale. Receive ordered samples of new varieties or new pesticides. Send samples of imported seeds intended for sale to the MOA for germination test. Send 4 l of samples of new pesticides to the MOA for trials. Perhaps opt to send samples of new varieties, intended as new introductions to the MOA for trials. Sell seeds of proven varieties after receiving positive results of germination test. Sell pesticides and other inputs. Distribute samples of new varieties and new pesticides to farmers and supervise their performance. Visit farmers every week to check on their needs; inspect farmers' fields for pests and advise on measures to be taken; advise farmers on dosage, storage conditions and methods of application for pesticides. Collect and evaluate results of performance of new varieties and pesticides. Repeat trials of new varieties and pesticides for another year. Identify superior and adapted new varieties or pesticides considered acceptable by farmers. Submit name and relevant information on new inputs (proven to be suitable in preceding activity) to the MOA for registration. Add new inputs to previous ones or replace some older inputs with new ones.	All containers of pesticides intended for sale must carry labels with information specified by the MOA. New varieties of seeds or new pesticides can be sold to farmers without awaiting results of trials. The company usually opts to wait 1-2 years after introducing samples for trials in farmers' fields. Only potatoes need to go through trials of 1-2 years before registration. Not all companies have programmes of supervision for trials of new seeds or pesticides. Some companies have their own farm on which they carry out trials of new varieties of new pesticides, in addition to trials in farmers' fields. Trials in farmers' fields cover a wide range of conditions and are thought by companies to be more reliable than trials carried out by the MOA. Supervision of trials, as well as advice, is undertaken by company 'extension agents'. Some new varieties may be entered by the MOA into trials. New inputs identified by company to be superior and accepted may or may not be compatible with the MOA results. Company relies on results of its own trials and experience in deciding which new inputs should be added.

Agricultural Research and Extension in Jordan: Issues, Constraints and Future Outlook

Agricultural research and extension in Jordan have been reorganised several times over the past 20 years. During the past decade, several attempts have been made to strengthen these facilities in order to provide modern and effective services for Jordanian farmers. But several issues and constraints have not yet been resolved. These are summarised in Table 11.9.

Agricultural research was one of the main topics discussed at the Jordanian Conference on Science and Technology held in 1978 in Amman. The Conference made several recommendations to the Government, including the establishment of a national unit within the National Planning Council (NPC) to plan, support and coordinate research. It also called for the formation of a National Committee for Science and Technology to plan, coordinate and evaluate research. In 1981 the NPC established a Department of Science and Technology with an annual budget of JD 150,000. The support activities of the NPC research need to be strengthened, particularly in defining the work of the National Committee which has not met since it was formed. The current (1981-85) Five-Year Development Plan includes a project to establish a national centre for research and extension.

Table 11.9: Summary of Issues and Constraints Facing Agricultural Research and Extension in Jordan (as of March 1984)

Issue or constraint	Remarks and manifestations
Mandate for research	Not clear, since there is no single identified national institution with a clear mandate to plan, organise, implement and evaluate research and ensure dissemination of its results. There are three bodies at present involved in research: a) FOA, University of Jordan b) DARE, in the MOA (rain-fed areas) c) Para-departmental project of the MOA and USAID in the Jordan Valley In addition, the University of Yarmouk is planning a number of animal and plant production projects in rain-fed areas of the country.
Organisation of research	The 1981-85 Five-Year Development Plan called for an autonomous national agricultural research centre; but no steps as yet have been taken to establish it. According to the Plan, such a centre is supposed to have a national mandate both for research and for the application of the results of the research.
Aggregate resource allocations for research	This is less than 1 per cent of AGDP, and fluctuates from one year to another, falling at times to less than 0.5 per cent of AGDP. This is much lower than targets called for by international organisations like the World Bank.
Continuity of service in research and extension	Uncertain in the MOA; many people move from one position to another within and outside the research and extension service.

Priority determination and extension works projects	There are no recognised criteria and there is no functional unit in charge of all bodies which implement research.
Fate of research results	Documentation and publication of results are not mandatory in the FOA. There is no systematic release of results to farmers. No vehicles to develop research results are available besides a limited facility to reproduce wheat seeds.
Mobility of research and extension works:	Very weak in the MOA as manifested by poor coverage of certain geographic areas.
Commodity coverage in research and extension	Distorted, since there are no recognised criteria for resource allocation. Pest problems keep appearing in epidemiological form for certain commodities.
Technical services, such as soil testing, plant pest diagnosis and water quality	Very weak, with no clear mandate for any particular institution. The Royal Scientific Society (RSS) is planning to establish a soil-testing service to fill the vacuum created by the inefficient handling of this service by others.
Role of private sector in technology transfer and extension	Although active and efficient in the Jordan Valley, there is no coordination or recognition of this role by the Government.
Role of farmers in identification, formulation and priority determination of research and extension	Very weak; no leverage for farmers in public institutions which implement research and extension services.
Incentives for research and extension workers	No such system exists. Promotion in the MOA is not based on achievement. Promotion criteria in the FOA are skewed to publications, with no clear recognition of the impact of results on farming improvement.

References and Other Works Consulted

Chamber of Commerce and Chamber of Industry, Records of company registration, Amman.

Daghestani, F. A., S. Qasem and B. Sakat (eds), 1978, 'Science and Technology for Development', paper presented at Science and Technology Policy Conference, Royal Scientific Society, Amman.

DARE (Department of Agricultural Research and Extension), Annual Reports, *1974–82*, Ministry of Agriculture, Amman.

Faculty of Agriculture, University of Jordan, *Annual Reports, 1974–82*, Amman.

Ghosi, Thulkifel and M. A. Arafat (eds), 1977, *Accomplishment of Agricultural Research in Twenty-Five Years, 1951–76,* Department of Agricultural Research and Extension, Ministry of Agriculture, Amman.

Gotsch, Carl, 1976, *Wheat price policy and demand for improved technology in*

Jordan's rainfed agriculture, Discussion paper No. 2. Studies of dryland agriculture, Ford Foundation, Amman.

El-Hurani, M. H., 1975, *Economic analysis of the development of the wheat subsector of Jordan*, unpublished PhD dissertation, Iowa State University, Ames, Iowa.

MOA (Ministry of Agriculture), Records of Company Licensing Section, Amman.

Oram, P. and Bindlish V., 1981, 'Resource allocations to national agricultural research: trends in the 1970s', Joint Report of International Food Policy Research Institute and International Service for National Agricultural Research, the Hague.

Qasem, Subhi, 1982, 'Problems and opportunities for strengthening national agricultural research systems in the Middle East', paper in Proceedings of Conference on The Role of International Associations in Strengthening National Agricultural Research, Joint Report of the Conference of International Service for National Agricultural Research (ISNAR) and the International Agricultural Development Service, ISNAR publications, No. C4, The Hague.

——, Sami Sunai and Hassan Gharabeh, 'Research in the agricultural sector', paper at the National Seminar of Science and Technology, Amman, 1977.

Sabet, Kamal, 1976, 'Organisation and administration of agricultural research in the Near East countries, Country Reports: Jordan', FAO (UN Food and Agriculture Organisation), Near East Regional Office, Cairo.

PART V

POLICY FORMATION

12 AGRICULTURAL POLICY FORMATION AND ADMINISTRATION

Subhi Qasem

Government and Administrative Structure

Legislative Authority

Legislative authority in Jordan is vested in the King and in the National Assembly. The latter is composed of two Houses: the Upper House (Senate) and the House of Representatives (Parliament). Members of the Senate, numbering 30 at present, are appointed by the King, while Members of Parliament are elected by the people. Parliament was dissolved in 1974 because of the particular political conditions following the Israeli occupation of the West Bank of Jordan in 1967. It was reconvened in early 1984.

According to the Constitution, the highest legislative act is the 'law'. All Bills of Law must pass through the following steps:

1. The Prime Minister submits Bills of Law to the House of Representatives, which has the power to ratify, reject or amend them.
2. Ratified or amended Bills are forwarded to the Senate.
3. Bills accepted by the National Assembly are then presented to the King for final confirmation through the issuing of a Royal Decree.

The second highest legislative act is the by-law (*nitham*). Each by-law must be based on a confirmed law and should not contradict it in any way. By-laws are confirmed by a Royal Decree from the King upon the recommendation of the Prime Minister. The Council of Ministers is responsible for the approval of all by-laws. Formulation of by-laws can come from Ministers, heads of institutions, or councils created by law. Other administrative procedures are organised through regulations or orders. Regulations are issued either by the Prime Minister, a Minister or a head of an institution as specified in the by-laws. Orders issued by the King, the Prime Minister, and occasionally by Ministers or other heads of institutions. The King's Orders, written or oral, do not absolve the Ministers or any other official from liability.

Most government affairs are administered through by-laws and regulations. Most financial affairs, especially taxes, are administered through laws. The administration of agricultural affairs will be discussed in more detail in the following paragraphs.

Executive Authority

Executive authority in Jordan is vested by the Constitution in the King, who exercises such authority through the Prime Minister and Ministers who make up the Council of Ministers. The Council of Ministers is responsible for the administration of all external and internal affairs of the state. The major administrative structural unit of the state is the Ministry, which is usually organised into Departments. All government administrative structures came under Ministries or Ministers until 1958, when the Constitution was amended to allow organisations other than Ministries, and individuals other than Ministers, to be in charge of administering some government affairs as specified by law.

At present there are 20 Ministries, in addition to the Prime Minister's Office which houses the Prime Minister (Chief Executive under the King) and his staff plus one Minister for Prime Ministerial affairs. Each Ministry is headed by a Minister assisted by an Under-Secretary. The Minister is responsible for the administration of all affairs of his Ministry, dealing with those within his jurisdiction and referring others to the Prime Minister. The Prime Minister either rules on matters referred to him by Ministers or presents them to the Council of Ministers. His authority is usually defined in each law or by-law, depending on the case.

In addition to the Ministries, there are a variety of institutions which are parastatal, and have autonomous or semi-autonomous status.

Organisational Structure

National affairs are organised within a number of structures, as outlined below.

Ministries

At present there are 20 Ministries:

- Agriculture
- *Awqaf* and Islamic Affairs
- Communications
- Culture and Youth
- Defence
- Education
- Finance
- Foreign Affairs
- Health
- Information
- Interior
- Justice
- Labour, Reconstruction and Development
- Municipal, Rural Affairs and Environment
- Public Works
- Social Development
- Supply
- Tourism and Antiquities
- Trade and Industry
- Transport

Each Ministry is organised into major structural and functional units called Departments, some enjoying more autonomy and flexibility than others. Some department heads or directors report to the Minister concerned; some report to the Under-Secretary. Ministries are created by means of laws, while Departments are set up through by-laws. Major administrative rules and procedures of Ministries are determined by by-law, and those of Departments by regulation, or sometimes by Order.

Councils, Authorities, Corporations and Organisations, and Associations or 'Syndicates'
The affairs of each institution are organised either by law or by-law, some enjoying more autonomy than others. Some are independent of Ministers, while others enjoy more autonomy and flexibility than Departments of Ministries, but are not independent of Ministries. There are some 27 such institutions in the country, the main ones being as follows.

Executive Institutions Independent of Ministers but Reporting to the Prime Minister
— Agricultural Credit Corporation
— Alia Corporation (Royal Jordanian Airlines)
— Central Bank of Jordan
— Government Employment Diwan or Office
— Jordan Cooperative Organisation
— Jordan Valley Authority
— National Planning Council
— Natural Resources Authority
— Water and Sewage Authority

Coordinating Councils Chaired by the Prime Minister or Ministers
— Higher Agricultural Council
— Higher Council of Health
— Higher Education Council
— Education Council

Other Executive Institutions
— Aqaba Port Authority
— Electricity Authority
— Household Water Corporation
— Jordan Television Corporation

Associations Created and Organised through Laws or By-Laws
— Professional Associations or Syndicates of: agricultural engineers, dentists, engineers, labourers, lawyers, medical doctors, and pharmacists.
— Jordan Valley Farmers' Association (JVFA)

Institutions Autonomous from Council of Ministers, Created by Law and Organised and Ruled through By-Laws
— University of Jordan
— University of Yarmouk

Ministries and Other Institutions Concerned with the Agricultural Sector

There are 11 Ministries and/or institutions whose functions are concerned with agriculture. Table 12.1 summarises the main functions and/or services provided to the agricultural sector by each.

Higher Agricultural Council (HAC)
Created in response to the need to coordinate management of institutions concerned with the agricultural sector, the HAC is composed of 16 members presided over by the Prime Minister. (See Annexe 12.1 for details.) The objectives and concerns of the HAC are as follows:

1. Formulation of agricultural policies related to agricultural production, utilisation of agricultural resources, taxation and customs exemption of agricultural commodities, as well as agricultural production inputs.
2. Coordination of the work of all institutions concerned in the service and development of the agricultural sector, including approval of developmental programmes and plans.

Table 12.1: Ministries and Other Institutions Concerned with the Agricultural Sector and their Main Roles

Ministry or institution	Main agricultural role
Higher Agricultural Council	— Formulation of agricultural policy — Coordination of work of institutions concerned in service or development of agriculture — Suggestions for legislative acts
Ministry of Agriculture	— Regulatory function of agricultural services as well as importation of production inputs — Provision of research and extension services — Provision of veterinary services — Implementation of afforestation projects and range protection — Import-export regulatory function for agric-cultural food commodities
Ministry of Supply	— Price control '*formulation*' of supply policy for all consumer goods including agricultural ones — Regulation of price control for all food commodities consumed in Jordan — Provision of cereal seeds for farmers
Jordan Valley Authority	— Identification, formulation and implementation of integrated social and economic development programmes of the Jordan Valley and Ghors, south of the Dead Sea — Management and distribution of irrigation water in the Jordan Valley
Agricultural Credit Corporation	— A parastatal institution which provides short-, medium- and long-term credit for farmers
Jordan Cooperative Organisation	— Organisation, management and promotion of the cooperative movement in Jordan, including farming cooperatives Provision of seasonal credit, agricultural services and agricultural inputs to members
National Planning Council	— Key role in formulation of comprehensive national social and economic development plans, including the agricultural sector — National focal point in negotiating and securing all loans, technical assitance funds and other financial resources for national development — Key role in priority determination and resource allocations for development projects at national level
Natural Resources Authority	— Development of water resources and irrigation projects in all areas outside the Jordan Valley and lowlands south of the Dead Sea — Formulation of national water policy
Ministry of Trade and Industry	— Responsibility for regulating imports of all goods, including agricultural inputs
Faculty of Agriculture, University of Jordan	— Sole higher education institution in the country which provides trained manpower at BSc and MSc level, as well as research services.

3. Suggesting amendments to existing laws and by-laws related to agriculture,. as well as the enacting of new ones.

Ministry of Agriculture (MOA)

The role and functions of the MOA are spelled out in the Agricultural Law enacted and confirmed in 1973 (see Annexe 12.2 for details). The 1973 Law is comprehensive and has replaced 14 previously enacted laws dealing with one aspect or another of the agricultural sector. It gives both the Minister and the MOA wide-ranging responsibilities to regulate, monitor and service the agricultural sector. Because there are several other institutions established to serve and develop natural resources, including agricultural resources, the impact of the MOA is not only weakened but is ambiguous in many ways. How the MOA operates in determining agricultural policy matters will be discussed below.

Ministry of Supply (MOS)

The main objective of the MOS is to formulate supply policy for all consumer goods, including agricultural ones. In addition, the Ministry is responsible for securing the necessary supplies in appropriate quantities, types and prices. The most important role of the MOS concerning agriculture is that it can control the prices of both locally produced goods and imports of food commodities at prices and in quantities it deems necessary to achieve its objective. If the MOA is concerned with the producer, the MOS is concerned with the consumer of agricultural goods. Two further aspects of the MOS mandate which concern agriculture are:

1. Importation of cereal seeds in coordination with the MOA and the Agricultural Credit Corporation (ACC) and the distribution of such seeds to farmers for planting.
2. Participation in the formulation of import-export policy for all types of consumer goods. Most of these functions overlap with the responsibilities of the MOA, with no clear provision for demarcation except that the MOS should coordinate its work with the MOA. The administrative mandate and methods of operation of the MOS are organised through By-laws 23 and 24 issued in 1974 and emanating from Article No. 120 of the Constitution.

Jordan Valley Authority (JVA)

The JVA's main concern is the development of the Jordan Valley through a comprehensive and integrated social and economic plan. The Authority's mandate includes the development of land and water resources, as well as the distribution and control of water used by farmers in the valley.

The work of the JVA was extended in 1980 to include the development of all land and water resources in the area known as Southern Ghors and extending from south of the Dead Sea to Aqaba. The mandate of the JVA, its organisation and operation have all been spelled out in Law No. 18 issued in 1977 as 'The Law of Jordan Valley Development'. The President of the JVA reports directly to the Prime Minister and enjoys wide-ranging authority and responsibilities considered to be more flexible than those enjoyed by Ministers.

Agricultural Credit Corporation (ACC)

The ACC is an autonomous institution, both financially and administratively, which was re-organised into its present form in 1963 through Law No. 12. Responsibility for the ACC's general policy and major decisions rests with the

Board of Directors headed by the Director-General and either other members, one of whom represents the MOA. The mandate of the ACC is to offer financial loans to individuals, organisations and groups of farmers working in agricultural production and the processing of agricultural products. The ACC offers three types of loans as follows:

1. Short-term loans repaid over a period of less than one year and limited to the purchase of agricultural production inputs.
2. Medium-term loans repaid within 1 to 10 years. These loans are given to finance machinery purchases, the establishment of fruit-tree orchards and animal production farms, digging water wells, the establishment of irrigation canals and reservoirs, and the construction of farm houses and farm stores.
3. Long-term loans repaid in 10-15 years. Such loans are offered to support major irrigation schemes shared by five or more farmers, agricultural industrial projects, land reclamation for orchard establishment in rain-fed areas and the purchase of agricultural land for small farmers or for those who purchase land to consolidate holdings and prevent fragmentation.

Jordan Cooperative Organisation (JCO)
The JCO was reorganised through the Law of Cooperation No. 20 issued in 1971. The JCO is listed among institutions concerned with agriculture because its main objective is to supervise and finance many cooperatives whose major function is agricultural production. The work of the JCO has expanded in recent years since it started offering seasonal loans in the form of agricultural production inputs or services. Inputs include seeds, pesticides and fertilisers, while services include renting machinery for ploughing, seed planting, fertiliser and pesticide application, harvesting and seed screening. The JCO has become involved in marketing agricultural produce both inside and outside Jordan. It is run by a Director-General who heads the Council of Administration and is responsible to it. One of the members of the Council represents the MOA and is appointed by the Minister.

National Planning Council (NPC)
The NPC was established in 1971 and was legally granted the mandate to formulate the National Plan for Social and Economic Development. The NPC formulates plans in cooperation and coordination with relevant institutions of the state. The objectives and mode of operative of the NPC are presented in Annexe 12.3. The Council has grown into a powerful institution because it is the national focal point for negotiating and securing loans and technical assistance funds to finance all national development projects. The NPC is also instrumental in the identification of development projects, the priority determination of such projects, and their financing from domestic and foreign sources. It is concerned with agriculture in its role as a planning agency, as well as in its active participation in priority determination and resource allocation for projects planned for the development of the agricultural sector. Prior to implementation, all planned projects within the national plan must be approved by the NPC, its Board of Directors and the Council of Ministers.

Thus far, the NPC has been responsible for launching three National Social and Economic Plans as follows:

First Plan	1973-75	(three years)
Second Plan	1976-80	(five years)
Third Plan	1981-85	(five years)

In the Third Plan, Agriculture, Water and Irrigation and Cooperatives are included as independent sectors.

Natural Resources Authority (NRA)
The NRA is a government institution established in 1968 under Law No. 12. Its main objective is to formulate and implement policies and projects for the development of water, minerals and rocks. The NRA is concerned with agriculture through its mandatory powers to formulate a national water policy for approval by the Council of Ministers, and cooperate with the MOA in the development of agricultural projects in reclaimed lands brought under irrigation.

The NRA's present role concerning agriculture is to develop water resources, both surface and underground, to control and monitor underground water utilisation, and to manage water and irrigation schemes in cooperation with the MOA. All such projects are located outside the Jordan Valley and Southern Ghors, which come under the mandate of the JVA. The NRA has a Board of Directors presided over by the Prime Minister. Others on the Board include the Vice-President and Director-General of the NRA, and representatives of several ministries and institutions, including the NPC and the MOA.

Other Institutions
There are four other institutions concerned with agriculture. These are

- The Ministry of Trade and Industry
- The Faculty of Agriculture, University of Jordan
- The Jordan Valley Farmers' Union
- The Association of Agricultural Engineers

All four are represented on the HAC and hence participate in the formulation of agricultural policy.
Ministry of Trade and Industry (MTI): The Ministry plays a regulatory role in importing agricultural inputs and in the formation of public and private shareholding companies serving agriculture. Its regulatory role is mainly in issuing licences of imports once they are cleared technically by the MOA.
Faculty of Agriculture (FOA): This is the only institution of higher education in the country which provides trained manpower at the BSc and MSc level, as well as research services.

Levels at which Policy Objectives Are Set and Resources Allocated

Who Decides on Policies
We presented above an outline of government and parastatal institutions concerned with the agricultural sector. The MOA is the oldest of them; all the others were created during the period from the late 1950s to the late 1970s. As these institutions became established, the legislative acts defining objectives and organisation granted them roles related to the agricultural sector which either overlapped with or replaced many of those previously assumed to have been under the umbrella of the MOA. The multiplicity of institutions concerned with agriculture made it difficult to pinpoint which one had a clear mandate to formulate agricultural policy. The creation of the HAC was seen by the Government as a necessary vehicle to draw all relevant institutions into one forum to formulate agricultural policy. Indeed, one of the clear objectives of the HAC is to formulate policies covering major aspects of agriculture. The confirmation and implementation of formulated policies are left to the individual

institutions concerned. Who decides on such policies and at what level is distributed over many national bodies. If a policy issue is regulated through a law, then the final decision requires the approval of the National Assembly; if it is regulated through a by-law, the final decision requires the approval of the Council of Ministers. The following is an outline of national bodies empowered by Constitution, law or by-law to rule on policy issues on various aspects of agriculture:

Policy issue	National body legally empowered to make decisions
Size of agricultural holding	National Assembly and Royal Decree
Taxation or fees of sale purchase, partitioning or inheritance of agricultural land	National Assembly and Royal Decree
Formulation of policies of land use	Higher Agricultural Council
Policies of utilisation of government land	National Assembly and Royal Decree
Formulation of national water policy	Natural Resources Authority
Confirmation of national water policy	Not clear[1]
Formulation of national agricultural development plans	National Planning Council
Confirmation (approval) of agricultural development plans	Higher Agricultural Council
Formulation of food supply policies, including agricultural commodities	Ministry of Supply
Confirmation of food supply policies, including agricultural commodities	Not clear under present laws[1]
Formulation of agricultural production policies	Higher Agricultural Council
Confirmation of agricultural production policies	Not clear[1]
Formulation of policies of utilisation of agricultural resources	Higher Agricultural Council
Confirmation of policies of utilisation of agricultural resources	Not clear[1]
Formulation of policies of taxation and customs exemption for agricultural commodities and production inputs	Higher Agricultural Council
Confirmation of policies of taxation and customs exemption for agricultural commodities and production inputs	National Assembly and Royal Decree
Formulation of agricultural credit policies	Agricultural Credit Corporation
Price policies of agricultural commodities produced locally or imported	Ministry of Supply
Price policies of agricultural inputs	Ministry of Agriculture
Price policies of cereal seeds	Ministry of Supply
Policies in soil and range conservation	Ministry of Agriculture

Note: [1] Confirmation of policies: Whenever formulation of policies is assigned to an institution without specifying who confirms it, it is assumed that if policy is formulated through a law or by-law, then confirmation is made through constitutional procedure. However, present laws does not specify the form in which policy should be formulated, hence which particular bodies are responsible for confirmation is not clear.

Policy Objectives

This section deals with how policy objectives are set, who decides on them, and who decides on the selection of instruments for implementing objectives. Policy objectives can be roughly classified as long-term and short-term. While it is difficult to distinguish between these two categories, for the purposes of this chapter we list all those which are enacted by law as being long-term objectives.

How Objectives are Set and Who Decides on Them: There are two main places where objectives of policy issues appear: in the legislative act (law or by-law) issued by the relevant body to adopt a policy; and in the National Social and Economic Development Plans. Examples of the first are the Law for Planning, the Law for Agricultural Credit and Law of Cooperatives (Cooperation) and the By-law of Supply. In each of these acts the objectives are set out. Naturally, those who approve the legislative act are the same as those who decide what the objectives are.

Secondly, with reference to the National Development Plans, development objectives and the means of achieving them are spelled out for the major sectors. We have indicated above that all contents of the National Development Plan are finally approved by the Council of Ministers. For the purposes of this chapter, objectives set out in the Development Plan represent the policy of the Government which approves that Plan. It is understood that all development objectives must conform with major policies that are formulated through legislative acts which constitute laws. However, there are several policies which have not been formulated and approved, despite the fact that the legislative acts pertaining to the institutions responsible for formulating such policies have been approved. Naturally, this presents gaps in policies that are known to planners or development agencies. Further details will be presented below under 'Present Policy Measures and their Objectives'.

How Instruments are Selected for the Implementation of Policy Objectives: Here again, two procedures are followed. First, the legislative act through which particular policy objectives are set includes articles for the establishment of the implementing instrument. One example is the Law of Planning, which both specified major planning objectives and called for the establishment of the NPC. The NPC has been operating in its present form since the confirmation of the Planning Law of 1971. Another example is the Law of Cooperatives, which called for the establishment of the JCO, which has been operating in its present form since 1971 when the Law was confirmed. Thus the level at which instruments are selected through the above procedure is the highest in the country, since confirmation of a law requires approval by the Council of Ministers and the National Assembly, and then the Royal Decree. This procedure is applied when Ministries, authorities, autonomous institutions, parastatal organisations or parastatal corporations are established.

The second level through which instruments are selected is a decision from the Council of Ministers through the formulation and approval of a by-law. All by-laws should be confirmed by Royal Decree. Such instruments take the form of a Department within an existing Ministry. Article 120 of the Constitution gives the Council of Ministers the power to organise all Ministries, both structurally and functionally, into Departments. Some Departments may enjoy more autonomy or flexibility in their work through being controlled by the Minister concerned or his Under-Secretary. The majority of Departments, however, are linked to the central administration of the Ministry concerned. An example of a semi-autonomous Department is the Department of Land and Survey of the Ministry of Finance. This Department enjoys wide-ranging authority and implements state laws and policies on the size of landholding, land sale, land purchase and leasing of leasing of state land to various users.

All Departments are organised into two or more Divisions which specialise in the implementation of certain functions assigned to the Department. Divisions are created by decision of the Minister concerned. The function of agricultural research, for example, is assigned to a Division within the Department of Agricultural Research and Extension (DARE) of the MOA.

Allocation of Funds

There are two types of financial allocation. The first is the recurrent budget of the institution, which includes salaries and operating costs; and the second is the capital expenditure, or what is usually referred to as the development allocations investment.

Recurrent Expenditure: The steps involved in deciding on such allocations are as follows. First, individual institutions present an annual budget to a sub-ministerial committee for discussion, with the Director of the Government's Budget Department playing a key role in the review. Budgets are usually prepared along guidelines issued annually by the Prime Minister. The biggest changes take place at this stage. Second, tentatively approved budgets for each institution (usually a Ministry) are integrated into one national budget and sent to the Prime Minister for review at the Council of Ministers. Further changes may take place at this stage. Third, the national budget approved by the Council of Ministers is then presented to the National Assembly for discussion and approval. Few, if any, changes are usually made at this stage. Finally, a Royal Decree is issued confirming the national budget approved by the National Assembly. The national budget is issued as a law for that year.

Development Allocations: Estimated costs of development projects are originally defined in the National Plan for Social and Economic Development (see Annexe 12.3 for mode of approval). The sources of finance are indicated in the Plan for each project. Sources may be either from the Government, from outside (loans, assistance, grants, etc.) or from both. Each year a review is held to decide on the final annual allocation for each project. These are included in the proposed budgets for each institution under the heading 'development projects'. The steps involved in approving such allocations are the same as for the recurrent budget. The only difference is the composition of the review team at the first step, where the NPC plays another key role in deciding on allocations.

General Remarks

Budgets of autonomous institutions like universities are not included in the process described above. The board of trustees of each univerisity approves the final annual budget.

Allocations for some projects implemented beyond one year are approved in total in the first year but reviewed annually for modifications. Final allocations for each year are included in the national budget.

Institutions through which Agricultural Policies are Administered

There are two forms in which agricultural policies are made and administered:

1. Policy-formulating institutions with no administrative responsibilities: such institutions are exemplified by the HAC, which is a policy-formulating

body. Implementation of all policies is left to the Ministries, authorities or other insitutions concerned.

2. Policy-formulating institutions with responsibilities for administering policies: The MOS is responsible for formulating policies on supplies of various consumer goods. Once policies are approved, the said Ministry will administer such policies. When policies are determined by law, implementation is either assigned to existing institutions, each with its own mandate, or a new institution is created to administer the policies. At present a law concerning water resources is in the process of enactment; it calls for the establishment of one authority responsible for policy formulation and policy administration of all water resources in the country. The following are examples of how some policies are administered:

Policy	Institution(s) Responsible for Administration
Size of land holdings	Department of Land and Survey, Ministry of Finance
Income tax	Department of Revenues, Ministry of Finance
Customs exemption of inputs or commodities	Department of Customs, Ministry of Finance
Export-import of agricultural commodities	Ministry of Agriculture and Ministry of Supply
Price control of agricultural commodities	Ministry of Supply
Agricultural credit	Agricultural Credit Corporation and the Jordan Cooperative Organisations
Agricultural land use	Ministry of Agriculture
Conservation of land resources	Ministry of Agriculture
Regulatory functions for imported inputs, quarantine services, technical services, etc.	Ministry of Agriculture
Agricultural production models for rain-fed and irrigated areas	Ministry of Agriculture
Price control of agricultural inputs	Ministry of Agriculture
Water resources development in rain-fed areas	Natural Resources Authority
Water resources development in the Jordan Valley	Jordan Valley Authority
Agricultural land development in the Jordan Valley and land distribution	Jordan Valley Authority

Geographical Coverage

All institutions administering policies have national coverage except the Jordan Valley and the lowlands south of the Dead Sea. The MOA has regional offices throughout the main governorates. Activities of the MOS are concentrated in the main cities, while the JCO has offices throughout the country.

Agricultural Policy Objectives and Instruments Used for Implementation: Issues, Current Status and Constraints

Although the formulation of policies related to the agricultural sector has been assigned to specific institutions through laws and by-laws, most of these institutions have not yet carried out such tasks. In fact, comprehensive policies for any agricultural sub-sector simply do not exist.

There are, however, several policy directives and/or decisions which appear in legislative acts (laws) concerned with agriculture. There are also other policy decisions which have been taken by the HAC, each dealing with a particular aspect of agriculture. Although one cannot consider such decisions, whether individually or collectively, as formulating integrated policy for any of the agricultural sub-sectors, we shall still discuss these decisions of directives for the purposes of this chapter.

Another group of directives or objectives has been spelled out in past and present development plans for the agricultural sector. Since the plans are approved by the Council of Ministers, they do constitute policy objectives, although perhaps of a short-range nature. These and other decisions related to the agricultural sector are presented below.

Policies Related to Agricultural Land and Landholdings

Issues Involved

Fragmentation of Landholdings: This is one of the most important constraints facing the agricultural sector in Jordan. Such fragmentation has had negative results because the size of most landholdings in rain-fed areas does not constitute an economic unit of production which can support the average farmers and his family. Existing laws give an individual the right to own land, whose size can be as small as 1,000 m^2, anywhere in Jordan. Within city or village limits it may be only 250 m^2. Another aspect of this constraint are the laws which govern the partitioning of land into single holdings with defined borders and location. The minimum size of landholding which can be partitioned into a separate holding is 1 ha in areas outside city or village limits. The results of such laws allow for common ownership of one single piece of land. The number of owners of a piece of land with an area of 1 ha (10 du) can be as many as 10. Common ownership of land comes as a result either of inheritance under Islamic law or by purchase. Islamic *Sharia* decrees that land owned by an individual passes to his inheritors after his death. The number of inheritors may vary, but as the years go by it grows. In 1975 the agricultural census reported the distribution of landholdings, which can be summarised as follows:

Size of holding (du)	No. of holdings	% of total
Less than 10	12,374	24
10–29	12,226	24
30–49	7,634	15
50–199	14,123	28
200–499	3,359	7
500+	1,075	2
Total number	50,791	100
Total area	3,904,031 du	—
Average size	77 du	—

The situation since 1975 has worsened, leading to further fragmentation. The *Annual Reports* of the Department of Land and Survey indicate that as many as 8,830 partitioning transactions were completed during the period 1976-81, as follows:

Year	No. of landholdings before partitioning	No. of landholdings after partitioning
1976	2,657	7,033
1977	2,445	8,313
1978	2,898	9,853
1979	3,028	10,295
1980	3,160	10,744
1981	3,480	11,832
TOTAL	17,668	59,070

The majority of these transactions were the result of land sales to new owners. Investors have turned to land purchasing for speculation because of the great escalation in land prices from year to year. About 60 per cent of these transactions were within city or town limits, meaning that the size of new landholdings was from 1 to 4 du. The remaining 40 per cent were outside city or town limits, so that new landholdings were 10 du or over.

Regional, City and Village Planning. In the last eight years the construction of houses, industrial plants and other commercial facilities has increased many times over in the main cities and towns of Jordan. As a result, older city limits have grown at the expence of agricultural land. Consequently, the price of agricultural land around villages, towns and cities has increased many times in the last 10 years. An example of the increase in the price of land over one year, as recorded by sale transactions carried out by the Department of Land and Survey, is as follows:

Area	Average price per du (1,000 m²)		
	1980 (JD)	1981 (JD)	% increase in 1981 over 1980
Amman	13,090	28,666	218
Naour (near Amman)	1,225	4,218	336
Abu Alanda	2,425	17,794	733
Badran	215	738	343

Source: Department of Lands and Surveys, 1981.

As a result of this price increase, owners have viewed land as a commodity for speculation rather than for production. Another result has been increased public pressure on both governorate and municipal councils to expand city or town limits so that more agricultural land can be included within them. Once these lands are included, further partitioning is allowed and prices increase many times over. The problem of what is agricultural land and what is not has become a major constraint, because much of the land has been purchased by investors who are more interested in reselling it at a higher price than in farming it.

Land Use: The laws of Jordan afford great freedom to landowners in the way they can utilize their land. Article 6 of Law No. 49, 1953 (published in the *Official Gazette* No., 1135 1/2/1953 under the title 'The Law of Utilisation of Non-transferable Properties') reads as follows:

> Every person who has land registered in his name has the right: to sell, to rent, to mortgage, to lend, to cultivate, to benefit from anything that grows in it without work, to cut or plant land with any tree, to cultivate land wih any crop for his benefit, to take down or construct on it a house, a store, a factory or any other building he needs for farming, to use its soil for brick making, and to sell what is in it of soil or rocks. The owner cannot build a village on his land.

The Law implies that all these rights are exercised within other existing laws of the country.

The Law of Agriculture No. 20, 1973, gives the MOA the authority to:

1. Determine regions in which certain crops can or cannot be cultivated
2. Regulate crop rotation at village or any other level
3. Determine the percentage of area allocated for the cultivation of specific crops out of the total area of each landholding
4. Determine cropping patterns

In addition, one of the responsibilities of the HAC is to formulate policies on cropping patterns as well as on ways of utilising land.

It may appear that there is no contradiction in all the above laws and regulations concerning land use. However, since the farmer has the final decision on what and where to plant, the enforcement of any law or regulation requires incentives as well as preventives. Farmers may be forced by present laws *not* to plant a certain crop in a certain area, but such laws do not force farmers to plant if they choose not to. Thus far, the MOA has been advocating certain cropping patterns and has not enforced its authority through by-laws which specify fines or other legal measures against violators. This situation is a result of a lack of coordination between legislation and implementation. It also reflects inconsistent views among various individuals as to how far the MOA should go to enforce cropping patterns. In the absence of such enforcement, the MOA and other institutions promote certain patterns through the preferential treatment of those farmers who apply recommended patterns; this includes providing credit, distributing subsidised inputs and providing assistance in the form of food materials. However, regulations call for fining farmers who exceed the approved area in some crops like tobacco. But this is a special case, because it pertains to a crop on which the Government charges fees on cropped area.

Policy Measures Taken

Jordan Valley Development Plan (1973): An excellent working programme to develop the lands of the Jordan Valley was launched in the Plan. It calls for the development of water resources, land resources, health facilities, roads, communications, education facilities, agricultural commodity marketing facilities, agricultural commodity processing facilities, electricity services and housing. It also calls for regional development of towns and villages, preventing farmers from building villages on agricultural land, and for alternative locations for farm villages with public service facilities. The most important aspect of the Plan which concerns us here is the redistribution and/or consolidation of landholdings in the Valley. All this development was the result of government policy spelled out in the Law of Jordan Valley Development, No. 18, 1977. This Law called for the redistribution and consolidation of landholdings receiving water from new irrigation schemes. The formula for the consolidation of, or redistribution to, old or new owners is detailed in the Law, and may be summarised as follows:

Old sizes of holdings	New sizes allocated to old owner
Less than 40 du	Allocated as a whole with provision to increase size to 40 du whenever possible
40-50 du	Allocated as a whole to owner
51-100 du	First 50 du are allocated plus 25 per cent of remaining area
101-500 du	First 62 du are allocated plus 17 per cent of area above 100 du
501-1,000 du	First 130 du are allocated plus 12 per cent of area above 500 du
1,001 du and above	200 du

Note: Holdings given to new owners (i.e. not those owning land before in the Valley) should be 40 du of Class A or B land or 50 du of Class C land.

The Law decrees that once the size and location of landholdings are defined for the owner, fragmentation (partitioning) is not permitted for any reason unless the size of the holding is over 80 du. The size of any new holding resulting from partitioning after sale or inheritance should not be less than 40 du. The Law, however, does not overcome the problem of there being several owners to each holding—which occurs according to Islamic inheritance rules once the individual owners die. The area of 40 du which was chosen as a minimum size was based on the premise that it constitutes a unit of land which gives sufficient returns to support an average family. Such a matter depends on many variables and warrants economic input-output studies.

The institution responsible for the development of the Jordan Valley and the implementation of the Law concerning it is the JVA. The targets for consolidation and redistribution of landholdings specified by law have not yet been reached. Legal, social and economic constraints have been encountered, plus the fact that water has still not been secured for all land in the Valley. The distribution of size of landholdings in the whole Jordan Valley in 1981 was as follows:

Size of landholding	No.	% of total
Less than 10 du[1]	956	15
10–29 du	2,421	38
30–199 du	2,810	45
200 du and above	150	2
	6,328	100

Note: [1] Many of these landholdings were there before the Plan was devised. They lie above the level of the irrigation canal and hence do not receive water by law. However, with the construction in 1983 of the Wadi al-Arab Dam, which is supposed to give water to these holdings on a permanent basis, the consolidation of as many of these holdings as possible will take place within the next few years.

Source: DOS 1981.

The Council of Ministers' Decision to Limit Land Fragmentation in Rain-Fed Areas: In response to increasing fragmentation (partitioning) of land reported by the Department of Land and Survey, and with the objective of freezing further fragmentation, the Council of Ministers took the following decision in April 1981: 'Until laws regulating land partitioning are modified, lands outside city, town and village limits will be partitioned to holdings not less than 60 dunums each.' This means that only holdings of 120 du or more can be partitioned, which is a major departure from previously existing laws which allowed partitioning of holdings of 20 du or above. It must be noted again here that such a decision will still not prevent common ownership of land. The decision became effective in April 1981. The Government is considering submissions for further modification of present laws regarding minimum size of landholdings (non-partitioned) and a great debate is taking place in related circles regarding such modifications.

Taxes and Fees on Land: Non-irrigated lands outside city, town or village limits are not subject to tax. Irrigated lands are taxed at the rate of JD 1.5 per annum for every dunum.

Fees for registration of land in the names of new owners as a result of sale are charged at the rate of 10 per cent of the value of land. An additional tax of 10 per cent of the cost of fees is also in effect. Fees for partitioning land among owners amount to only 1 per cent of the value of partitioned land.

Agricultural Land Utilisation: The present laws of Jordan do not make it mandatory for owners of agricultural land to utilise it. Although it is implied that owners should use land for agricultural production, statistics have shown a sharp increase in unfarmed agricultural land in recent years. In rain-fed areas, much of the marginal land is left unfarmed in drought years. This is indeed the normal practice. But in recent years there is increasing evidence that land is not farmed because it belongs to absentee owners who purchased it recently only in order to re-sell it later at a higher price. In 1974 the total area under crops was 3.8 mn du. If this is considered a base year, the total area cultivated in subsequent years has decreased, as shown in the following:

Year	Total area cultivated (mn du)	% of base
1974 (base year)	3,89	100
1975	2.36	60
1976	2.54	65
1977	2.35	60
1978	2.54	65
1979	2.04	52
1980	2.70	69
1981	2.38	61

Land and Soil Conservation: One of the objectives of the current 1981-85 Five-Year Development Plan is to protect land and soil resources, and to develop them along the lines of optimum utilisation. This comes as a response to the continuing deterioration of soil as a result of erosion in hilly areas, water courses, lands surrounding dams and rangelands. It also comes as a response to increasing soil salinity in irrigated areas in the Jordan Valley, Southern Ghors and other arid areas irrigated by underground water like Wadi Dhuleil. The Plan calls for the implementation of the following projects:

Project	Projected allocations (JD mn)	Implementing institution and remarks
Afforestation	8.4	Ministry of Agriculture, Implementation is slow.
Development of Zarqa River basin	18.6	Ministry of Agriculture, Preparation of project is underway.
Terracing hilly lands and fruit-tree planting	12.5	Ministry of Agriculture and private sector with aid from World Food Programme. Implementation is slow compared to the volume of work needed.
Drainage of soils in the Jordan Valley	3.5	Jordan Valley Authority. Implementation is slow.
Rangeland protection	0.1	Limited studies in fenced areas through recurrent budget of Ministry of Agriculture and another project implemented by cooperative organisation.
Desert soil dams	0.6	Jordan Valley Authority. Allocation is too low for work required throughout the country. Implementation is slow.

Notes: There are no set criteria on which resources are allocated. Annexe 12.3 summarises the mode of planning of projects for a five-year period.

The figures shown in the Plan are projected allocations, but actual allocations may change according to availability of funds. Each year, the

Higher Development Committee, which is composed of the Ministers of Finance, Trade and Industry, and Supply, and the Presidents of both the Central Bank of Jordan (CBJ) and the NPC, considers various projects, looks at ensured resources and recommends further allocations for each project to the Council of Ministers. If funds are available through grants, a technical assistance programme or a loan, then the project has a better chance of receiving allocations. For example, projected allocations of JD 12.5 mn for terracing hilly lands and fruit-tree planting are divided between two sources: around $US 18 mn from the World Food Programme (WFP) and an equal sum from the Government. The WPF commitment arrives in the form of food materials and the Government allocates a sum of about $US 3 mn each year.

Land Use and Cropping Patterns: The declared objective in the current Five-Year Plan, as well as in decisions taken by the HAC, is to prepare a national cropping pattern plan for both irrigated and rain-fed agricultural land. The land use and cropping patterns are to be developed along the following lines or objectives:

1. Utilisation of Jordan's unique climatic conditions, such as in the Jordan Valley, for the production of high-value cash crops
2. Promotion of the production of strategic commodities like wheat throughout Jordan
3. Promotion of fruit-tree planting in hilly areas
4. Allocation of steep lands (25 per cent or more) to afforestation
5. Non-cultivation of all lands with 200 mm of annual rainfall, and allocation of all such lands (not under irrigation) to natural ranges
6. Afforestation and terracing of hilly and steep lands in watersheds, water courses and river basins. Planting of terraced areas with fruit trees
7. Promotion of cereals, legume production and summer vegetables in rain-fed areas receiving 300 mm or more annual rainfall and less than 9 per cent in gradient.

Although the 1973 Law of Agriculture gives the Minister of Agriculture the authority to define land use and cropping patterns for all agricultural lands in Jordan, no steps have yet been taken to translate such authority into a plan of action. Policy formulation to define land use plans and cropping patterns was granted to the HAC. Although the subject has been discussed several times in the Council, no policies have yet been formulated.

Which cropping patterns to choose and how to enforce their adoption have proved to be complex questions. In the Jordan Valley, for example, the soil and temperature are suitable for the cultivation of a wide variety of crops, such as wheat, corn, cucurbits, tomatoes, potatoes, eggplants, peppers, beans, cruciferous vegetables, citrus, olives, bananas and, in some regions, grapes. But the water supply is the limiting physical factor. Each crop has different water requirements. While no definite information exists on the exact amount of water needed for each crop, the following may be listed from highest to lowest levels of water required: bananas, citrus, corn, tomatoes, eggplants, potatoes, cucurbits, beans, crucifers and wheat.

All these crops require supplementary irrigation, since rainfall in the Valley is low, averaging annually from about 300 mm in the north to less than 100 mm in the south. But the water available for irrigation is limited: there is not enough water to plant all the Jordan Valley with bananas or citrus. Farmers, however, both from experience and in response to economic factors, prefer to plant these crops. A dunum of citrus may give a net return of more than JD 200 a year, while a dunum of wheat will not give more than JD 20 a year. The Government would like the farmers to plant wheat because it is considered a strategic commodity and it requires less water. Taking into account both the limited water supply and the farmers' desire to plant more citrus, the JVA concluded that about 25 per cent of Jordan Valley land could be planted with citrus. This figure was based on the water available for permanent irrigation during the summer. The quandary was: who should plant and where?

Until the 1960s, landowners had aquired water rights which allowed them to plant a large percentage of their holdings with citrus. As the JFA developed water resources and built extensive water transportation schemes which served about two-thirds of the land in the Valley, the difficulties of keeping track of the percentage of each piece of land and what it should be planted with became enormous. As a result, the Authority in more recent years has defined the amount of water each farmer can receive. This rationed water supply differs from one area to another, based on previous water rights. Landowners with no previous water rights receive amounts of water relative to the size of their holdings. Under this new arrangement, farmers are free to utilise their rations as they deem fit. This situation is not thought to be satisfactory, since farmers take chances and plant crops which may need more water than that provided by their rationed supplies.

Rain-fed areas with a rainfall of 400 mm and above are suitable for wheat, legumes, summer vegetables, tobacco, olives, grapes and stone fruits, among other crops. The economic net returns on each of these crops are different. Some need more capital investment than others. If the Government forces farmers to plant a specific crop, then it must ensure competitive returns. Government policies have not been crystalised in this field. At one time, the Government declared that it would buy all the wheat produced by the farmers at prices higher than those on the world market. However, the flat land which is good for wheat is also good for stone fruits, summer vegetables and tobacco. Farmers know that stone fruits and other crops give them higher returns. As a result, fewer farmers are opting to plant wheat.

Policies Related to Water Resources

Every Jordanian, whether in the public or the private sector, is well aware of how vital water is for his welfare. Water resources are so limited in Jordan that the Government has been prompted to grant priority to the development, conservation and efficient use of such resources.

Issues Involved

Multiplicity of Institutions Concerned with Water Resources: In addition to private citizens who claim water rights to surface and underground water, there are six public institutions, each of which has clauses in its laws to deal with water development and consumption in one way or the other. These institutions are as follows:

Name of Institution	Area of concern
City and town municipalities	Distribution and sale of water within city limits
Drinking Water Corporation	Development and transportation of drinking water outside Amman city limits
Water and Sewage	Development, transportation and distribution of water

Authority	for household and industrial usage within Amman city limits
Natural Resources Authority	Development and management of water resources for agricultural purposes outside the Jordan Valley and Southern Ghors
Jordan Valley Authority	Development, management and distribution of water resources for agricultural and household usage in the Jordan Valley and Southern Ghors
Ministry of Health	Quality control of drinking water

The geographical coverage of these institutions overlaps and several institutions use water from the same sources.

Priorities of Usage: Although the responsibility for formulating water policy has been given to the NRA, no such policy exists to date. One of the prime objectives of such a policy would be to determine who has rights to water, how it should be developed and for what purposes, and the priorities of water use for households, industry and agriculture.

Present Policy Measures and their Objectives
Establishment of one National Water Authority: In response to the major difficulties facing Jordan in making ends meet between existing and potential water resources on the one hand and the nation's needs for water on the other, the Government decided early in 1984 to establish the National Water Authority (NWA) which is now in operation. The new Authority was created by a special law and has the following objectives:

1. Formulation of national water policy
2. Development of water resources
3. Distribution and utilisation of water
4. Conservation of water resources, including quality control
5. Optimisation of water usage in all sectors

The NWA has taken over the functions of two institutions dealing with water, the Water and Sewage Authority (WSA) and the Drinking Water Corporation. Within the next five years, the new Authority will also take over the work and functions related to water which are presently handled by the NRA and the JVA.
Development of Water Resources: The objectives of the present Five-Year Development Plan are to improve and support water institutional reorganisation, to develop water resources, to optimise water utilisation in all sectors, and to conserve water resources. To attain these ends the Plan has called for the implementation of projects at a cost of JD 521.7 mn over the five-year period 1981-85; this constitutes 15.8 per cent of total investment allocations of the Plan. Major projects are as follows:

Project	Total allocations (JD mn)	Implementing institution and/or comments
Dam construction and expansion	32.1	Jordan Valley Authority: one completed, two others at planning stage
Second stage of Jordan Valley irrigation scheme	200.0	Some stages being implemented; some still at planning phase
Other irrigation schemes	42.5	Some being implemented, while others still at planning stage
Water resource development for households, and its transportation mainly in Amman Governorate	109.0	Various municipalities and authorities
Water and sewage in other governorates of Jordan	137.1	Various municipalities and authorities

Policies Related to Farmers' Organisations

Farmers are the least organised group of producers in Jordan. All professionals are organised in 'syndicates' or associations: businessmen are organised through chambers of commerce in various cities; industrialists through chambers of industry; and all labourers except farm labourers are organised in the labour 'syndicate' or association. Farmers, irrespective of the size of their holdings, are not organised through any chamber of their own. One of the issues facing the agricultural sector is the definition of who is a farmer. Such a definition does not exist.

Traditionally, neither farmers nor Government have been active in initiating a farmers' organisation. In more recent years farmers have become more interested in organising themselves and the Government has become more responsive to their demands. Such developments have manifested themselves in the creation of the Jordan Valley Farmers' Union (JVFU) and the Jordan Valley Farmers' Association (JVFA) in the Jordan Valley and farmers' cooperatives. These two groups are organised through laws and are supported by the Government. In 1983 the poultry producers banded together and obtained a ruling from the MOS to be the sole agency for organising the supply and marketing of eggs throughout Jordan. Although this may not represent policy, it is still a positive step on the part of the Government to recognise the interests of egg producers as a group within the general policy of commodity supply of the country.

Policies Related to Production Inputs

Exemption from Customs Duties: Customs duties on imported goods constitute a major proportion of government revenues, in addition to being a mechanism to regulate import-export policies. All agricultural inputs (except spare parts and a few other minor items) used in agricultural production, including machinery, are exempted from customs duties. Since these exemptions are regulated by law, they are considered to be, and indeed are, an expression of policy which supports the promotion of national agricultural production.

No Subsidy to Production Inputs: The Government, through its various institutions, has been steadily moving towards the implementation of a policy of no government subsidies to agricultural production inputs. These include fertilisers, vegetable seeds, cereal seeds, pesticides, implements, tractors, machinery and the majority of fruit-tree seedlings. The MOA provides seedlings of olives, grapes, citrus and some stone and pome fruits at prices lower than those in the private sector. This programme is part of a World Food Programme (WFP) initiative to develop hilly rain-fed areas and is administered by the MOA.

However, the Government does support organisations like cooperatives and the JVFA which compete with the private sector in providing agricultural production inputs to farmers. Government support comes in the form of loans to provide working capital for such organisations. The idea behind such support is to have parallel competitive markets for production inputs in order to keep prices down.

Policies Related to Priority Determination of Developmental Projects in Agriculture

Capital investment in water, irrigation and agriculture constituted only 6.4 per cent of total expenditure in the 1976-80 Development Plan. In the current Plan (1981-85) both agriculture and irrigation receive more attention. Expenditure in previous Plans and aggregate allocations in the current one for the major sectors are as follows:

Sector	Actual expenditure under 1976-80 Plan (JD mn)	(% of total)	Allocations in 1981-85 Plan (JD mn)	(% of total)
Agriculture, water & irrigation	78.4	6.4	755.2	22.92
Industry, mining & electricity	416.1	34.1	922.2	27.94
Tourism	33.5	2.7	65.7	1.99
Transport	253.4	20.5	545.5	16.53
Communications	22.8	1.9	106.8	3.24
Housing	257.7	21.3	308.1	9.34
Construction & other services	160.0	13.1	596.5	18.0
TOTAL	1,221.9	100	3,300	100

Source: NPC 1981-85.

Policies Related to the Promotion of Agricultural Practice

Government polices include additional measures and services intended to help small and other farmers to reduce farmers' costs and increase their revenues. Such measures are scattered about in the various laws but can be summarised as follows:

Exemption of Agricultural Income from Tax: All income generated by individuals through agricultural production activities is exempt from income tax. This policy is enacted through income tax laws currently in effect.

Provision of Credit at Interest Rates Lower than those Charged by Commercial Banks: The ACC provides short-, medium- and long-term credit to farmers at rates of 6-8 per cent a year. Present credit rates of commercial banks are 10-12 per cent a year. In addition, the JCO provides short-term credit to its members at rates of 6-8 per cent a year.

Promotion of Wheat Production through the Purchase of Produce at Prices Higher than those on the World Market: The MOS is responsible for such purchases and offered JD 110 per t in 1983. The world market price for wheat in 1983 was about JD 70 CIF Amman.

Strengthening of Agricultural Research and Extension Activities: The Government recognises the need to strengthen agricultural research and extension activities in order to help producers overcome the technical constraint of production. To this end, the present Five-Year Plan calls for the establishment of a national research centre with a unit to serve the Ghors at a projected cost of JD 5.93 mn over the Plan period. The unit in the Ghors has started functioning but the centre is still under consideration.

Policies Related to Supply and Demand of Agricultural Food Commodities

Jordan is one of the countries in the region where domestic food production is much lower than demand. Deficits vary from one commodity to another throughout the year as well as during certain months of each year. Perhaps the most difficult problem of the agricultural sector is to devise policies concerning the supply of food commodities to meet domestic demand. Before discussing policy measures related to this issue it may be pertinent to present some of the major issues facing Jordan in this area.

Issues Involved

Sources of Supply: At present, there are three main sources of supply for food commodities. These are:

1. Production on the West Bank of Jordan
2. Domestic production on the East Bank of Jordan
3. Imports

The Government has decided not to stop the flow of foodstuffs produced on the West Bank (occupied by Israel since 1967) to the East Bank. It is not our intention here to discuss the complexities of this issue, but rather to present it as one of the variables which Jordan has to deal with when considering food supply policies. Although the Jordanian authorities control the upper limits of the type and quantity of food commodities flowing in from the West Bank, they have little control over the type and quantities which actually flow in below the ceiling.

Domestic production of food commodities on the East Bank is characterised by the following:

1. The final decision as to what, when, where and how much to produce is still in the hands of individual producers.
2. Production of food commodities fluctuates in type and quantity from one season to another and from one year to another. With certain food commodities such as eggs, production exceeds demand during some months of the year; this is also true for vegetables like tomatoes and cucumbers. For most other foodstuffs, however, production falls short of demand throughout the year.

Food imports have to fill these gaps in type, quantity and time available. It is easy to see that such a process will make various groups—for different reasons—dissatisfied with government intervention in this field.

'Transitional' Pressure on Demand: Demand for food commodities on the East Bank suddenly increased in the late 1960s, due to the expulsion there of hundreds of thousands of inhabitants from the West Bank. Although the biggest surge took place after 1967, the movement has continued at varying intensities.

To complicate matters further, there are at least 500,000 Jordanians (mostly from the West Bank) who work abroad and visit Jordan during the summer months. All these groups of people increase the demand for food in a way not normally encountered in other countries.

Roles of the Private and Public Sectors and Price Control: All domestic production of food is in the hands of the private sector, while imports are shared between the private sector and the MOS. The objective of the MOS is to provide consumers with food at the lowest possible profit margin (or no profit at all), which is not the case with the private sector. This has meant that the Ministry's role in providing food commodities at a low price has grown in recent years. Virtually all basic foodstuffs, such as wheat, rice, sugar, meat and vegetable oil, are imported by the MOS. Retail distributors obtain these commodities from the MOS with fixed marginal profits. Private sector importers claim they cannot compete with the Ministry. MOS objectives are to ensure that foodstuffs are available in type and quantity, and at time and prices, suited to consumer needs and means. Some of these objectives—or perhaps all of them—are not necessarily compatible with the interests of domestic producers or private sector importers. Where the MOS draws the line between being consistent with government policy in promoting domestic production on the one hand, and with its mandate to secure food for consumers on the other, is still unclear.

The MOS intervenes almost daily in determining food prices, especially for fruit and vegetables. Farmers and other producers complain that such interventions are not in their interest. They claim that prices set by the Ministry are not always realistic and do not ensure a fair profit margin on production costs. What strengthens their claims is that agricultural production inputs are not subsidised, and their prices are not controlled. Another complaint by producers concerning price controls is the fact that the MOS tries to keep prices low by importing food from countries (mainly Comecon countries) where farmers do enjoy a government subsidy. Such practices provide unfair competition for Jordanian producers. Producers conclude that MOS practices constitute negative incentives for domestic production.

While the concern of both the MOS and producers is justified, there is a need to reconsider MOS practices in price control and importation to ensure fair prices for producers. Such prices should take into account the cost of domestic production, a reasonable profit margin for farmers, contingency costs for risks taken by farmers and—most important of all—a reduction in marketing and transportation costs to ensure fair farm-gate prices.

Policy Measures and Objectives

Steps to Minimise the Deficit in Agricultural Balance of Trade: The current Five-Year Plan has projected targets to increase the value of AGDP by 43 per cent in 1985, i.e. to increase agricultural income from JD 60 mn in 1980 to JD 86 mn in 1985 at 1980 prices. Production targets call for an increase in total production of major food commodities as follows:

Commodity	Average production	(000 tons)	% change
Field crops	109	142	30
Vegetables, including watermelons	376	600	60
Olives	24	32	33
Grapes	26	40	54
Citrus	30	50	67
Other fruits	20	25	25
Dairy products	39	45	15
Meat	7	15	114
Poultry	25	36	44
Eggs (mn)	280	450	61

To achieve such targets, the Plan has called for the implementation of organisational measures, some 30 development projects, and the strengthening of public institutions (both government and parastatal) to serve agriculture better. These actions and measures are summarised in Annexe 12.4.

MOA Intervention in the Import-Export of Agricultural Food Commodities: The undeclared policy of the Government is to protect domestic production from competition by imports. The interventions can be summarised as follows:

1. To stop importing commodities considered to be in good supply from domestic production
2. To permit imports of commodities in short supply to meet domestic demand
3. To permit exports of commodities thought to be in excess of domestic demand
4. (Rarely) to prevent exports of commodities thought to be in short supply from domestic production

The effectiveness of such policy practices is weakened by several factors, including:

1. A weak information base on the projected type and quantity of commodities from domestic sources
2. Price controls practised by the MOS
3. Import practices of the MOS, with their minimal coordination with the MOA
4. Fluctuations in type and quantity of commodities available from domestic production
5. The failure of concerned institutions and the private sector to reach a stable formula for a cropping pattern based upon domestic and regional market demands

There is no doubt that this area constitutes one of the weakest links between the supply and demand of food commodities. To achieve government policy objectives in the promotion of domestic production, there is an acute need for stronger institutional capacities and better coordinated and integrated measures.

Improvement of Farmers' Incomes: One of the declared policy objectives set by the HAC is that farmers should receive the highest possible percentage of the wholesale price of the food commodities they produce. Another objective is to improve and stabilise marketing facilities, information organisation and procedures to ensure the highest possible income for farmers. (HAC decisions, 1979).

Establishment of the Jordanian Company for the Marketing and Processing of Agricultural Products (JCMPAP): In early 1984 the Government established a public share-holding company with a capital of JD 12 mn. Fifty-one per cent of the shares are to be owned by parastatal institutions, namely the JVFA Association, the ACC and the JCO. The other 49 per cent are open to the public, including farmers. The establishment of the Company is a reflection of government policy to improve organisation and strengthen facilities,. information and procedures for marketing agricultural commodities. The objectives of the Company explicitly promote the role of the private sector and farmers in marketing, as well as providing a source of short-term financing to farmers as an alternative to the middleman presently handling the marketing of most domestic production.

The decision to establish the JCMPAP was taken by the HAC in mid-1983. The Company will take over all marketing facilities established by the Government throughout Jordan.

The MOS and Supply Policies: The MOS was established in 1974 to formulate and administer the application of a general supply policy for Jordan. Although no such policy has yet been formulated, the Ministry has made several interventions in the supply of agricultural commodities through price control and importation. The goals set in the Five-Year Plan for the MOS are as follows:

1. To provide basic food supplies in sufficient quantities and at prices which reflect real cost; and to supply other items at reasonable prices
2. To ensure a minimum level of food security
3. To set up standard specifications for food items and ensure effective control and enforcement
4. To protect the consumer from malpractices and fraud
5. To encourage the production of food items
6. To improve the level of the baking and flour-mill industries
7. To introduce new patterns of consumption which would reduce costs to the public while at the same time ensuring adequate nutritional intake

The MOS is currently implementing the following projects as part of the Five-Year Development Plan.

Project	Allocations (000 JD)
Flour mill	1,900
Cold storage warehouses	2,280
Expansion of grain silos	15,500
Feed plant	3,500
Automated bakeries	1,800
Ordinary warehouses	600
Poultry slaughter-house, plus upgrading of existing slaughter-house in the Governorate of Amman	2,400
Feed concentration plant	2,000
TOTAL	30,980

Policies Related to Agricultural Production

As already indicated, there are no integrated and comprehensive policies related to agricultural production. But several individual decisions and practices exist which throw light on the general direction of policies. Agricultural production in this context covers five main aspects: who produces agricultural commodities;

what should be produced and how much; where they should be produced; government intervention to optimise such production; and government intervention to maximise net returns of production operations to producers. In the following, we shall attempt to present decisions, principles and interventions related to agricultural production which have policy implications.

Who Produces Agricultural Commodities: All agricultural land and animal farms (except experimental farms and nurseries) are privately owned. Declared policy decisions, as well as the laws of the country, emphasise the role of the private sector in agricultural production. The state of parastatal institutions are not involved in any commercial agricultural production. Land owned by the state which is developed by public institutions for purposes of crop production is distributed to farmers as new owners. Land owned by the Government and proved to have potential for crop production, especially through irrigation, is leased to individual citizens. Such individuals will eventually be given the title deeds of ownership once it is proven by a special committee that improvements have been undertaken to make the land productive. This is probably the most clear-cut of all aspects related to agricultural production.

What Should be Produced and How Much: This is the most difficult aspect of all. It has not been resolved or organised either to promote the interests of producers or to follow policy guidelines set by the Government. The present policy guidelines may be summarised as follows:

1. The type and quantity of commodities produced should be compatible with domestic consumption needs as well as with existing and potential needs of regional and other markets outside Jordan.
2. Production operations should follow the principle of free competition and the production of those commodities which ensure highest net returns under Jordanian conditions of soil, water, climate, capital, labour and inputs.
3. Production should be based on production patterns recommended by the public institutions concerned. Such patterns are based on the physical limitations of production resources.
4. Production boards for special commodities or groups of commodities should be established, charged with organising production and marketing aspects of these commodities. Boards are to be composed of producers, as well as representatives of relevant public and private institutions, and may establish special funds for crop insurance and price stabilisation.
5. The JVA should administer water rationing, leaving it to the farmers to decide which crops to cultivate.
6. There should be minimal government intervention, except in setting production patterns and in support of higher prices of certain commodities selected by the Government, like wheat.

Where Commodities Should Be Produced: The following guidelines are presently followed:

1. Poultry, dairy, sheep and other animal farms should comply with city, town and regional planning regulations. All farms of permanent structure should be licensed by both the MOA and the relevant local municipal authorities. In recent years the application of such rules has become stricter.
2. The public institutions concerned should promote ecological and topographic zones suitable for crop production. Present conditions leave farmers virtually free to decide whether or not to follow recommended land use patterns. Credit, and other government services, promote the production of fruit trees in hilly areas, leaving cereals, legumes and other seasonal crops to flat lands in rain-fed areas.

Government Interventions to Optimise Productivity: Perhaps the greatest gap between the Government's declared objectives and actions on the one hand, and the effectiveness and impact of such actions on the other, appears in this aspect of agricultural production. We have already discussed government interventions and policies on this matter, including issues related to land, water resoures, infrastructural facilities in rural areas and production inputs. However, one of the biggest gaps is in government commitment to provide strong technical services, including research to solve technical constraints, the effective dissemination of information, soil- and water-testing services and pest diagnosis services. Government policy thus far has been that most of these services should be provided to farmers free of charge, with the remainder at a minimal or subsidised fee. So far the performance of public-sector institutions in this field has been far below the necessary levels in both quality and quantity. Most farmers have resorted to other sources to obtain such services. Those who suffer most in this process are the small farmers. The reorganisation of such services, from the standpoint of sources of finance, structure, function, autonomy and size and quality of allocated resources, is vital to achieve government objectives and to meet farmers' needs.

Positive policy measures

1. The Government follows a policy of no direct subsidy to production inputs. At the same time, it encourages parastatal organisations to compete with the private sector in providing agricultural inputs to farmers.
2. State laws exempt income generated through agricultural production from income tax.
3. Parastatal organisations provide credit to farming operations and agricultural development projects at rates lower than those provided by commercial banks.
4. The MOA rations the importation of commodities which compete with domestically produced ones.
5. The MOS supports the prices of some commodities considered vital to food security, such as wheat.
6. The Government attempts to provide marketing facilities, like grading and packing plants. Such facilities have been limited to the Jordan Valley.
7. The Government supports public share-holding companies dealing with marketing, with government participation in working capital but with autonomous management and operation.

Policy disincentives and omissions

1. The MOS controls the ceiling of retail prices to consumers for agricultural food commodities without ensuring fair farm-gate prices to farmers.
2. MOS importing practices are not always coordinated with domestic supplies and constitute disincentives to farmers.
3. Farmers are provided with weak information services about the demand and marketing potential of domestically produced commodities.
4. Municipalities charge wholesale fees on commodities marketed within city limits which are all borne by farmers but do not provide marketing services or facilities in return.
5. No effective control exists on fees charged to farmers by wholesalers of agricultural commodities.
6. No insurance is given to farmers for risks or environmental disasters.
7. There is a multiplicity of institutions intervening in supply and marketing.

Annexe 12.1
Composition of the Higher Agricultural Council*

Date of establishment: 1979, through By-law No. 22 based on Article 120 of the Constitution.

Composition: Prime Minister as President and membership of:

- Minister of Agriculture
- Minister of Trade and Industry
- Director-General, Agricultural Credit Corporation
- Vice-President, Natural Resources Authority
- Dean, Faculty of Agriculture, University of Jordan
- Director-General, Jordanian Company for the Marketing and Processing of Agricultural Products
- Two members from the private sector concerned with agriculture appointed for a three-year term, renewable, by the Prime Minister upon recommendation of the Minister of Agriculture
- Minister of Supply
- President, Jordan Valley Authority
- President, National Planning Council
- Director-General, Jordan Cooperative Organisation
- Under-Secretary, Ministry of Agriculture
- Director from the Ministry of Agriculture, appointed by the Minister of Agriculture

Functions and Responsibilities

1. To formulate agricultural policies in the areas shown below, within the general framework and objectives of development:
   i) Policy of animal and plant agricultural production
   ii) Policy of marketing of agricultural products and agricultural inputs, and of determining prices of each
   iii) Policy for the utilisation of agricultural resources: agricultural land, water, labour, capital and technical expertise
   iv. Policy of taxation and customs exemption for agricultural commodities and production inputs
2. To approve agricultural development plans and programmes within the framework and objectives of the National Development Plan.
3. To coordinate the work of institutions (departments, organisations and corporations) concerned with the agricultural sector in order to consolidate and unify efforts as well as to avoid duplication.
4. To submit of suggestions and amendments on laws and by-laws related to agricultural development.

Meetings of the Council: At least once every month with 10 members consitituting a forum.

* Source: Hashemite King of Jordan, By-law No. 22, *Official Gazette*, Feb 1979, Amman.

Annexe 12.2
Summary of Articles of the Agricultural Law*

Date and Number: Law No. 20, 1973, published in the *Official Gazette*, No. 2419, on 16 May 1973. The Law is comprehensive and replaced 14 other laws concerned with one aspect or another of agriculture, issued during the period 1927-72.

Contents:

Book One: Composed of 10 chapters and 130 articles concerned with agricultural production.

Book Two: Composed of 4 chapters and 54 articles concerned with animal wealth.

General: Composed of 14 articles concerned with regulating the marketing of agricultural and animal commodities.

Annexe: Fees and prices of agricultural services and/or inputs provided by the Ministry of Agriculture (MOA).

Book One: Crop and Fruit Production

Chapter 1: Five articles determining the role of the Minister of Agriculture and the MOA in the organisation of agricultural production. Major items include:
— The Minister of Agriculture is responsible for the determination of the national cropping pattern, crop rotation, date of crop planting and harvest, methods of production and cropping pattern within each farm.
— The MOA may establish production schemes of its own and may sell such land to farmers residing in the area of the scheme.

Chapter 2: Registration of crop varieties through a committee which determines type of crop and name of variety. All varieties must be registered before use.

Chapter 3: Production and distribution of foundation and planting seeds, including areas where they can be produced in the country. In addition, articles in this Chapter regulate:
— Stations where seeds are cleaned and screened.
— Seed testing procedures and methods. All seeds must be tested before sale or distribution.
— Import and export of seeds. The approval of the MOA is required for both imports and exports.
— The seed trade. The MOA is in charge of registration and determination of prices.

It must be noted that not all regulation apply to activities carried out by the MOA.

Chapter 4: Organisation of fruit-tree orchards and seedling production.

Chapter 5: Olive oil processing plants.

Chapter 6: Regulations through a committee for the use of fertilisers, including types, procedures of registration, and prices.

Chapter 7:
— Crop protection: regulation of private and general crop protection campaigns, including those for desert locusts.
— Pesticides: regulations through a committee, including testing, registration, exports, import distribution, sale and prices.
— Quarantine service: regulations related to the import and export of plant materials for propagation and consumption.

* Source: Hashemite Kingdom of Jordan, Law of Agriculture No. 20, *Official Gazette*, No. 2419, May 1973, Amman.

Chapter 8: Agricultural landholdings. Regulations related to the organisation and use of landholdings for agricultural purposes and records of information needed for each holding.
Chapter 9: Protection and conservation of fruit trees and crops from grazing or any other harm done by individuals or animals.
Chapter 10: Forestry and soil conservation. Regulations relating to the protection, management and utilisation of forested areas.
Chapter 11: Rangelands. Definition of range (all land with annual rainfall of 200 mm or below), methods and patterns of rangeland utilisation, management and development.

Book Two: Animal Production
Chapter 1: Export and import of animals. Regulations related to the organisation of the animal trade (including poultry) and animal slaughtering.
- Animal feed: regulations related to the determination of animal feed, feed manufacturing and processing, animal feed registration and trade.
- Protection of wild animals—including birds—and hunting regulations.
- Regulations on how and where to keep bees, their products and trade.

Chapter 2: Animal health
- Regulations related to the management and protection of animals from disease as well as the control of such diseases.
- Veterinary quarantine.

Chapter 3: Slaughtering of animals and animal skinning
- Specification and methods of slaughtering as well as requirements of slaughter houses.

Chapter 4: Fisheries. Regulations on methods of fishing as well as management and protection of marine life.

General Articles
Organisation of marketing of animal and plant commodities. Trade regulations for plant and animal commodities including determination of type, duration and quantities of exported or imported commodities. Regulation of the establishment and operations of central markets for trade in agricultural commodities.

Annexes
Annexe 1: Fees for agricultural services including spraying, quarantine, soil testing, registration and licences.
Annexe 2: Veterinary animal production fees, including fees for disease diagnosis, quarantine services and animal product testing.
Annexe 3: Fees on imported agricultural commodities and on services for bee-keeping.

Annexe 12.3
National Planning Council (NPC): Objectives and Modes of Operation*

Date of Establishment: Law No. 68, 1971.
Composition of Board of Directors: Prime Minister as President, President of NPC as Vice-President, with eight members as follows:
- Minister of Finance
- Minister of Trade and Economy

- Governor of Central Bank
- President, University of Jordan
- President, Royal Scientific Society
- Secretary-General, NPC
- President, Chamber of Commerce Union
- Two members from the private sector appointed by the Council of Ministers

Functions and Responsibilities:

1. To formulate long-range national plans for economic, social, cultural and manpower development. Also, to formulate comprehensive development programmes with defined priorities to be executed over determined periods of time.
2. To negotiate agreements with friendly countries and international agencies to secure funds and aid for financing various development projects. The NPC is empowered to implement studies to determine the feasibility of development projects.
3. To identify the manpower needs of Jordan for development programmes and to facilitate the training of such manpower.
4. To organise a system to monitor and evaluate development programmes.
5. To cooperate in all its activities with Ministries and institutions in charge of implementing development programmes.

* Sources: Law of NPC, No. 68, 1971.
National Planning Council, *Economic and Social Development Plan, 1981-85*, Amman.
Prime Minister's orders.
Proceedings of seminars.

6. The NPC may establish research and planning centres or institutes. It may also contract with public or private institutions to conduct studies or research needed for planning and development operations.
7. The NPC is the link (focal point) between all Ministries and foreign financial sources. Prior to their implementation, all development projects must be approved by the NPC, its Board of Directors and the Council of Ministers, in that order.

Mode of Planning Operations: the NPC is empowered to suggest to the Prime Minister the establishment of planning units in all Ministries and/or other public institutions. The following steps are followed in development plan formulation and approval:

1. Formulation of an *ad hoc* national planning committee usually made up of 20 members representing major economic, social and cultural sub-sectors. The committee's function is to monitor sectoral committees and to supervise and guide the formulation of the plan.
2. Formation of *ad hoc* committees at national level for major sectors. In the 1981-85 Development Plan, the following were recognised as major sectors:

- Monetary Policy and Banking System
- Cooperatives
- Mineralogy and Industry
- Energy and Electricity
- Transportation
- Culture and Youth
- Education

— Social Development
— Housing and Government Buildings
— *Awqaf*
— Agriculture
— Water and Irrigation
— Tourism and Archaeology
— Trade and Supply
— Communications
— Information
— Health
— Manpower and Work
— Municipal, Rural Affairs and Environment
— Science and Technology and Statistics

3. The Ministries of the Interior, of Foreign Affairs and of Defence do not represent a sector, while six areas not represented by Ministries are considered sectors for the purposes of planning. Each sectoral committee is composed of senior officials plus representatives of the private sector organisations or institutions concerned. The function of each sectoral committee is to prepare and submit to the national committee a 'plan' for development of the sector it represents. Sometimes seminars with the wider participation of individuals working in, or concerned with, the sector are held to discuss policy guidelines, priorities of projects, and the role of the private and public sectors in order to reach consensus. Agriculture was the subject of one such seminar in all three plans launched.
4. The plan formulated by the national committee is submitted to the Board of Directors of the NPC and to the Council of Ministers for final approval.

Annexe 12.4
Organisational Measures, Agricultural Projects, and Measures to Develop Resources as well as to Strengthen Institutions: 1981-85 Five-Year Development Plan

Developmental Goals:

1. To conserve agricultural resources, i.e. soil, water and natural vegetation, and develop their use in accordance with the ecological conditions of each agricultural area; and to prepare the corresponding land use maps.
2. To organise the afforestation of 300,000 du of state forestry land, giving priority to land within the main catchment areas and to land most subjected to soil erosion.
3. To plant 245,000 du with fruit trees, giving priority to sloping land and land within the main catchment areas, having a slope of more than 9 per cent.
4. To expand the production of broiler meat and eggs to meet local demand and increase exports. To establish cattle and sheep farms.
5. To organise, develop and protect the rangelands in order to increase their bearing capacity as well as to increase the production of meat.
6. To limit the production of the main field crops (wheat, barley and lentils) to areas most suitable for them, and to improve yields to achieve a 30 per cent increase in average production during the period of the 1981-85 Five-Year Development Plan.
7. To increase productivity in the agricultural sector and speed up the introduction of farm mechanisation.

8. To increase storage capacity for agricultural products to regulate consumption and marketing operations.
9. To strengthen collective and cooperative programmes in the sector, especially in the fields of production and marketing.
10. To integrate planning for agricultural development through comprehensive development planning at the regional level.
11. To support joint Arab action in the field of integrated agricultural production in order to achieve Arab food security.
12. To initiate participation by the Agricultural Credit Corporation (ACC) in the equity capital of companies formed to establish and manage agricultural enterprises, including those dealing with agricultural inputs.
13. To provide loans of JD 48 mn through the ACC during the period 1981-85 to finance the following categories of projects:
    - i) Land reclamation and development in rainfed areas, especially in the Zarqa catchment area and the highlands.
    - ii) Establishment of irrigation projects, installation of modern irrigation systems, and drilling and equipping of artesian wells.
    - iii) Establishment of livestock development projects, including cattle and sheep farms.
    - iv) Mechanisation of agriculture and development of food processing projects.
    - v) Construction of farm buildings, and establishment of nurseries for fruit-tree seedlings and vegetable transplants, as well as extending seasonal loans to the JVFA to enable it to supply its members with the necessary inputs.

Organisational Measures:

1. Establishing a Directorate, which reports directly to the Minister of Agriculture, to manage the development of the Zarqa catchment area.
2. Intensifying effort in the field of applied agricultural research, publication of the findings of such research, support of extension services, preparation of bulletins for guidance in agricultural development, and the formation of a committee for agricultural research with representatives from the agencies concerned.
3. Setting up three joint boards for livestock, fruit and vegetables. These boards would emanate from the Higher Agricultural Council and be responsible for formulating and following up on the implementation of policies relating to the production, marketing and consumption of these commodities.
4. Forming a permanent commission for agricultural credit to be entrusted with the tasks of reviewing and coordinating agricultural credit policies, developing lending criteria, assessing financial needs and securing the requisite funds.
5. Encouraging the establishment of mixed companies and cooperative societies specialising in plant and animal production to provide agricultural inputs, fruit-tree seedlings and vegetable transplants, and marketing services. Supporting parastatal organisations such as the Jordan Cooperative Organisation and the JVFA to enable them to provide agricultural inputs, in competition with the private sector.
6. Expanding the method of contract farming with individual farmers or groups of farmers, to produce agricultural products which are in demand, and developing this system to include the provision of agricultural inputs, support services, means of transportation and other facilities that will encourage farmers to enter contract farming.
7. Adopting marketing policies which will ensure the maintenance of existing markets and the identification of new markets for agricultural exports.

8. Formulating integrated programmes to deal with violations against state lands, and deciding upon a method of exploitation of marginal lands within environmental and climatic limitations.
9. Enacting the necessary legislation and taking the decisions required to deal with the problem of land fragmentation by consolidating ownership into suitably sized holdings.
10. Carrying out a comprehensive survey of agricultural land in order to demarcate the different ecological zones, and to determine the crops most suited to these zones.
11. Expanding the planting of fruit trees in the highlands through the provision of subsidies to farmers, so as to encourage them to increase and continue production.
12. Reviewing legislation which defines the responsibilities of government agencies dealing with agriculture in order to attain the highest possible level of efficiency in performance, coordination and integration.
13. Taking the necessary steps towards a unified Arab plan for agricultural integration and Arab food security, in cooperation with the Arab organisations and institutions concerned.
14. Amending the ACC Law to enable the Corporation to participate in the equity capital of companies established to initiate and manage agricultural projects and agricultural inputs.
15. Augmenting the resources of the ACC to meet the necessary financial requirements of agricultural development.

Summary of Agricultural Projects, 1981–85:

Project	Total allocation (000 JD)
Annual Afforestation	6,840
National Afforestation Programme	1,570
Development of Forest Tree Nurseries	1,330
Development of Zarqa River Catchment Area	18,665
Soil Conservation and Fruit-Tree Planting	12,550
Production of Fruit-Tree Seedlings	5,370
Development of Grain Production	500
Production of Vegetable Transplants	2,050
Production of Potato Seeds	725
Collective Agricultural Pest Control in the Jordan Valley	1,700
Development of Meat Production	8,600
Establishment of a Grand Parent Stock Poultry Farm	6,500
Production of Veterinary Serum and Vaccines	350
Improving Methods of Production and Marketing of Poultry Products	4,000
Establishment of a National Centre for Agricultural Research	3,000
Agricultural Research Unit for the Ghors	2,930
Agricultural Marketing Centres	3,115
Tomato Paste Factories	1,300
Refrigerated Warehouses	1,000
Feed Concentrate Plant	2,000
Feed Plant	3,500
Poultry Slaughter-house	2,000

Water and Irrigation Sector	
Wadi al-Arab Dam	16,000
Raising the King Talal Dam	12,000
Raising the Kafrein Dam	3,500
Desert dams	600
Jordan Valley Irrigation Project, Second Phase	200,000
Irrigation of Wadi Araba	6,000
Subterranean drainage	3,500
Private Sector Investments	133,370

References and Other Works Consulted

Department of Lands and Surveys, *Annual Reports, 1974-81* inclusive. Prepared and published annually by the Department of Land and Surveys. Average size of each Report is 80 pages. Each Report contains statistics on all transactions of land sale, partitioning, inheritance, mortgage and consolidation. All Reports are in Arabic; English translations started in 1980.)

DOS (Department of Statistics), *Agricultural Census, 1975*, Amman.

— *Agricultural Sample Survey in the Ghors, 1981*, Amman.

Higher Agricultural Council, *First Annual Report, 1979*, Amman, 23pp (in Arabic). (This includes By-law, membership, and all decisions taken by the HAC up to the end of 1979.)

— *Minutes of Meetings of HAC, 1982*, Amman.

Jordanian Lawyers' Association, *Compendium of Laws, By-laws and Related Supreme Court Acts Confirmed in Jordan*, Jordanian Lawyers' Association, 1977-, Amman. (Compendium of 16 volumes containing all laws and by-laws arranged in alphabetical order up to 1976.)

NPC (National Planning Council), *Social and Economic Development Plan, 1981-85*, Amman, 360 pp. (in Arabic and English).